Sustainable Land Use and Management

Sustainable Land Use and Management

Editors

Lu Zhang
Bing Kuang
Bohan Yang

MDPI

Basel • Beijing • Wuhan • Barcelona • Belgrade • Novi Sad • Cluj • Manchester

Editors
Lu Zhang
Faculty of Political Science
College of Public Administration
Central China Normal
University
Wuhan, China

Bing Kuang
Faculty of Political Science
College of Public Administration
Central China Normal
University
Wuhan, China

Bohan Yang
Faculty of Political Science
College of Public Administration
Central China Normal
University
Wuhan, China

Editorial Office
MDPI
St. Alban-Anlage 66
4052 Basel, Switzerland

This is a reprint of articles from the Special Issue published online in the open access journal *Sustainability* (ISSN 2071-1050) (available at: https://www.mdpi.com/journal/sustainability/special_issues/sustainable_land_use_and_management).

For citation purposes, cite each article independently as indicated on the article page online and as indicated below:

Lastname, A.A.; Lastname, B.B. Article Title. *Journal Name* **Year**, *Volume Number*, Page Range.

ISBN 978-3-03928-599-0 (Hbk)
ISBN 978-3-03928-600-3 (PDF)
doi.org/10.3390/books978-3-03928-600-3

Contents

About the Editors . . . . **vii**

Lu Zhang, Bing Kuang and Bohan Yang
Sustainable Land Use and Management
Reprinted from: *Sustainability* **2023**, *15*, 16259, doi:10.3390/su152316259 . . . . **1**

Shandong Niu, Xiao Lyu and Guozheng Gu
What Is the Operation Logic of Cultivated Land Protection Policies in China? A Grounded Theory Analysis
Reprinted from: *Sustainability* **2022**, *14*, 8887, doi:10.3390/su14148887 . . . . **5**

Guoqun Ma, Danyang Lv, Tuanbiao Jiang and Yuxi Luo
Can Land Transfer Promote Agricultural Green Transformation? The Empirical Evidence from China
Reprinted from: *Sustainability* **2023**, *15*, 13570, doi:10.3390/su151813570 . . . . **25**

Ying Tang and Menghan Chen
Impact Mechanism and Effect of Agricultural Land Transfer on Agricultural Carbon Emissions in China: Evidence from Mediating Effect Test and Panel Threshold Regression Model
Reprinted from: *Sustainability* **2022**, *14*, 13014, doi:10.3390/su142013014 . . . . **43**

Shuping Fan, Peng Li, Qi He, Jiaru Cheng, Mingfeng Zhang, Nan Wu, et al.
Study on the Spatial-Temporal Evolution of Land Use Ecosystem Service Value and Its Zoning Management and Control in the Typical Alpine Valley Area of Southeast Tibet—Empirical Analysis Based on Panel Data of 97 Villages in Chayu County
Reprinted from: *Sustainability* **2022**, *14*, 10057, doi:10.3390/su141610057 . . . . **59**

Min Zhou, Hua Zhang, Zixuan Zhang and Hanxiaoxue Sun
Digital Financial Inclusion, Cultivated Land Transfer and Cultivated Land Green Utilization Efficiency: An Empirical Study from China
Reprinted from: *Sustainability* **2023**, *15*, 1569, doi:10.3390/su15021569 . . . . **79**

Gül Aslı Aksu, Şermin Tağıl, Nebiye Musaoğlu, Emel Seyrek Canatanoğlu and Adnan Uzun
Landscape Ecological Evaluation of Cultural Patterns for the Istanbul Urban Landscape
Reprinted from: *Sustainability* **2022**, *14*, 16030, doi:10.3390/su142316030 . . . . **99**

Qi Yang, Zhonggen Sun and Hubiao Zhang
Assessment of Urban Green Development Efficiency Based on Three-Stage DEA: A Case Study from China's Yangtze River Delta
Reprinted from: *Sustainability* **2022**, *14*, 12076, doi:10.3390/su141912076 . . . . **125**

Gubu Muga, Shougeng Hu, Zhilan Wang, Luyi Tong, Zongnan Hu, Hui Huang and Shijin Qu
The Efficiency of Urban–Rural Integration in the Yangtze River Economic Belt and Its Optimization
Reprinted from: *Sustainability* **2023**, *15*, 2419, doi:10.3390/su15032419 . . . . **145**

Zian Lin, Yuanfa Ji and Xiyan Sun
Landslide Displacement Prediction Based on CEEMDAN Method and CNN–BiLSTM Model
Reprinted from: *Sustainability* **2023**, *15*, 10071, doi:10.3390/su151310071 . . . . **165**

Bingkui Qiu, Jinjiang Yao, Siyu Han and Zhe Zhu
Spatio-Temporal Variation of Habitat Quality for Bird Species in China Caused by Land Use Change during 1995–2015
Reprinted from: *Sustainability* **2022**, *14*, 10078, doi:10.3390/su141610078 . . . . **185**

Shewakena Aytenfisu Abab, Feyera Senbeta and Tamirat Tefera Negash
The Effect of Land Tenure Institutional Factors on Small Landholders' Sustainable Land Management Investment: Evidence from the Highlands of Ethiopia
Reprinted from: *Sustainability* **2023**, *15*, 9150, doi:10.3390/su15129150 . . . . **197**

Yu Lu and Ziheng Shangguan
Reassessing Resettlement-Associated Poverty Induced by Water Conservancy Projects in China: Case Study of the "Yangtze to Huai River Inter-Basin" Water Diversion Project
Reprinted from: *Sustainability* **2023**, *15*, 9477, doi:10.3390/su15129477 . . . . **221**

About the Editors

Lu Zhang

Lu Zhang, Associate Professor, Doctoral supervisor, School of Public Administration, Central China Normal University, and Visiting Scholar at Queensland University of Technology and University of North Carolina at Chapel Hill. His research areas mainly include sustainable land use development, cultivated land protection and food security. He has published more than 60 academic papers, provided a series of decision-making advisory reports on social and economic development for China's central and local governments, and chaired a number of national research projects.

Bing Kuang

Bing Kuang, Associate Professor and Master's supervisor, School of Public Administration, Central China Normal University. His research areas mainly include land use management, regional green development, and farmland protection policy analysis. He presides over the Key Program of the post-funded project of the National Social Science Foundation of China and the Youth Fund of the National Natural Science Foundation Natural Science Foundation of China (Natural Science Foundation of China) and other kinds of scientific research projects. He has published more than 60 academic papers, and an academic work.

Bohan Yang

Bohan Yang, Lecturer and Master's supervisor, School of Public Administration, Central China Normal University, and Joint-training Doctor of Huazhong Agricultural University and the Institute for Environmental Studies (IVM), Vrije University Amsterdam. He has presided over a National Natural Science Foundation youth project, a Humanities and Social Science Youth Fund of the Ministry of Education, and three other science projects. His research interests include ecological and environmental effects of cropland displacement, cropland system resilience, and land system modelling.

sustainability MDPI

Editorial

Sustainable Land Use and Management

Lu Zhang *, Bing Kuang and Bohan Yang

Faculty of Political Science, College of Public Administration, Central China Normal University, Wuhan 430079, China; kuangbing117@163.com (B.K.); bohan.yang@ccnu.edu.cn (B.Y.)
* Correspondence: zhanglu54522@ccnu.edu.cn

With the rapid development of urbanization and social economy, the utilization and protection of land have become one of the great social problems globally. Excessive urbanization has not only brought significant challenges to the sustainable utilization of urban land [1], but also imposed far-reaching, negative implications on farmland as well as ecological environment protection [2,3], as both urban and rural land are faced with overexploitation, and the harmony of the human–land system has yielded to discord. Unreasonable land-use planning and allocation are gradually reducing the efficiency and sustainability of urban land use, and also branching out the conversion scale of farmland to construction land. However, farmland reduction and urbanization not only give rise to ecological environmental issues, such as land degradation, environmental pollution, carbon emission increases, and so on, but also induce many social problems around land interests [4–6].

The implementation of Protection policies for cultivated land, as an important type of land use, is an essential element of sustainable land use and management. Meanwhile, cultivated land conservation is a recognized worldwide topic and is central to ensuring food security and maintaining social stability [7]. The first paper in this Special Issue first summarizes the current dilemmas of China's cultivated land protection at the theoretical level, and preliminarily depicts the external foundation of CLPP in view of China's topography and spatial distribution of cultivated land (Contribution 1). This paper uses CLPP texts as research samples based on grounded theory to construct an analytical framework.

From a practical point of view, land transfer, as an important means of farmland policy, is an important channel in sustainable land use and management. And optimizing land management is a promising approach to mitigating climate change [8]. Based on inter-provincial panel data from 2005 to 2020, the study examined the influence of land transfer on agricultural green transformation and its underlying mechanism by using a two-way fixed effect model and an intermediary effect model. The study found that land transfer substantially promotes agricultural green transformation and encourages the progress of agricultural technology (Contribution 2). On the basis of exploring the mechanism and effect of agricultural land transfer on agricultural carbon emissions, the correlation between agricultural land transfer and agricultural carbon emissions was tested so as to clarify the mechanism of agricultural land transfer affecting agricultural carbon emissions and its future trends (Contribution 3). Meanwhile, under the background of ecological civilization construction and the overall planning of land and space, the paper, taking Chayu County, a typical alpine valley area of southeast Tibet as an example and based on the remote sensing interpretation data of three periods in 2000, 2010 and 2020, employed the three-level spatial scale from the village level to analyze and calculate the regional ecosystem service value and their dynamic changes (Contribution 4). It is also important to examine theoretically and empirically whether and how Digital financial inclusion (DFI) can reinforce cultivated land green utilization efficiency (CLGUE) through the mediator of cultivated land transfer (CLT) under the background of food security, social stability and environmental protection (Contribution 5). The study explored the mediating mechanism between DFI and CLGUE from the perspective of CLT, and the results showed that there is regional heterogeneity in

Citation: Zhang, L.; Kuang, B.; Yang, B. Sustainable Land Use and Management. *Sustainability* **2023**, *15*, 16259. https://doi.org/10.3390/su152316259

Received: 15 November 2023
Accepted: 21 November 2023
Published: 24 November 2023

DFI in terms of promoting and improving CLGUE, with a more positive relationship in the main grain marketing areas.

In addition to agricultural land, rapid and excessive urbanization has also brought significant challenges for the sustainable use and management of urban land. The metropolitan area of Istanbul, with its rapid urbanization rate, has faced intense pressure regarding the sustainability of urban habitats. This paper provides an understanding of how urbanization changed the function of the spatial distribution of the urban mosaic by combining medium-resolution data with high-resolution satellite imagery, evaluating the overall landscape structure and elucidating the landscape functions in an urban environment based on the landscape structure (Contribution 6). Moreover, there is also a study on urban green development efficiency (GDE) adopting a three-stage DEA model for Yangtze River Delta cities (Contribution 7). The results showed that the GDE level showed heterogeneity in different cities. And the GDE was enhanced by increasing the proportion of the tertiary industry and the green area of built districts but weakened when the area of built districts (ABD) reflecting urban construction was expanded.

The following papers of this Special Issue examined the efficiency of sustainable green development from the perspective of urban–rural integration and explored policies and strategies for the sustainable land use and management in the context of natural geological disasters and social issues surrounding land interests. Under the constraints of scarce land resources and the need for high-quality economic and social development, one paper measured the efficiency of URID from the input–output perspective, taking into account the impact of carbon emissions; it also calculated the efficiency of URID and described the spatio-temporal characteristics in 73 cities within three major city clusters in the Yangtze River Economic Belt (YREB) from 2010 to 2019, and analyzed the input–output optimization strategies for URID within each of these major urban systems (Contribution 8). As a typical geological disaster, landslides also bring a great challenge to sustainable land use and management. The GRA–MIC fusion correlation calculation method was used to select the factors influencing landslide displacement, and the CNN–BiLSTM model was used for prediction. The experimental prediction results showed that the model proposed in this paper can be popularized and applied in areas with frequent landslides and provide strong support for disaster prevention and reduction and land use management (Contribution 9). The ecological impacts of land use change are also reflected on the quality of bird habitats. Habitat loss and degradation due to land use change and loss is a major threat to biodiversity worldwide (Contribution 10). Studies have shown an inverted U-shaped relationship between the intensity of LUC and the PGSH. This study could provide a reference for measuring the impacts of LUCC on bird species, enabling the protection of bird species and habitats that need it most.

In relation to land interests, one study assessed the extent to which Land Tenure Institutional Factors (LTIFs) influence on-farm Sustainable Land Management (SLM) investment in the highlands of Ethiopia through unbundling tenure security across a bundle of rights. The study strengthened the notion that security of tenure may be a necessary condition. And an in-depth analysis of the security of tenure categories across a bundle of rights is necessary to help formulate context-specific SLM policy and strategy incentivizing smallholders' on-farm SLM investment (Contribution 11). Land management issues are also embedded in displacement and resettlement-associated poverty caused by water conservancy projects (WCP). The study found that rural re-settlers were more resilient to forced majeure because land guarantees employment and food supply, allowing households avoidance of secondary livelihood destruction (Contribution 12).

Sustainable development is currently a hot topic that has attracted global concern, and the process of land use and management profoundly affects the realization of sustainable development goals [9,10]. This Special Issue gathered studies regarding sustainable land use and management from different research perspectives, aiming to contribute to the global challenges of the sustainable urban and rural development in the rapidly urbanizing world.

Conflicts of Interest: The authors declare no conflict of interest.

List of Contributions:

1. Niu, D.; Lyu, X.; Gu, G. What Is the Operation Logic of Cultivated Land Protection Policies in China? A Grounded Theory Analysis. *Sustainability* **2022**, *14*, 8887.
2. Ma, G.; Lv, D.; Jiang, T.; Luo, Y. Can Land Transfer Promote Agricultural Green Transformation? The Empirical Evidence from China. *Sustainability* **2023**, *15*, 13570.
3. Tang, Y.; Chen, M. Impact Mechanism and Effect of Agricultural Land Transfer on Agricultural Carbon Emissions in China: Evidence from Mediating Effect Test and Panel Threshold Regression Model. *Sustainability* **2022**, *14*, 13014.
4. Fan, S.; Li, P.; He, Q.; Cheng, J.; Zhang, M.; Wu, N.; Yang, S.; Pan, S. Study on the Spatial-Temporal Evolution of Land Use Ecosystem Service Value and Its Zoning Management and Control in the Typical Alpine Valley Area of Southeast Tibet—Empirical Analysis Based on Panel Data of 97 Villages in Chayu County. *Sustainability* **2022**, *14*, 10057.
5. Zhou, M.; Zhang, H.; Zhang, Z.; Sun, H. Digital Financial Inclusion, Cultivated Land Transfer and Cultivated Land Green Utilization Efficiency: An Empirical Study from China. *Sustainability* **2023**, *15*, 1569.
6. Aksu, G.A.; Tagil, S.; Musaoglu, N.; Canatanoglu, E.S.; Uzun, A. Landscape Ecological Evaluation of Cultural Patterns for the Istanbul Urban Landscape. *Sustainability* **2022**, *14*, 16030.
7. Yang, Q.; Sun, Z.; Zhang, H. Assessment of Urban Green Development Efficiency Based on Three-Stage DEA: A Case Study from China's Yangtze River Delta. *Sustainability* **2022**, *14*, 12076.
8. Muga, G.; Hu, S.; Wang, Z.; Tong, L.; Hu, Z.; Huang, H.; Qu, S. The Efficiency of Urban–Rural Integration in the Yangtze River Economic Belt and Its Optimization. *Sustainability* **2023**, *15*, 2419.
9. Lin, Z.; Ji, Y.; Sun, X. Landslide Displacement Prediction Based on CEEMDAN Method and CNN–BiLSTM Model. *Sustainability* **2023**, *15*, 10071.
10. Qiu, B.; Yao, J.; Han, S.; Zhu, Z. Spatio-Temporal Variation of Habitat Quality for Bird Species in China Caused by Land Use Change during 1995–2015. *Sustainability* **2022**, *14*, 10078.
11. Abab, S.A.; Senbeta, F.; Negash, T.T. The Effect of Land Tenure Institutional Factors on Small Landholders' Sustainable Land Management Investment: Evidence from the Highlands of Ethiopia. *Sustainability* **2023**, *15*, 9150.
12. Lu, Y.; Shangguan, Z.. Reassessing Resettlement-Associated Poverty Induced by Water Conservancy Projects in China: Case Study of the "Yangtze to Huai River Inter-Basin" Water Diversion Project. *Sustainability* **2023**, *15*, 9477.

References

1. Pan, H.; Page, J.; Shi, R.; Cong, C.; Cai, Z.; Barthel, S.; Thollander, P.; Colding, J.; Kalantari, Z. Contribution of prioritized urban nature-based solutions allocation to carbon neutrality. *Nat. Clim. Chang.* **2023**, *13*, 862–870. [CrossRef]
2. Krause, A.; Haverd, V.; Poulter, B.; Anthoni, P.; Quesada, B.; Rammig, A.; Arneth, A. Multimodel analysis of future land use and climate change impacts on ecosystem functioning. *Earth's Future* **2019**, *7*, 833–851. [CrossRef]
3. Hong, C.; Burney, J.A.; Pongratz, J.; Nabel, J.E.M.S.; Mueller, N.D.; Jackson, R.B.; Davis, S.J. Global and regional drivers of land-use emissions in 1961–2017. *Nature* **2021**, *589*, 554–561. [CrossRef] [PubMed]
4. Zheng, Q.; Ha, T.; Prishchepov, A.V.; Zeng, Y.; Yin, H.; Koh, L.P. The neglected role of abandoned cropland in supporting both food security and climate change mitigation. *Nat. Commun.* **2023**, *14*, 6083. [CrossRef] [PubMed]
5. Chen, K.; Wang, Y.; Li, N.; Xu, Y.; Zheng, Y.; Zhan, X.; Li, Q. The impact of farmland use transition on rural livelihood transformation in China. *Habitat Int.* **2023**, *135*, 102784. [CrossRef]
6. Zangue, Y.D.; Melot, R.; Martin, P. Diversity of farmland management practices (FMP) and their nexus to environment: A review. *J. Environ. Manag.* **2022**, *302*, 114059. [CrossRef] [PubMed]
7. Zhu, Y.; Wang, Z.; Zhu, X. New reflections on food security and land use strategies based on the evolution of Chinese dietary patterns. *Land Use Policy* **2023**, *126*, 106520. [CrossRef]
8. Sha, Z.; Bai, Y.; Li, R.; Lan, H.; Zhang, X.; Li, J.; Liu, X.; Chang, S.; Xie, Y. The global carbon sink potential of terrestrial vegetation can be increased substantially by optimal land management. *Commun. Earth Environ.* **2022**, *3*, 8. [CrossRef]

9. Lu, X.; Zhang, Y.; Lin, C.; Wu, F. Analysis and comprehensive evaluation of sustainable land use in China: Based on sustainable development goals framework. *J. Clean. Prod.* **2021**, *310*, 127205. [CrossRef]
10. Cao, M.; Tian, Y.; Wu, K.; Chen, M.; Chen, Y.; Hu, X.; Sun, Z.; Zuo, L.; Lin, J.; Luo, L.; et al. Future land-use change and its impact on terrestrial ecosystem carbon pool evolution along the Silk Road under SDG scenarios. *Sci. Bull.* **2023**, *68*, 740–749. [CrossRef] [PubMed]

sustainability

MDPI

Article

What Is the Operation Logic of Cultivated Land Protection Policies in China? A Grounded Theory Analysis

Shandong Niu, Xiao Lyu * and Guozheng Gu

School of Humanities and Law, Northeastern University, Shenyang 110169, China; 2010005@stu.neu.edu.cn (S.N.); 2010006@stu.neu.edu.cn (G.G.)
* Correspondence: lvxiao@mail.neu.edu.cn; Tel.: +86-187-6931-9508

Abstract: Cultivated land protection policies (CLPP) are essential for maintaining social stability, guaranteeing food security, and ensuring sustainable development. However, a mismatch exists between policy performance and the objectives that influence the implementation of CLPP, and the system mechanisms of CLPP must be revealed and explored. Based on the literature review, this paper summarizes the current dilemmas of China's cultivated land protection at the theoretical level, and preliminarily depicts the external foundation of CLPP in view of China's topography and spatial distribution of cultivated land. This paper uses CLPP texts as research samples based on grounded theory to construct an analytical framework. The results show that the operation logic of the CLPP is founded on situation–structure–motivation–action–space–outcome. Accordingly, systematic analysis and in-depth understanding of the operation logic of CLPP will help to re-examine the profound relationship between policy text and implementation effect from such perspectives as transnational, trans-regional, and multi-scale. It also helps to reveal the hidden scientific value of spatiotemporal pattern for cultivated land protection, and serve the formulation and implementation of relevant policies in the future. Under the background of the new era of ecological civilization, it is urgent to enhance the operational effectiveness of the CLPP, identifying the focus of policy implementation, and scientifically formulating the CLPP is of great significance to its success.

Keywords: cultivated land protection policies; operation logic; current dilemmas; grounded theory

Citation: Niu, S.; Lyu, X.; Gu, G. What Is the Operation Logic of Cultivated Land Protection Policies in China? A Grounded Theory Analysis. *Sustainability* **2022**, *14*, 8887. https://doi.org/10.3390/su14148887

Academic Editors: Lu Zhang, Bing Kuang and Bohan Yang

Received: 13 June 2022
Accepted: 19 July 2022
Published: 20 July 2022

Publisher's Note: MDPI stays neutral with regard to jurisdictional claims in published maps and institutional affiliations.

1. Introduction

Cultivated land is a critical subject, and its protection is central to ensuring food security and maintaining social stability [1]. The projections show that feeding a world population of 9.1 billion people in 2050 would require raising overall food production by 70% between 2005 and 2050 [2]. Since 1960, the amount of cultivated land per capita in the world has decreased from 0.41 ha to 0.21 ha [3]. Furthermore, it is estimated that 5 to 10 million ha of cultivated land is abandoned every year as a result of soil degradation and the construction of buildings and infrastructure (Food and Agriculture Organization of the United Nations, FAO, 2009). In 2019, the Intergovernmental Panel on Climate Change (IPCC) noted in its report entitled "Risk Management and Decision-making in Relation to Sustainable Development" that climate and land changes result in compound risks to food systems, human and ecosystem health, and livelihoods. Control of land degeneration and the sustainable use of cultivated land plays an important role in reducing soil erosion, eliminating hunger, and coping with climate change [4,5]. In particular, the 2030 Agenda (Transforming our World: The 2030 Agenda for Sustainable Development) also established the "land degradation neutral world" to reset or reduce the level of soil degradation. This is a significant challenge for scientists and policymakers worldwide who seek to protect cultivated land [6] and achieve sustainable development of agricultural and socio-economic production while ensuring food security [7,8].

Protecting cultivated land is a recognized worldwide topic. Internationally, cultivated land protection forms and priorities have their own characteristics, but they have certain commonalities in scientific planning, legal protection, government regulation, and market operation. The cultivated land protection policy of the United States is mainly macro-control oriented. First, it is to formulate a complete legal system for cultivated land protection, that is, to plan and protect cultivated land from the aspects of planning, taxation, requisition, etc.; then it is to implement the land use control system and the land development right system [9,10]. Canada mainly guides cultivated land protection through the planning system, including laws and regulations, land use planning, and restrictions on the right to agricultural land development [11]. In the face of high urbanization, Britain still retains a large number of cultivated land thanks to its focus on the implementation of the land development right system [12,13]. After the 1960s, many laws have been issued to strengthen the planning and management of land resources in Britain. Comparatively speaking, in France, the cultivated land protection policy puts more emphasis on centralization, and the central government exercises the land legislative power. Private cultivated land must be used for agricultural management, and it is not allowed to abandon cultivation, inferior cultivation, or to build houses on cultivated land [14]. In Germany, the ecological compensation policy and the ecological index trading mode ensure land use and ecological balance, which is a policy worthy of reference [15]. Given its limited land resources, Japan has not only established a sound legal system for the protection of cultivated land, but also paid special attention to the creation of new urban agriculture [16]. In addition, South Korea also formulated a series of laws to reasonably protect and develop agricultural land resources [17]. In general, compared with the above-mentioned countries, the particularity of China's cultivated land protection policy lies in basic national conditions and the complex relationship between the multiple subjects of cultivated land protection.

Food security in China faces many challenges [18,19]. China is a developing country with a large population, less cultivated land per capita, less high-quality cultivated land, and less cultivated land reserve resources than developed countries [20,21]. Cultivated land protection in China began in the 1980s and is the most stringent cultivated protection system in the world [22,23]. However, in reality, cultivated land protection has not achieved the expected effect of the policy [24,25]. According to the monitoring data of the Ministry of Natural Resources [26], cultivated land decreased by 354,700 ha, 388,000 ha, and 336,500 ha in 2013, 2014, and 2015, respectively. By 2015, China's per capita cultivated land area was only one-third of the world average. In 2015, the national high-quality cultivated land area was 3.9738 million ha, accounting for only 2.9% of the total national cultivated land area. The quality of cultivated land in some areas has continued to deteriorate [27]. In particular, the performance of cultivated land in terms of quality and ecological protection is relatively low [21,28,29] because rapid urban expansion has led to cultivated land degradation [30,31], cultivated land conversion [32,33], cultivated land abandonment, and the decline of cultivated land fertility [34]. In addition, excessive intensive use of cultivated land has caused various problems, such as overloaded agricultural ecosystem operations and increasing environmental pressure [35,36], which has seriously undermined the sustainable development of agriculture in China [37]. Because of this, based on the development situation of cultivated land use and protection, the Chinese government formally put forward a series of policy objectives and measures to strengthen cultivated land protection from the three dimensions of quantity, quality, and ecology in 2012 [38,39].

Is the effect of the CLPP consistent with the expected goal of three-dimensional protection [40]? The evolution of the CLPP has been systematically analyzed [41,42], including the policy performance [43]. A literary analysis of the CLPP literature reveals two main points. First, relevant scholars carried out a performance evaluation of CLPP based on the macro and meso scale and found that the implementation of a land use control system and basic cultivated land protection policy yielded positive results for cultivated land protection [44,45]. At the same time, the CLPP plays an important role in promoting food production [46], maintaining the quantity, and reducing illegal activities. Second, some scholarly research shows that the performance of the CLPP is disappointing, and the cultivated land control system does not effectively constrain the demand and supply of cultivated land conversion [47,48]. Additionally, the land policy to control the growth of construction land and prevent the loss of cultivated land has proven a two-fold failure [49]. These arguments provide a basis for optimizing the path of cultivated land protection in China. As public policy, the historical logic and comprehensive attribution of the CLPP and the complexity of the policy's external environment cannot fully explain the cause and effect whereby cultivated land protection is not consistent with the goal. Whether the system mechanism of the policy is clarified is a key factor affecting the policy effect. In the context of China's land and space governance in the new era, what are the practical problems faced by the cultivated land protection policy? How do the elements of policy system interact? Is there any inherent logic law for induction? Therefore, a clear policy can explain the operation logic and help construct the theoretical framework of the CLPP.

This paper focuses on these issues to explore the operational logic and construct the theoretical framework of CLPP. It also supports current discussions of the systematic law of policy operation and goes on to identify the focus of policy implementation. This paper is organized as follows: In Section 2, we briefly review the relevant literature regarding cultivated land protection to introduce the current dilemma. In Section 3, we describe the research methods and data sources. In Section 4, using 62 texts of cultivated land protection policies as research samples, we perform coding analysis and construct an analytical framework based on grounded theory to explore the operational logic of CLPP. In Section 5, we discuss the operational logic of cultivated land protection and analyze the results. In the last section, we provide both conclusions and policy implications.

2. Literature Review

2.1. Mismatch between Cultivated Land Resource Value and Cultivated Land Protection Path

The comprehensive value of cultivated land resources includes economic, production, ecological, and social value. The root cause of cultivated land loss is that the total value of cultivated land resources and their reasonable distribution are ignored [50]. The non-market value of cultivated land such as ecosystem service function, ensuring national food security, maintaining social stability, and providing basic living security for farmers attracts much attention [51]. Researchers found that the excessive loss of cultivated land conversion accounted for 44.73% of the total cultivated land conversion area because the non-market value of cultivated land from 1989 to 2006 was ignored [52]. According to studies [53], the ratio of economic value, ecological value, and social value of cultivated land is approximately 1:2:3, indicating that the current market value of cultivated land resources is far from fully reflective of the value of cultivated land resources. However, the value of cultivated land has not been widely studied by the community. On the one hand, as the first behavior subject of cultivated land protection, farmers' one-sided cognition of cultivated land value, low comparative benefits of cultivated land's social and economic value [54], and obvious externality of cultivated land protection affect farmers' enthusiasm for protecting cultivated land [55]. However, these factors also bring a series of problems, such as a deteriorating cultivated ecological land environment due to the excessive pursuit of economic benefits of cultivated land. The lack of comprehensive value cognition of cultivated land results in a disparity between requirements and the supply of cultivated land multi-functions [56]. In reality, China's topography are three ladder distributions. According to the spatial distribution of cultivated land in 2017, the cultivated land quantity in eastern area is more than that in Western area, and the cultivated land quantity in areas with "light–temperature–water–soil" suitable conditions such as southeast coastal area are gradually decreasing (Figure 1a). The economic, production, ecological, and social value of cultivated land in different regions are spatiotemporal heterogeneity. Additionally, insufficient attention to the ecological attributes of cultivated land with a large amount of chemical investment in cultivated land use weakens the cultivated land ecological function. Therefore, some scholars have proposed focusing on the non-market value and ecological value of cultivated land [57], the value of social responsibility [58], and the value of social welfare [59] to squeeze the benefit space of cultivated land conversion. Overall, the value of cultivated land has experienced a development process that has grown in scope. The cognition of cultivated land comprehensive value is still in the connotation deepening and standard quantification stages. A scientific and comprehensive evaluation system of cultivated land value is yet to be developed; the benefit distribution is unbalanced, which leads to a deviation in the path to cultivated land protection.

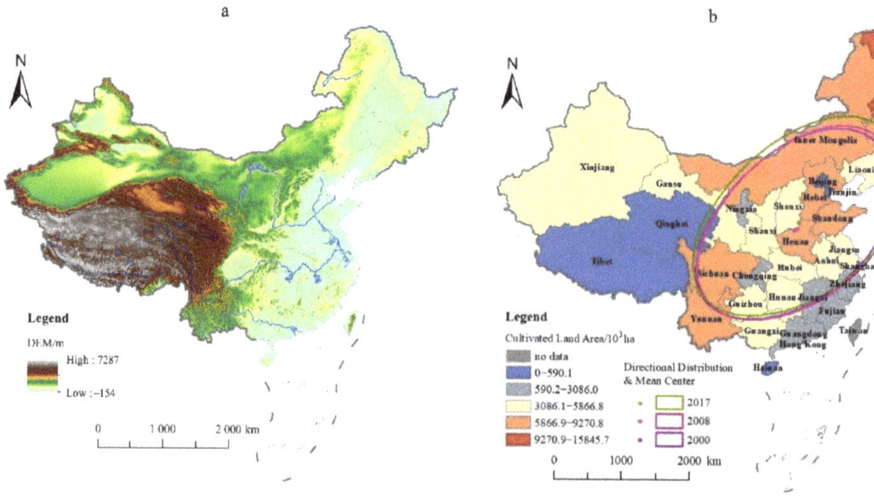

Figure 1. Chinese topography and spatial distribution of cultivated land. (**a**) represents the topographic map of China; (**b**) represents the spatial distribution of cultivated land area in China.

2.2. Imbalance between Cultivated Land Protection Policy Objectives and Stakeholders

Cultivated land protection is a spatial allocation process in which multiple stakeholders participate in responsibility rights [60], and its stakeholders include the central government, local governments, and farmers [61,62]. However, the ambiguity of land property rights is the root of the interest game among multiple stakeholders involved in cultivated land protection in China [63]. The spatial allocation of cultivated land protection is both a complex process and an unbalanced relationship, which leads to a mismatch between the goals and achievements of central and local governments [64]. The central government assumes ensuring all social welfare as the basic goal, local governments focus on regional economic development, and farmers pay more attention to the production efficiency of cultivated land. The interest game among multiple stakeholders is a key reason for the failure of the cultivated land protection policy [65]. A principal–agent relationship is formed between central and local governments in the process of responsibility and rights distribution [66]. However, the rights and obligations of each subject are not clear, which can easily lead to responsibility shifting. The central government (the client) is the policy-maker and must coordinate social, economic, and ecological development holistically. Local government (the agent) follows the logic of maximizing the benefits of local economy and political achievements, and its responsibilities as an intermediate principal are far greater than its rights. However, local governments are not all compensated accordingly, which easily leads to abuse of power by local governments and land violations. Farmers often fail to realize their rights and interests due to insufficient participation in public decision making. Finally, these actions have made a profound influence upon the replacement of policy objectives because of an imbalance of power, responsibilities, and interests.

2.3. Structural Contradiction between Restraint Mechanism and Incentive Mechanism of Cultivated Land Protection

Many scholars believe that a problem to be solved is establishing restrictive and incentive mechanisms to stimulate the enthusiasm of multiple stakeholders in cultivated land protection [31]. At present, cultivated land occupation and destruction frequently occur, the law enforcement mechanism is not perfect, protection responsibility is not clearly defined [67], and the cultivated land protection legal framework has not achieved legislative goals. In particular, China's cultivated land resources are mainly distributed in the eastern monsoon area, and the high-quality cultivated land concentration areas are highly overlapped with the economically developed regions. Simultaneously, China's cultivated land quantity gravity center is gradually migrated northward (Figures 1b and 2), combined with agricultural non-point source pollution, planting structure adjustment, natural disaster and ecological conversion, which will bring new challenges to China's cultivated land protection in the new era. For example, due to the rapid urbanization occupying a large number of high-quality cultivated land, the existing cultivated land protection restrictive and incentive mechanisms are difficult to adapt to the practical needs of the development of ecological civilization in the new era, which may further aggravate the resource mismatch pattern of "South-to-North Water Transfer Project" and "North-to-South Grain Transport Project", thus forming the endogenous contradiction of establishing restrictive and incentive mechanisms. In addition, governments at all levels bear responsibility for protecting cultivated land. The current administrative restrictive mechanisms fail to adjust the interest game of governments at all levels and cannot meet the actual needs of cultivated land protection [60,68].

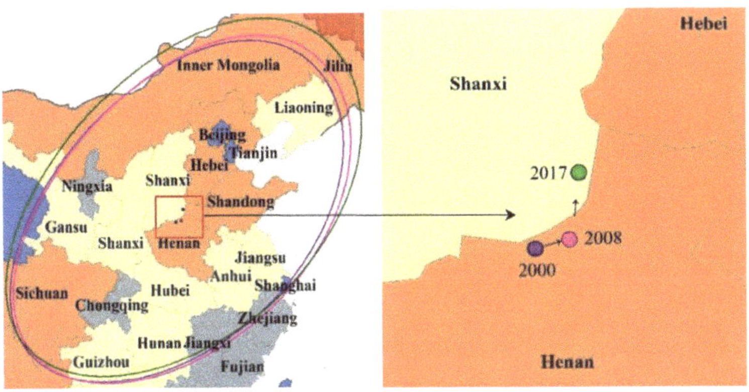

Figure 2. The migration trajectory of cultivated land quantity gravity center in China from 2000 to 2017.

2.4. There Are Deficiencies in the Compensation Mechanism for Cultivated Land Protection

In recent years, the government and theory fields have explored the transformation of reliance on restrictive systems to an emphasis on both incentives and restraints, and the compensation mechanism of cultivated land protection is considered an important component [69]. The multi-functional spillover of cultivated land and the low cost of non-agricultural conversion are the main reasons for the loss of cultivated land [67]. Moreover, the profit losses in the process of cultivated land and non-grain conversion greatly reduce the enthusiasm of relevant stakeholders in cultivated land protection. Therefore, the construction of an economic compensation mechanism to compensate for the interests of multiple stakeholders in the process of cultivated land conversion could play a role in balancing the interests of relevant stakeholders in cultivated land protection [68]. However, the compensation mechanisms of cultivated land protection remain in the theoretical analysis and method exploration stages [70]. Currently, there is a lack of systematic research on many aspects, including the compensation subject, object, and compensation standards and modes [55]. There are still some problems in the operation path and specific links of the regional compensation mechanism. In particular, regarding compensation for cultivated land protection, factors such as economic production conditions, resource endowment, and ecological environment are ignored. Additionally, the compensation mechanism for cultivated land protection needs to be improved in terms of complementing cultivated land from other areas and ecological compensation.

As a whole, the institutional system and policy context of cultivated land protection have basically taken shape, but the awareness of active protection at the farmer level has not been formed in society as a whole. Construction occupation continues to promote extensive use of cultivated land through a series of forms of transformation. Land comprehensive consolidation tends to increase land use indicators and there is not enough to prioritize improving the quality of cultivated land. At the economic and social level, the enthusiasm and initiative of various stakeholders for cultivated land protection are still in a passive state, that is, "no push, no go," and soil pollution in different regions has not been effectively curbed. Although there are many research perspectives, content, and related concepts, the logic is the same. The CLPP must answer a key question: how can we effectively implement the trinity protection mode?

3. Research and Data Source

3.1. Research Method

Grounded theory is a qualitative research method, and its core idea is to first collect data and then encode the data level by level, refine the concepts, and compare them repeatedly without hypothesis by classifying and coding the original text materials (information fragments) [71,72]. Multiple concepts (concise definitions) reflecting real data are formed by repeatedly summarizing and comparing the key information points to gradually break through the stereotypes formed by the existing research. Thus, subjective path dependence among researchers is avoided (eliminating bias).

Therefore, what follows is the content with common characteristics (common convergence). However, the reliability and validity of coding should be ensured while mining the core categories of text data (theoretical saturation). Figure 3 shows the basic research steps needed for this paper based on grounded theory methodology.

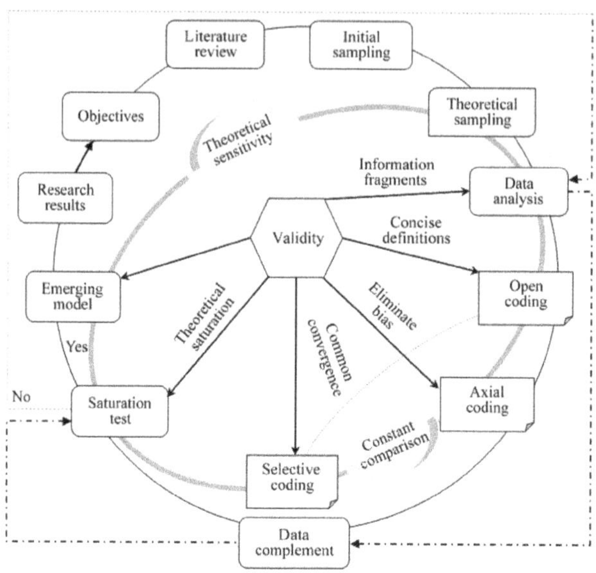

Figure 3. The grounded theory research method and process.

3.2. Data Sources

The data in this paper mainly include two forms of policy text and statements by President Xi Jinping on cultivated land protection. This paper considers the CLPP from the founding of new China to today as the research object through the direct retrieval of the official websites of the CPC (the Communist Party of China) Central Committee and relevant state ministries and commissions. The relevant content of the research literature and the existing policy texts are retrieved retrospectively to obtain published policy texts related to cultivated land protection. To ensure that the information content of the policy is consistent with the theme, and that the texts are accurate, the following principles are followed: (1) the main content or part of the policy is directly related to cultivated land protection; (2) the nature of this policy is legislative documents within the scope of the law or other normative documents, such as binding opinions, measures, and notices formulated by the Party Central Committee, the State Council, and other state organizations.

4. Data analysis

4.1. Open Coding

Open coding is a process of cutting and giving concepts to words, sentences, paragraphs, and the whole text in the original materials. This process requires that the original materials be systematically classified, but the hidden concepts behind the materials be deeply excavated and combined with the research situation for an in-depth analysis of research problems (Figure 3). This method explores the key concepts hidden behind the original materials in the context of this research. For the open coding, the 71 data texts are read word by word followed by the principles of coding independence, openness, temporality, and revisability, which are constantly compared and summarized, and 94 initial concepts are condensed. According to the connotation and extension of each initial concept in the research context, 45 categories are formed, as shown in Table 1.

Table 1. Open coding and categorization.

Original Data	Labeling	Conceptualization	Categorization
In November 2013, when President Xi Jinping visited Shandong province, he said, "we should add wings to science and technology in agriculture and lay emphasis on increasing production and efficiency, combining good seed and good law, combining agricultural machinery with agronomy, and coordinating production ecology. We should promote the integration of agricultural technology, mechanization of labor processes, production and operation informatization, legalization of safety and environmental protection, and speed up construction of the technical systems required by the development of safe agriculture with high yields, high quality, high efficiency, and ecology."	p1 We should speed up the construction of technology systems to meet the requirements of high yield, high quality, high efficiency, ecological, and safe agricultural development.	P1 High yield, high quality, high efficiency, ecological, and safe	PP1 Technical systems
In December 2013, President Xi Jinping delivered a speech at the central rural work conference, "The fundamental guarantee for national food security is cultivated land, and it is the lifeblood of grain production." Farmers can be non-agricultural, but cultivated land cannot be non-agricultural. If the cultivated land is not farmed, we will have no land to live on.	p2 Farmers can be non-agricultural, but cultivated land cannot be non-agricultural.	P2 Cultivated land cannot be non-agricultural.	PP2 Lifeblood

Table 1. *Cont.*

Original Data	Labeling	Conceptualization	Categorization
Circular on strengthening land management and stopping unauthorized occupation of cultivated land (the CPC Central Committee and the State Council,1986) suggested initiatives to "strengthen land management and resolutely stop the illegal occupation and abuse of cultivated land," urgent circular on forbidding development zones and urban construction from occupying cultivated land and abandoning it (General Office of the State Council, 1992); circular on strengthening the management of various development zones (General Office of the State Council, 2003); urgent circular on suspending examination and approval of various development zones. The General Office of the State Council,2003, put forward "strictly control the loss of cultivated land"; circular on "no building houses in rural areas" (Ministry of Natural Resources, Ministry of Agriculture and Rural Areas, 2020)	a6 Strengthen land management and resolutely stop the illegal occupation and abuse of cultivated land.	A6 Unauthorized occupation of cultivated land is banned	AA3 Management control (A6, A10, A21, A72, A73)
Circular on resolutely stopping the "non agriculturalization" of cultivated land The General Office of the State Council proposed that the permanent basic cultivated land that has been included in the core reserve of nature reserves should be included in the ecological conversion and be withdrawn in an orderly manner.	a82 Ecological returning of cultivated land	A82 Ecological returning of cultivated land	AA37 Balance and coordination

4.2. Axial Coding

The principal axial coding aims to analyze each category in the context of research and socio-cultural background. It not only eliminates the gap between theory and practice, but also improves the explanatory power of theory to social phenomena or behaviors. The connotations of the relationship between the main category and primary category are shown in Table 2.

Based on the qualitative analysis of NVivo12.0 software, this paper adapts human–computer cooperation to manage and code the data texts of CLPP, which will guarantee the reliability and consistency of the coding reach the qualified level. There are four basic problems that have been refined during literature review and study of policy texts, but the systematic research framework has not been set up before the coding process. The problems are as follow: What is the social foundation of CLPP? What is the institutional environment of its development? What kind of governance structure has been formed by CLPP? How does the policy system affect the allocation of cultivated land resources? On the basis of theoretical analysis and comparison of policy text materials, the coding analysis of the original data has been completed. For the principal axial coding process (Figure 3), this paper forms 22 main categories, including external environment, governance philosophy, internal conditions, stakeholders participation, rice bowl theory, red-line consciousness, bottom-line thinking, institutional rules, system construction, propaganda and guidance, support systems, internal core, external concurrence, composite space, new strategy, capacity increase, peasants' subject positions, sustainability, pattern optimization, reform and innovation, improve the system, and ecological efficiency. Based on the dimension situation–structure–motivation–action–space–outcome, we define the logical relationship between primary categories in Table 2.

Table 2. Open coding and categorization.

Dimension	Main Category	Primary Category	Connotation
Situation	External environment	AA1 Consciousness awakening	Farmers have their own land; there is clear ownership of rural land property rights.
		AA2 Clear concept	We should establish the basic national policy of cultivated land protection and pay attention to the protection and rational use of cultivated land.
	Governance philosophy	PP3 Panda theory	Cultivated land is the most valuable resource in China, and as vital as the protection of the giant panda.
	Internal conditions	AA20 Production environment	It is necessary to classify soil organic matter and improve soil quality.
		AA22 Factor input	New agricultural inputs, such as new fertilizers, low toxicity and high efficiency pesticides, multi-functional agricultural machinery, and degradable agricultural film should be developed.
Structure	Stakeholders participate	AA34 Multiple stakeholders participate	Encourage government and social capital cooperation (PPP) mode, guide rural collective economic organizations, farmers, and new agricultural operators.
Motivation	Rice bowl theory	PP4 Keeps the rice bowl	Chinese people need to put their rice bowls in their own hands and hold their own food.
	Red-line consciousness	PP5 Keep red line	Keep the red line of cultivated land protection firm.
	Bottom-line thinking	PP2 Lifeblood	Cultivated land is the lifeblood of food production.
		AA27 Bottom line thinking	We should stick to the four bottom lines: no change in the nature of public ownership of land, no breaching the red line of cultivated land, no reduction in grain production capacity, and no damage to farmers' interests.
Action	Institutional rules	AA3 Management control	Using mandatory policy tools to strengthen land management, stop unauthorized occupation of cultivated land, stop "non-agricultural" cultivated land use.
		AA4 Constraint incentive	Permanent basic cultivated land protection, spatial planning, three-line delineation.
		AA6 Command control	Strictly control incremental and classified management, economical and intensive utilization of land use.
		AA8 Technology	Topsoil stripping, national cultivated land reserve resources survey and evaluation, land and resources remote sensing monitoring "one map", conservation tillage.
		AA10 Supervision and inspection	Supervision and assessment, local government responsibility, and natural resources supervision.
		AA11 Land use regulation	Land use regulation institution.
		AA12 Law responsibility	The crime of destroying cultivated land should be established.
		AA13 Index control	Cultivated land index, construction land index, agricultural land conversion index.
		AA18 Economic measure	Land reclamation fees, cultivated land occupation tax
	System construction	AA14 Broaden channels and control total amount	Quality improvement, combination of compensation and improvement, and improvement from drought to water; attract social capital and financial capital to participate in land consolidation and high standard cultivated land construction.
		AA16 Capacity reserve	A reserve of cultivated land quantity, paddy field, and production capacity should be established.
		AA25 Land consolidation+	Relying on the cultivated land protection mechanism driven by land consolidation technology innovation, a land consolidation mechanism dominated by the government, dominated by farmers and participated by the society will be formed.

Table 2. *Cont.*

Dimension	Main Category	Primary Category	Connotation
	Propaganda and guidance	AA5 Propaganda and guidance	Strengthen the propaganda of cultivated land protection.
	Supporting system	AA7 Balance of occupation and compensation	Land requisition and compensation.
		AA17 Joint responsibility system	Responsibility system of cultivated land protection, off-office auditing of cadres.
		AA24 Economic compensation incentive	Compensation mechanism of cultivated land protection.
		AA29 Ecological compensation	Ecological compensation system for forest, grassland, wetland, and soil and water conservation.
Space	Internal core	AA9 Quantity security	Cultivated land reserved.
		AA23 Ecological elasticity	Ecosystem protection of cultivated land.
		AA26 Quality tapping potential	High standard construction of basic cultivated land, prevention, control, and remediation of heavy metal contaminated cultivated land.
	External concurrence	AA30 Rotation fallow	Rotation and fallow of cultivated land.
		AA32 Game competition	Construction occupation and agricultural structure adjustment.
		AA37 Trade-off coordination	Ecological deterioration, disaster-damaged area, contaminated zone.
	Composite space	AA35 Key protected areas	Grain production function zone and important agricultural product production protection zone.
	New strategy	AA19 Rural revitalization strategy	All the income from adjustment is used to consolidate the achievements of poverty alleviation and support rural revitalization.
		AA21 International trade adjustment	Make use of the international agricultural product market and agricultural resources to effectively adjust and supplement the domestic food supply.
		AA28 Urban-rural integration	Break the institutional barriers that hinder the free flow and equal exchange of urban and rural elements.
Outcome	Capacity increase	AA15 Comprehensive production capacity	Steadily improve the comprehensive grain production capacity.
	Peasants' subject positions	AA31 Peasantry's inclination	Respect peasantry's inclination and implement them in a safe and orderly manner.
		AA33 Farmers' interests	Farmers' interests will not be damaged.
	Sustainability	AA36 Continue increasing productivity	Comprehensively enhance the capacity of sustainable yield increase of cultivated land.
	Pattern optimization	AA38 Quantity-ecological- quality all in one	New pattern of special protection of permanent basic cultivated land with strong protection, intensive and efficient management, and strict supervision.
	Reform and innovation	AA39 Improve means	Improve the balance in the management of cultivated land occupation and compensation.
	Improve the system	AA40 Policy system	The permanent basic cultivated land management and control system, cultivated land protection system, and balance policy system need to be further improved.
	Ecological efficiency	PP1 Technical system	Accelerate the construction of the technology system to meet the requirements of high yield, high quality, high efficiency, ecological, and safe agricultural development.

4.3. Selective Coding

After continuous comparison of the main categories (Figure 3), it is clear that the core category of this paper is "the operational logic of CLPP". Around this core category, we derive six dimensions of situation, structure, motivation, action, space, and outcome. This paper defines theory as the theoretical model of cultivated land protection operation logic, as shown in Figure 4.

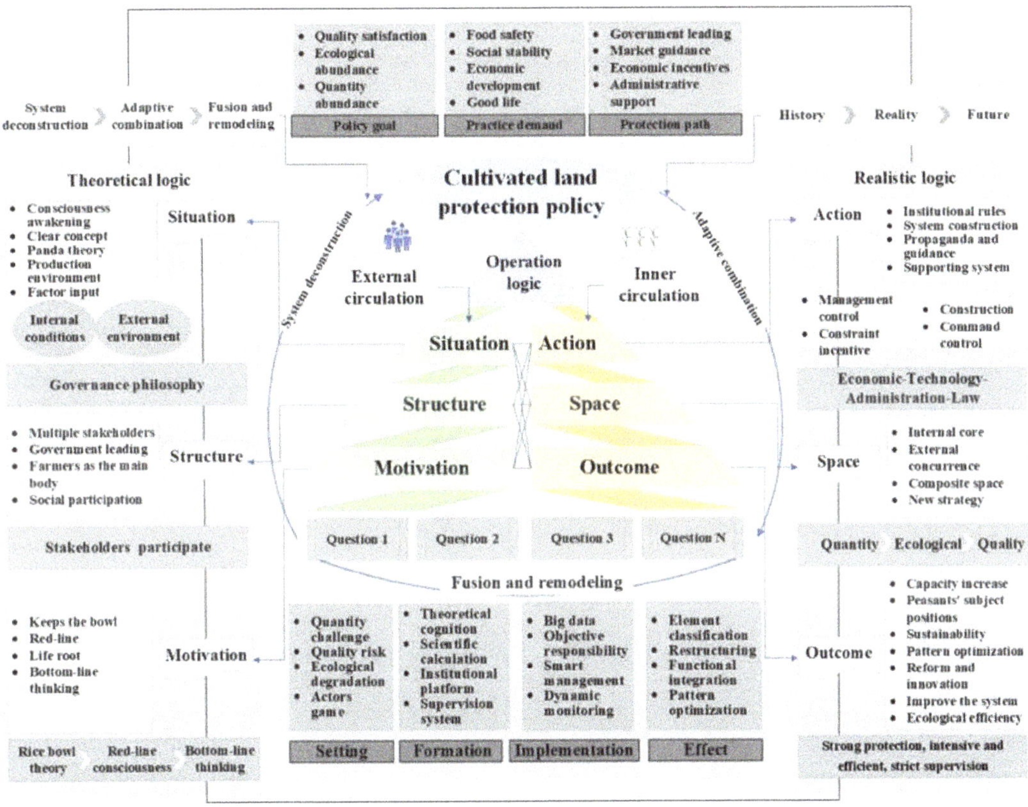

Figure 4. The theoretical model of the CLPP operation logic in China.

Figure 4 shows that "Situation" refers to all types of environment in the CLPP process, including natural, social, economic, and other institutional environments; "Structure" refers to the allocation of resources by different actors in the process of cultivated land protection, and the structural relationship between different actors will also affect the process of policy operation; "Motivation" refers to the goal tendency or internal driving force of cultivated land protection policymaking; "Action" refers to the selection and use of relevant policy tools by relevant stakeholders in the process of cultivated land protection; "Space" refers to the spatial division of CLPP implementation; and "Outcome" refers to the specific efficiency or result of CLPP implementation.

4.4. Theoretical Saturation Test

Theory saturation means that the newly collected data cannot make a new contribution to the theoretical construction, which is used to identify the decision to stop sampling (Figure 3). We recoded and classified the remaining 20 CLPP texts and found no new concepts and categories and no new changes in the relationship between concepts and categories. This shows that the theoretical model shown in Figure 4 has passed the theoretical saturation test and has strong practical explanatory power.

5. Discussion

The cultivated land protection system has been implemented for more than 40 years in China. While it has achieved positive results, there are also some new contradictions and problems. To further promote the theoretical and practical innovation of cultivated land protection systems, problem awareness and goal orientation should be intensified, and there should be increased efforts to solve the challenges and difficulties of the new era. A breakthrough in theoretical innovation and the research on the essential stipulation of objectives and directions, thinking and ideas, elements and structures, mechanisms, and paths is required. Relying on regional cultivated land resource endowments integrates sustainable development, green development, and other elements in ideas, concepts, and implications. We emphasize the need to change "two-dimensional protection" into "three-dimensional protection", "passive protection" into "active protection", and "external push-pull" into an "endogenous driving force". The mechanisms and path of cultivated land protection aim to achieve sustainable development and food security at the macro level, coordinated regional development and flexible space of cultivated land protection at the meso level, and innovation of cultivated land use and stimulation of endogenous power at the micro level. From a practice-deepening perspective, simplistic, representational, and sportive policy implementation should be avoided in favor of building a dynamic and balanced virtuous circle mechanism using "government and market" and "responsibility and incentives". The combination of "macro–meso–micro" common direction, which contains the characteristics green, sustainability, initiative, prevention, and control, will span the whole process of cultivated land protection and be cast in the work of coordinating the spatial connection between the quantity structure and quality toughness of cultivated land. Tunneling through the environmental Kuznets curve should guide the conceptualization of the essential provisions for constructing long-term mechanisms for cultivated land resource use and protection. The following is mainly based on the subject–action–space dimension that explains the operation logic of cultivated land protection in China.

5.1. Stakeholders Logic

Under the complex institutional change environment from agricultural civilization to industrial civilization and then to ecological civilization, and from a planned economy to a market economy, the main structure of cultivated land protection is formed with government leadership. Farmers compose the main body, and there is social participation motivated by red-line consciousness and bottom-line thinking (Figure 4). All parties follow the inclusive interest theory logic of Olsen's collective action in the profit-seeking game. The central government should comprehensively coordinate social, economic, and ecological aspects and formulate relevant norms at the macro level. Local governments face the contradiction between non-agricultural economic benefits and protection against economic costs, which leads to weak implementation. However, farmers are selective when protecting cultivated land based on farming benefits. From the perspective of the logic mechanism of CLPP, the long-term existence of weak actionability and limited thinking and cognition are mutually superimposed, which creates a strong incentive for multiple stakeholders to fall into Olsen's collective action. That is, limited livelihood capital is constrained by the ability to take action, which is difficult to effectively transform in the operations of the socio-economic system but also shows a decreasing trend due to continuous loss. Accordingly, it will also inhibit the enthusiasm of the main protectors and accelerate the reverse cycle between cultivated land protection and actionability. First, the cultivated land protection system has the regulatory dimension of national compulsion and implementation guaranteed by national compulsion. Its highly differentiated organizational structure provides a stable organizational and institutional environment for the implementation of CLPP. The pressure of local financial and political promotion, high-pressure accountability performance appraisal mechanisms, and inefficient policy supervision mechanisms under fiscal and tax decentralization provide an institutional driving force for the local government to block the implementation of CLPP. Second, the weak values in the normative dimension inhibit the public's constraint on the implementation of CLPP, which provides a favorable social normative environment for the generation of a policy implementation block. Moreover, since the reform and opening up of China, the cognitive framework with economic development as the main task prevents local government policy executors from correctly recognizing the importance of cultivated land protection but provides a suitable cognitive environment for the formation of policy implementation blocks.

5.2. Action Logic

The action logic of cultivated land protection is to reasonably intervene in the system operation of cultivated land protection by means of economic and technical tools, laws and regulations, administrative supervision, and other means, on the premise of adhering to the red line of cultivated land and ensuring food security to achieve Pareto optimization of resource elements (Figure 4). This optimizes natural resources, market elements, and social levels and then builds a transaction system oriented by the functional integration of cultivated land use systems if required. Similarly, by focusing on optimizing the spatial patterns of cultivated land protection, a solid foundation could be laid to improve the sustainable production capacity of grain and promote the sustainable and intensive development of cultivated land use. For example, for disaster prevention and control, coupling the motivation to seek advantages and avoid the disadvantages of cultivated land use with agricultural-induced production substitution behaviors and narrowing the income gap between urban and rural areas will reduce the pressure of cultivated land to promote national economic growth. These steps are critical to promoting the components of cultivated land ecosystems from damage to recovery. It is also critical to improve the technical level and efficiency of integrated prevention and control of biological disasters by increasing crop diversification [68]. Specifically, on the one hand, in terms of green development, it is possible to create, construct, and improve green initiatives and conditions of market operations in poor areas by promoting the allocation of production and service resources, such as the biological seed industry, heavy agricultural machinery, smart agriculture, and green inputs. On the other hand, operational mechanisms should be constructed, such as land use regulation, balance of occupation and compensation, constraints and incentives, and economic compensation as a guarantee based on greater visibility of high-quality green development. There should be guidance, encouragement, and support for new agricultural operators who integrate modern agricultural production holistically creating rural collective economic organizations. Farmers and new agricultural business entities should be more active in protecting cultivated land. Furthermore, through these actions, a joint force to protect cultivated land and promote the use and protection of cultivated land resources with high-quality economic and social benefits should be formed.

5.3. Space Logic

Space is the carrier of the subject and action. According to the standard of spatial scale, cultivated land use systems can be divided into internal core, external concurrence, composite space, and new strategy. From quantity protection to both quantity and quality, and then to adhere to the trinity pattern of quantity control, quality management, and ecological management and protection, cultivated land protection in China has gradually formed an internal core space. That core space is composed of a stable quantity safety zone, a quality potential tapping zone, and a sustainable ecological elastic zone. In particular, external competition and cooperation space is composed of a game competition to resolve use conflicts and balance. Coordination emphasizes complementary integration, rotation, and fallow space to emphasize recuperation. The new strategic space is composed of rural revitalization to coordinate the development, an urban–rural integration space emphasizes organic interaction, and a key development space relies on the composite ecosystem (Figure 5). With the continuous progress of ecological civil construction, the space and function of cultivated land face a contradiction of imbalance and insufficiency, which is mainly reflected in the spatial conflict between the pollution and damage of cultivated land and the conflict between the production function and ecological service function of cultivated land. For instance, the excessive application of chemical fertilizers and pesticides leads to three-dimensional pollution of soil and water biogenesis, the mismatch between single-crop cultivation and multi-dimensional ecological water resources efficient use mode, and the excessive neglect of the role of cultivated land forest network construction. In the future, under the requirements of ecological civilization construction, there will be 25.13 million ha of cultivated land, accounting for 18.5% of the total cultivated land in China [41]. This land must be managed by adjusting to local conditions and using fallow rotation to reconcile regional ecological risks. Therefore, we should internalize the positive externality of the internal core space through intensive management and ecological production; shape the external competition and cooperation space of population, resources, environment, economy, ecology, and policy; and expand the new strategic space of resource-saving, environment-friendly, and ecological conservation efforts.

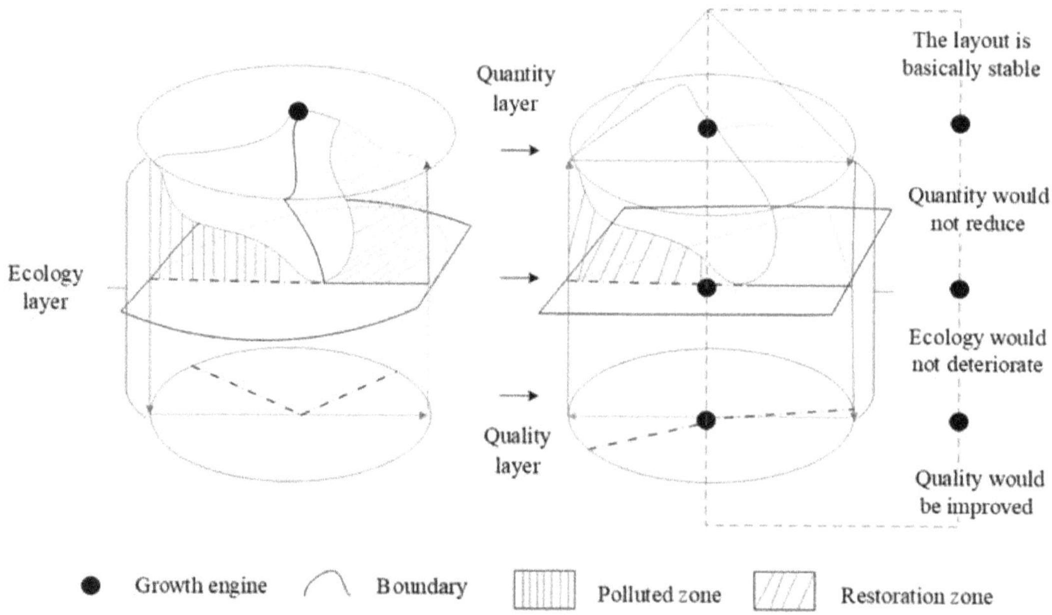

Figure 5. The theoretical model of the CLPP operation logic in China.

5.4. Systematic Integration Logic of Stakeholders–Action–Space

From the perspective of the occurrence mechanism, the three dimensions of subject, action, and space are interwoven, embedded, and coupled in each link of the input–conversion–output–feedback in CLPP operation. In the internal core space, external competition, cooperation space, and new strategic space are the ultimate goals of cultivated land protection. The logical starting point is the relationship between the types and structures of cultivated land use in different spaces, which sometimes evolve into mutually restraining variables (Figure 4). For example, having more organic carbon and nitrogen in the internal core space can usually increase food production and help achieve the goals of food security and climate change mitigation [2]. However, nitrogen management of cultivated land in the external competition and cooperation space and the new strategic space becomes particularly complex because the natural endowment of regions is geographically different, and various types of soil will convert part of the input nitrogen into N_2O, thus increasing the burden of greenhouse gases [4]. The main body of cultivated land use in different spaces chooses the corresponding management behavior based on their cognition, and the cultivated land protection action formed in this process is not unified but is closer to the trade-off synergy of spatio-temporal differentiation. Therefore, the Chinese government has begun to implement a series of measures to promote the transformation of China's grain production. With these measures, the objectives of cultivated land protection are expanded. One measure is to ensure the safety of agricultural product supply and the other is to protect agricultural resources and the ecological environment. Therefore, the integrated logic of subject–action–space of cultivated land protection is not the cultivated land resource itself but that the external environment is composed of the social system, economic development, and cultural consciousness. It is particularly obvious that the use and protection of cultivated land contains a huge opportunity cost, and responsibility and incentives are an indispensable regulatory measure to guide and regulate the behavior of new agricultural operators. At the same time, ecological governance and comprehensive compensation mechanisms play a substantial role in balancing the relationship between market supply and demand, the scarcity of agricultural products, and the benefits of ecological restoration [21]. We should explore the third-party governance model represented by social capital, particularly agricultural machinery services for small farmers and the organization and management of agricultural enterprises.

In short, there is a complex relationship between the multiple stakeholders of cultivated land use and protection (Figure 6). Their actions are mainly affected by the external factors of cultivated land protection. The conflict and bridging of different spaces have become an inexhaustible driving force for the development of a cultivated land protection system. In the future, the spatial distribution of cultivated land should be optimized according to the light and heat suitability of soil and water to agricultural production to highlight the targeted effect of green management of cultivated land. Based on the mutual coupling of various spatial types, ecological management measures should be implemented accurately to achieve the status of a basically stable layout, no reduction in quantity, no degradation of ecology, and improvements in quality to accelerate the spatial layout of cultivated land quantity, quality, and ecology in a coordinated and sustainable development direction. Meanwhile, the green power of multiple stakeholders, the adjustment of actions, spatial resilience, and multi-functional integration should be strengthened. The main responsibility of cultivated land protection is overall protection, system restoration, and comprehensive management. In addition, it is essential to integrate big data, artificial intelligence, and other modern technologies with operation processes to provide technical support for the information supply of sustainable use of cultivated land and reshape the cultivated land protection system in the new era.

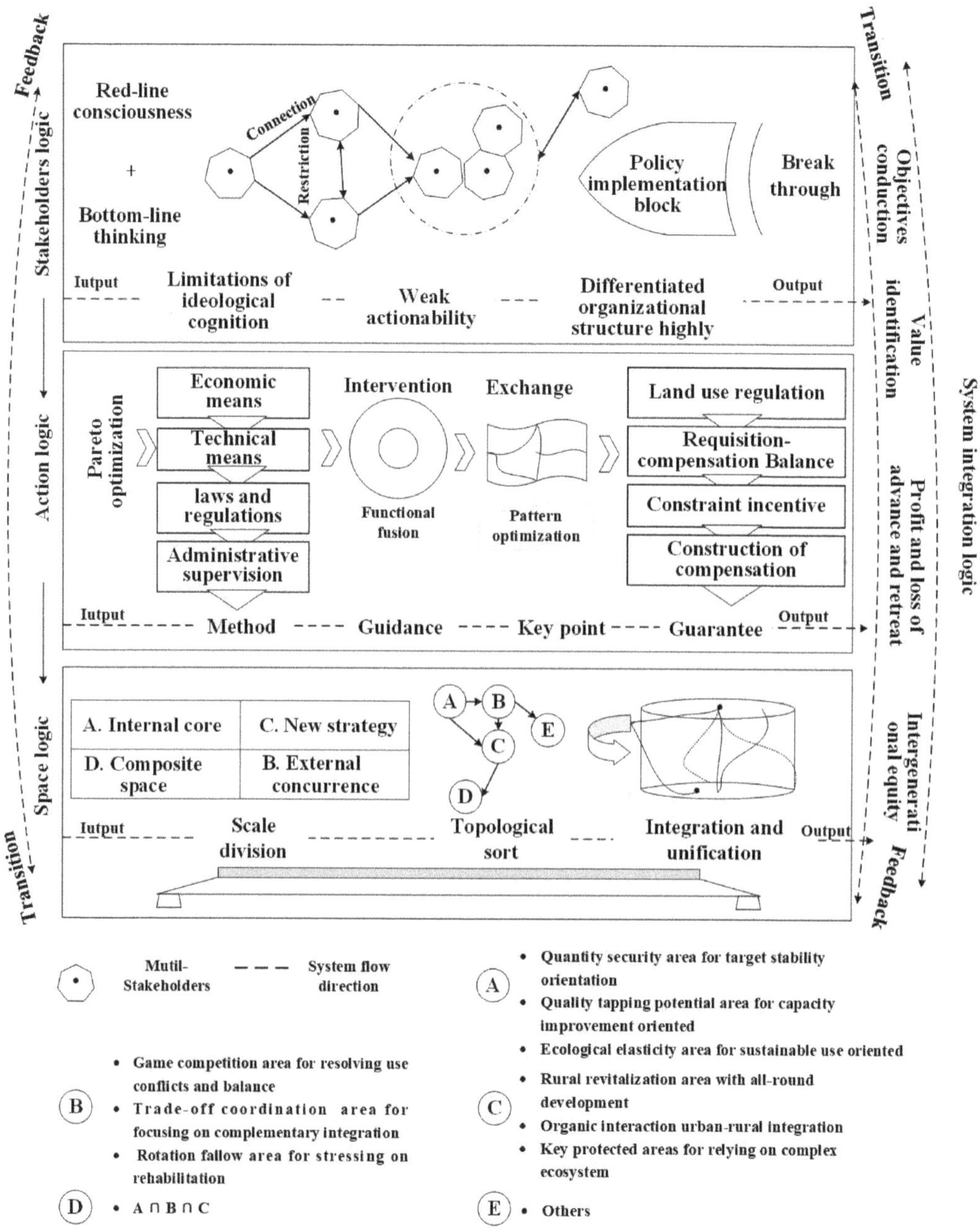

Figure 6. The theoretical model of the CLPP operation logic in China.

6. Conclusions

This paper takes the policy texts of CLPP as the research object by grounded theory and analyzes the major dilemmas and operation logic of CLPP. We show that the logic operation of the CLPP is based on situation–structure–motivation–action–space–outcome. More attention should be paid to the logical operation of the CLPP in China, which we explain in detail by analyzing the relationship among stakeholders, action, and space. The research conclusions help to identify the mechanisms of CLPP and to clarify the factors affecting the implementation of protection behavior. The research conclusion also sheds light on the obstacles faced by cultivated land protection in the process of ecological transformation and can guide improvements in protective measures. The key findings are as follows:

(1) The basic logic of the CLPP operation is to take the ultimate goal of cultivated land protection as the logical starting point and red-line consciousness and bottom-line thinking as the motivation. Based on a structure of government leadership with farmers as the main body and with social participation, this policy takes Pareto optimization of resource elements as the main direction. Multiple measures, such as the economy, technology, laws and regulations, and administrative supervision stimulate the functional integration of cultivated land use system. Then, relying on the internal core space, external competition and cooperation space, composite space, and new strategic space, the spatial pattern of cultivated land protection is optimized. The three dimensions of subject, action, and space are intertwined, embedded, and coupled in the input–conversion–output–feedback, and the conflict and bridging of different spaces become the inexhaustible driving force for the development of a cultivated land protection system. Therefore, we believe that the key to guaranteeing the effectiveness of CLPP in the future lies in solving the contradiction between theoretical abstraction and practical execution. Accordingly, we should distinguish the policy types and implementation methods of command control, economic incentive, and publicity guidance. In different stages of economic and social development, the optimization and combination of multiple policy tools should be reasonably used to ensure the effect of cultivated land protection. Moreover, in order to reduce the negative externality of cultivated land occupation, we should appropriately increase the comprehensive cost of converting cultivated land into construction land, and improve the efficiency of optimal allocation of land resources through land marketization measures. At the same time, land marketization measures should also be taken to improve the efficiency of optimal allocation of land resources.

(2) CLPP is a comprehensive system of human development and natural protection information, which integrates administration, the economy, technology, and culture. In the practice of national agricultural regionalization protection, the theory of cultivated land use and protection is consolidated. The CLPP continues to maintain continuity, stability, and sustainability, and plays a supporting role in China's socialist modernization. The value and importance of CLPP in this era are reflected in the practice of the new development stage, new development concepts, and new development patterns of cultivated land protection. The completion of the goals and tasks of cultivated land protection does not mean the end of the system, but that China will continue to implement the world's most stringent cultivated land protection system. The evolution process of CLPP is the result of the game of multiple stakeholders, which shows significant path dependence characteristics. Therefore, how to use policy implementation to effectively improve the self-enthusiasm of stakeholders has become the key to the innovation of cultivated land protection system in the future. In particular, we should find a safety coefficient interval to balance the cultivated land protection and construction needs of CLPP, and coordinate the interest demands and bureaucratic structure of different subjects. Some pension policies, low interest loan policies, preferential taxes, and other policy compensation should be explored in the institutional framework of cultivated land protection. In addition, we should strengthen agricultural production technology, agricultural product marketing, and other supporting measures to improve the enthusiasm of agricultural managers.

(3) CLPP should be based on the connotation of cultivated land and its protection objectives, and then implement adaptive governance for different forms of cultivated land use. Some factors such as the allocation of land use indicators and their marketization should also be fully considered to ensure the authority and applicability of the

policy. Simultaneously, we should promote the legislation of cultivated land protection from the aspects of legal concept, control methods, compensation means, target responsibility, which will be beneficial to improve the systematization and integrity of the legal system related to cultivated land protection. Furthermore, the cultivated land protection system needs to cope with the transformation of cultivated land use brought about by climate change, smart agriculture, and food system transformation, and it must become more inclusive and sustainable in the process of ecological governance. The system can support the higher productivity levels of economic growth, such as sustainable intensification of cultivated land use. Scientific and technological innovation and technology integration play various roles in the implementation of cultivated land protection systems, which can create extensive efficiency. In addition, accurate assessment of human needs, seed quality, cultivated soil, and agricultural product trading will be the basis for effective protection of cultivated land.

Author Contributions: Conceptualization, S.N. and X.L.; methodology, S.N. and G.G.; software, S.N. and G.G.; formal analysis, X.L.; investigation, S.N.; data curation, S.N.; writing—original draft preparation, S.N. and X.L.; writing—review and editing, S.N. and X.L.; visualization, S.N. and G.G.; All authors have read and agreed to the published version of the manuscript.

Funding: This paper won the support of the National Natural Science Foundation of China (Approval Number: 42071226). Fundamental Research Funds for the Central Universities (Grant/Award Number: N2214001, N2114006).

Data Availability Statement: Data supporting the results of this study can be obtained by contacting the authors.

Conflicts of Interest: The authors declare no conflict of interest.

References

1. Gottero, E. Identifying vulnerable farmland: An index to capture high urbanisation risk areas. *Ecol. Indic.* **2019**, *98*, 61–67. [CrossRef]
2. Dorward, A.R. Agricultural labour productivity, food prices and sustainable development impacts and indicators. *Food Policy* **2013**, *39*, 40–50. [CrossRef]
3. Linner, H.; Messing, I. Agricultural land needs protection. *Acta Agric. Scand.* **2012**, *62*, 706–710. [CrossRef]
4. Greiner, R. Optimal farm management responses to emerging soil salinisation in a dryland catchment in eastern Australia. *Land Degrad. Dev.* **2015**, *8*, 71–93. [CrossRef]
5. Ye, S.; Song, C.; Shen, S.; Gao, P.; Zhu, D. Spatial pattern of arable land-use intensity in china. *Land Use Policy* **2020**, *99*, 104845. [CrossRef]
6. Sokolow, A.D. Federal policy for preserving farmland: The farm and ranch lands protection program. *Publius J. Fed.* **2010**, *40*, 235–256. [CrossRef]
7. West, P.C.; Gerber, J.S.; Engstrom, P.M.; Mueller, N.D.; Brauman, K.A.; Carlson, K.M.; Cassidy, E.S.; Johnston, M.; MacDonald, G.K.; Ray, D.K.; et al. Leverage points for improving global food security and the environment. *Science* **2014**, *345*, 325–328. [CrossRef]
8. Zabala, A. Land and food security. *Nat. Sustain.* **2018**, *1*, 335. [CrossRef]
9. Bunce, M. Thirty years of farmland preservation in North America: Discourses and ideologies of a movement. *J. Rural. Stud.* **1998**, *14*, 233–247. [CrossRef]
10. Klein, L.R.; Hendrix, W.G.; Lohr, V.I.; Kaytes, J.B.; Sayler, R.D.; Swanson, M.E.; Elliot, W.J.; Reganold, J.P. Linking ecology and aesthetics in sustainable agricultural landscapes: Lessons from the Palouse region of Washington, USA. *Landsc. Urban Plan.* **2015**, *134*, 195–209. [CrossRef]
11. Magnan, A. The financialization of agri-food in Canada and Australia: Corporate farmland and farm ownership in the grains and oilseed sector. *J. Rural. Stud.* **2015**, *41*, 1–12. [CrossRef]
12. Newman, L.; Powell, L.J.; Wittman, H. Landscapes of food production in agriburbia: Farmland protection and local food movements in British Columbia. *J. Rural. Stud.* **2015**, *39*, 99–110. [CrossRef]
13. Nixon, D.V.; Newman, L. The efficacy and politics of farmland preservation through land use regulation: Changes in southwest British Columbia's Agricultural Land Reserve. *Land Use Policy* **2016**, *59*, 227–240. [CrossRef]
14. Zangue, Y.D.; Melot, R.; Martin, P. Diversity of farmland management practices (FMP) and their nexus to environment: A review. *J. Environ. Manag.* **2022**, *302*, 114059. [CrossRef]
15. Nitsch, H.; Osterburg, B.; Roggendorf, W.; Laggner, B. Cross compliance and the protection of grassland–Illustrative analyses of land use transitions between permanent grassland and arable land in German regions. *Land Use Policy* **2012**, *29*, 440–448. [CrossRef]
16. Hashiguchi, T. Japan's Agricultural Policies after World War II: Agricultural Land Use Policies and Problems. In *Social-Ecological Restoration in Paddy-Dominated Landscapes*; Springer: Tokyo, Japan, 2014. [CrossRef]

17. Choi, G.H.; Lee, D.Y.; Jeong, D.K.; Kuppusamy, S.; Lee, Y.B.; Park, B.J.; Kim, J.H. Perfluorooctanoic acid (PFOA) and perfluorooctanesulfonic acid (PFOS) concen-trations in the South Korean agricultural environment: A national survey. *J. Integr. Agric.* **2017**, *16*, 1841–1851. [CrossRef]
18. Li, H.; Wu, Y.; Huang, X.; Sloan, M.; Skitmore, M. Spatial-temporal evolution and classification of marginalization of cultivated land in the process of urbanization. *Habitat Int.* **2017**, *61*, 1–8. [CrossRef]
19. Gu, B.; Zhang, X.; Bai, X.; Fu, B.; Chen, D. Four steps to food security for swelling cities. *Nature* **2019**, *566*, 31–33. [CrossRef]
20. Liu, Y.; Qiao, L. Innovating system and policy of arable land conservation under the new-type urbanization in China. *Econ. Geogr.* **2014**, *34*, 1–6.
21. Kong, X.B. The Connotation and realization path of ecological governance of cultivated land protection in China. *China Land Sci.* **2020**, *36*, 1–11.
22. Han, H.; Zhang, X. Exploring environmental efficiency and total factor productivity of cultivated land use in China. *Sci. Total Environ.* **2020**, *726*, 138434. [CrossRef] [PubMed]
23. Wang, Y.; Jiang, Z.; Wang, J. The differences of institutional performance of farmland protection in China. *J. Public Manag.* **2010**, *7*, 21–30+123.
24. Tan, S.; Zhang, H. Performance evaluation on the policies of cultivated land protection in China from the perspective of quantity protection. *China Popul. Resour. Environ.* **2010**, *20*, 153–158.
25. Chen, M.; Liu, T. Improving the effectiveness of the protection of farmland in the new era of China. *Res. Agric. Mod.* **2018**, *39*, 1–8.
26. Shen, J.; Li, C.; Mi, G.; Li, L.; Yuan, L.; Jiang, R.; Zhang, F. Maximizing root/rhizosphere efficiency to improve crop productivity and nutrient use efficiency in intensive agriculture of China. *J. Exp. Bot.* **2013**, *64*, 1181–1192. [CrossRef] [PubMed]
27. Zhang, H.; Tan, M.; Kong, X.; Xu, Y.; Xu, E.; Shang, E. Research on the strategy for improving cultivated land quality in China. *Strateg. Study CAE* **2018**, *20*, 16–22. [CrossRef]
28. Lu, X.; Shi, Y.; Chen, C.; Yu, M. Monitoring cropland transition and its impact on ecosystem services value in developed regions of China: A case study of Jiangsu province. *Land Use Policy* **2017**, *69*, 25–40. [CrossRef]
29. Lu, X.; Qu, Y.; Sun, P.; Yu, W.; Peng, W. Green transition of cultivated land use in the Yellow River Basin: A perspective of green utilization efficiency evaluation. *Land* **2020**, *9*, 475. [CrossRef]
30. Liu, X.; Zhao, C.; Song, W. Review of the evolution of cultivated land protection policies in the period following China's reform and liberalization. *Land Use Policy* **2017**, *67*, 660–669. [CrossRef]
31. Xie, H.; Huang, Y.; Choi, Y.; Shi, J. Evaluating the sustainable intensification of cultivated land use based on emergy analysis. *Technol. Forecast. Soc. Chang.* **2020**, *165*, 120449. [CrossRef]
32. Chen, Z.G.; Zhang, X.; Huang, X.J.; Chen, Y. Influence of government leaders' localization on farmland conversion in Chinese cities: A "sense of place" perspective. *Cities* **2019**, *90*, 74–87. [CrossRef]
33. Ustaoglu, E.; Williams, B. Determinants of urban expansion and agricultural land conversion in 25 EU countries. *Environ. Manag.* **2017**, *60*, 717–746. [CrossRef] [PubMed]
34. Li, S.; Li, X. Progress and prospect on farmland abandonment. *Acta Geogr. Sin.* **2016**, *71*, 370–389.
35. Lyu, X.; Niu, S.; Li, Z.; Huang, X.; Zhong, T. Present situation and trends in research on cultivated land intensive use in China. *Trans. Chin. Soc. Agric. Eng.* **2015**, *31*, 212–224.
36. Xiao, Y.; Wu, X.; Wang, L.; Liang, J. Optimal farmland conversion in China under double restraints of economic growth and resource protection. *J. Clean. Prod.* **2017**, *142*, 524–537. [CrossRef]
37. Zuo, L.; Zhang, Z.; Carlson, K.M.; MacDonald, G.K.; Brauman, K.A.; Liu, Y.; Zhang, W.; Zhang, H.; Wu, W.; Zhao, X.; et al. Progress towards sustainable intensification in China challenged by land-use change. *Nat. Sustain.* **2018**, *1*, 304–313. [CrossRef]
38. Niu, S.; Fang, B. Cultivated land protection system in China from 1949 to 2019: Historical evolution, realistic origin exploration and path optimization. *China Land Sci.* **2019**, *33*, 1–12.
39. Niu, S.; Lu, X.; Gu, G.; Zhou, X.; Peng, W. Sustainable intensification of cultivated land use and its influencing factors at the farming household scale: A case study of Shandong province, China. *Chin. Geogr. Sci.* **2021**, *31*, 109–125. [CrossRef]
40. Liu, D.; Gong, Q.; Yang, W. The evolution of farmland protection policy and optimization path from 1978 to 2018. *Chin. Rural. Econ.* **2018**, *12*, 39–53.
41. Qi, X.; Zang, Z.; Huang, X. The contradiction of cultivated land protection in the new era and its innovative countermeasures. *China Land Sci.* **2018**, *32*, 9–15.
42. Xie, H.; He, Y.; Choi, Y.; Chen, Q.; Cheng, H. Warning of negative effects of land-use changes on ecological security based on GIS. *Sci. Total Environ.* **2020**, *704*, 135427. [CrossRef]
43. Kuang, B.; Han, J.; Lu, X.; Zhang, X.; Fan, X. Quantitative evaluation of China's cultivated land protection policies based on the PMC-Index model. *Land Use Policy* **2020**, *99*, 105062. [CrossRef]
44. Zhong, T.; Huang, X.; Chen, Y. Arable land conversion effects of basic farmland protection policy. *China Popul. Resour. Environ.* **2012**, *22*, 90–95.
45. Guo, L.; Yan, J. Study on the decoupling of cultivated land occupation by construction from economic growth in China. *China Popul. Resour. Environ.* **2007**, *17*, 48–53.
46. Lyu, X.; Huang, X.; Chen, Z.; Tang, J.; Zhao, Y. Evaluation on the grain production performance of the cultivated land protection policy in China. *Resour. Sci.* **2010**, *19*, 82–87.
47. Qu, F.; Chen, J.; Chen, W. Theoretical and empirical study on the land conversion economic driving forces. *J. Nat. Resour.* **2005**, *20*, 231–241.

48. Zhong, T.; Mitchell, B.; Scott, S.; Huang, X.; Li, Y.; Lu, X. Growing centralization in China's farmland protection policy in response to policy failure and related upward-extending unwillingness to protect farmland since 1978. *Environ. Plan. C Politics Space* **2017**, *35*, 1075–1097. [CrossRef]
49. Liu, T.; Liu, H.; Qi, Y. Construction land expansion and cultivated land protection in urbanizing China: Insights from national land surveys, 1996–2006. *Habitat Int.* **2015**, *46*, 13–22. [CrossRef]
50. Yu, F.; Cai, Y. A new insight of cultivated land resource value. *China Land Sci.* **2003**, *17*, 3–9.
51. Chen, H.; Lu, Y. Preservation compensation accounting for farmland based on opportunity costs and Markov chains: A case study on Xuzhou City. *Resour. Sci.* **2015**, *37*, 17–27.
52. Li, X.; Qu, F.; Jiang, H.; Jiang, D. Research on target of Chinese cultivated land protection based on excessive loss measurement and elimination. *China Land Sci.* **2008**, *22*, 4–11.
53. Cao, Z.; Hao, J.; Liang, L. The value accounting of cultivated land resources in Huang-huai-hai Regio. *J. Arid. Land Resour. Environ.* **2009**, *23*, 5–10.
54. Niu, H.; Wang, K. Measurement and analysis of cultivated land protection externalities under different sample schemes. *Resour. Sci.* **2017**, *39*, 1227–1237.
55. Feng, Y. Rural households' cognition of arable land value and protection willingness in economically developed areas of south China—A case study of Panyu district, Guangzhou. *Chin. J. Agric. Resour. Reg. Plan.* **2013**, *34*, 51–57.
56. Song, X.; Wu, Z.; Ouyang, Z. Changes of cultivated land function in China since 1949. *Acta Geogr. Sin.* **2014**, *69*, 435–447.
57. Li, G.; Qiu, D.; Wang, L.; Wang, P. Impacts of difference among livelihood assets on the choice of economic compensation pattern for farmer households farmland protection in Chongqing city. *Acta Geogr. Sin.* **2012**, *67*, 504–515.
58. Fang, B.; Wang, B. Study on value compensation for social responsibility of cultivated land based on the level of regional economic development. *Geogr. Res.* **2011**, *30*, 2247–2258.
59. Yao, L.; Zhao, M.; Xu, T. Social benefits under land conservation policy: A choice experiment for non-market valuation. *Issues Agric. Econ.* **2017**, *38*, 32–40+1.
60. Duan, Y.; Shen, J. Farm land protection regulation in China based on the bargaining among multiple Stakeholders. *Urban Dev. Stud.* **2010**, *17*, 107–112.
61. Zhu, H. The behavior tendency and game relation about protectors of cultivated land resource security in China. *China Popul. Resour. Environ.* **2009**, *19*, 82–87.
62. Xu, H. Cultivated land protection: Game analysis of farmers, local government and central government. *Reform Econ. Syst.* **2011**, *4*, 65–68.
63. Liu, S. The changes in collective land rights system and agricultural performance—A review of research on agricultural land system during China's forty years' reform. *J. Agrotech. Econ.* **2019**, *1*, 4–16.
64. Guo, G.; Wu, Q. On system obstruction of cultivated land protection in China: From the viewpoint of principal—Agent theory. *China Land Sci.* **2008**, *22*, 49–55.
65. Zhou, X.; Li, X.; Song, W.; Kong, X.; Lu, X. Farmland transitions in China: An advocacy coalition approach. *Land* **2021**, *10*, 122. [CrossRef]
66. Zhang, Y.; Zhang, J.H.; Tian, Q.; Liu, Z.H.; Zhang, H.L. Virtual water trade of agricultural products: A new perspective to explore the Belt and Road. *Sci. Total Environ.* **2018**, *622–623*, 988–996. [CrossRef]
67. Jiang, G.; Kong, X.; Zhang, F.; Li, C.; Zheng, H. Analyzing the mechanism of economic compensation for farmland protection. *China Land Sci.* **2009**, *23*, 24–27.
68. Song, G.; Liu, Q.; Wang, Y. Study on compensation mechanism for cultivated land protection in northeast major grain producing areas based on the value of farmland development rights. *China Land Sci.* **2014**, *28*, 58–64.
69. Fang, B.; Qi, X.; Wang, Q. Theoretical framework and estimating the value balance between the occupation and reclamation of cultivated land within China. *China Land Sci.* **2013**, *27*, 12–17.
70. Ou, M.; Wang, K.; Guo, J. Research progress on ecological compensation mechanism of farmland protection. *Res. Agric. Mod.* **2019**, *40*, 357–365.
71. Connelly, L.M. Grounded theory. *Medsurg. Nurs.* **2013**, *22*, 124–127. [CrossRef]
72. Zhang, W.; Zhang, M.; Zhang, W.; Zhou, Q.; Zhang, X. What influences the effectiveness of green logistics policies? A grounded theory analysis. *Sci. Total Environ.* **2020**, *714*, 136731. [CrossRef] [PubMed]

sustainability

Article

Can Land Transfer Promote Agricultural Green Transformation? The Empirical Evidence from China

Guoqun Ma [1,2], Danyang Lv [1], Tuanbiao Jiang [1,2,3,*] and Yuxi Luo [1,2,4,*]

1 School of Economics and Management, Guangxi Normal University, Guilin 541006, China; mgq0709@mailbox.gxnu.edu (G.M.); l2309566100@163.com (D.L.)
2 Pearl River-Xijiang River Economic Belt Development Institute, Guangxi Normal University, Guilin 541004, China
3 Center for Southwest Urban and Regional Development, Guangxi Normal University, Guilin 541004, China
4 Guangxi Key Laboratory of Landscape Resources Conservation and Sustainable Utilization in Lijiang River Basin, Guangxi Normal University, Guilin 541004, China
* Correspondence: 452697151@mailbox.gxnu.edu.cn (T.J.); yluogxnu@mailbox.gxnu.edu.cn (Y.L.)

Abstract: As an important means of farmland policy, whether land transfer can promote agricultural green transformation is worthy of further study; however, related research is relatively rare. Based on the inter-provincial panel data from 2005 to 2020, this paper examines the influence of land transfer on agricultural green transformation and its underlying mechanism by using a two-way fixed effect model and an intermediary effect model. This study reveals significant findings as follows: (1) Land transfer substantially promotes agricultural green transformation. (2) Energy consumption is a major contributor to the growth of agricultural carbon emissions; however, land transfer can mitigate this by reducing energy consumption. (3) Land transfer can promote agricultural green transformation by fostering agricultural technology progress. (4) Further analysis reveals that land transfer in economically developed areas and the southeastern side of the "Hu-Huanyong Line" significantly enhances agricultural green transformation. Based on these findings, this paper suggests promoting land transfer while considering regional differences. Additionally, attention should be directed towards reducing energy consumption and encouraging agricultural technology's progress.

Keywords: land transfer; energy consumption; carbon emissions; agricultural technology's progress

Citation: Ma, G.; Lv, D.; Jiang, T.; Luo, Y. Can Land Transfer Promote Agricultural Green Transformation? The Empirical Evidence from China. *Sustainability* **2023**, *15*, 13570. https://doi.org/10.3390/su151813570

Academic Editors: Lu Zhang, Bing Kuang and Bohan Yang

Received: 9 August 2023
Revised: 4 September 2023
Accepted: 10 September 2023
Published: 11 September 2023

1. Introduction

Global warming poses a significant threat to sustainable economic development, becoming a shared challenge for all nations [1]. Consequently, reducing carbon emissions has garnered substantial attention worldwide [2]. Among the contributors to global carbon emissions, the agricultural sector alone accounts for 14% [3]. To reduce the negative impact of greenhouse gases, it is necessary to reduce carbon emissions from the agricultural sector. In countries such as China, agricultural carbon emission reduction has witnessed heightened attention due to the detrimental impacts of the traditional extensive production model, characterized by high investment and pollution emissions, on the agricultural environment [4]. Since 2005, the Chinese government has introduced land transfer policies and measures, the main purpose of which is to promote large-scale and intensive agricultural production, optimize the allocation of agricultural factors, and improve the agricultural production environment. In addition, land transfers have also been highly valued by other countries. For example, the governments of France and Vietnam use land transfers to improve agricultural management and the utilization efficiency of elements [5,6]. Based on the "Opinions on Innovating Institutional Mechanisms to Promote the Green Development of Agriculture", the "National Strategic Plan for Quality Agriculture (2018-2022)", the "14th Five-Year National Agricultural Green Development Plan", and other relevant documents, it can be believed that the core of agricultural green transformation lies in energy conservation and emission reduction; therefore, the specific performance can be investigated from two dimensions: the agricultural energy consumption and the agricultural carbon emissions. Theoretically, land transfer can reduce the fragmentation of the land and improve the efficiency of mechanical utilization and the use of fossil energy [7], thereby promoting agricultural green transformation. The average annual growth rate of the ratio of land transfer area to total cultivated land area in China from 2005 to 2021 is 13.7% (data sources: National Rural Economic Situation Statistics, China's Rural Management Statistical Annual Report and 2019 Statistical Annual Report on China's Rural Policies and Reforms), while energy consumption and carbon emissions decreased by 7.06% and 7.00%, respectively (data sources: China Energy Statistical Yearbook and China Rural Statistical Yearbook). By reducing agricultural carbon emissions and enhancing the agricultural production environment [8], land transfer becomes a critical driver for China's agricultural green transformation [9].

Land transfer can enhance farmers' work efficiency and factory utilization through moderate-scale operations, thereby promoting agricultural green transformation [10]. However, it is important to note that land transfer does not necessarily guarantee the agricultural green transformation [11]. This is due to the increasing opportunity cost of agricultural production compared to the benefits of farming, leading agricultural producers to pursue non-agricultural industries with higher returns for their own or household income [12]. In situations where labor is insufficient, some agricultural households with unproductive land may choose to abandon their farmland [13], while others may compensate for the lower workforce by increasing chemical inputs and utilizing agricultural machinery and equipment [14]. These practices, although ensuring farmland productivity, can hinder agricultural green transformation [15]. Therefore, it is essential to examine the impact of China's land transfer policy, which the government has implemented and intends to continue implementing for an extended period, on the requirements of the era of agricultural green transformation.

Existing literature primarily focuses on the influence of agricultural production agglomeration [16] and agricultural insurance [17] on agricultural green transformation. Surprisingly, there is limited investigation into the impact of land transfer on agricultural green transformation. Some scholars have pointed out that an imperfect land transfer system and an underdeveloped market contribute to the shift of land from a "carbon sink" to a "carbon source", leading to high carbonization in agricultural development [18]. Additionally, an increased scale of land transfer may hinder the carbon reduction effect associated with expanding land operation scale [19]. Conversely, other scholars argue that land transfer promotes moderate land scales and facilitates agricultural green transformation through scale production and knowledge spillover effects [20]. For example, Hu et al. (2023) and Wang et al. (2023) found that land transfer can significantly inhibit agricultural carbon emissions [5,21]. Gao et al. (2023) analyzed the data of 46 prefectures in Japan based on a structural equation model and found that land transfer could effectively improve the efficiency of land resource allocation, inhibit land abandonment, and contribute to sustainable land development [22]. In fact, land transfer is not only a focal point in agricultural green transformation but also a key driver for agricultural technology progress [23]. Some scholars pointed out that land transfer can promote agricultural technology progress through agricultural organization reform and management mode transformation [24]. For example, Rada et al. (2018) found that land transfer facilitates the adjustment of regional industrial structures and enables efficient land management, thereby promoting agricultural technology progress [25]. Some scholars found that agricultural technology progress can improve energy efficiency [26] and resource utilization rate, which can significantly promote agricultural green transformation [27,28] For example, Yang and Li (2017) found that agricultural technology progress influences the marginal replacement rate between different factors, leading to improvements in mechanized operations, work efficiency, and energy factor utilization, ultimately impacting agricultural green transformation [29]. In addition, there will be regional heterogeneity in the impact of land transfer on agricultural green transformation. For example, Luo et al. (2020) found that the establishment of major grain-producing areas was conducive to reducing agricultural carbon emissions [30]. Using the data of G7 countries, Ibrahim et al. (2023) found that economic growth would intensify carbon emissions [31].

Based on previous studies, it was found that the existing literature mainly has the following shortcomings: (1) Fail to consider the effect of land transfer on agricultural green transformation; (2) Fail to consider the mediating effects of energy consumption and agricultural technology progress; (3) Regional heterogeneity is not taken into account. Theoretically, the agricultural technology progress resulting from land transfer holds several positive externalities that contribute to the agricultural green transformation [32]. Therefore, incorporating land transfer, agricultural technology progress, and agricultural green transformation into the same theoretical framework is essential for further exploring the internal mechanisms by which land transfer impacts agricultural green transformation. In light of this, the primary focus of this paper is on the relationship between land transfer and agricultural green transformation. The paper aims to achieve the following: (1) From the perspective of land transfer, using a two-way fixed effect model to explore its impact on agricultural green transformation; (2) Using the mediation effect model to explore the internal mechanism of land transfer affecting agricultural green transformation; (3) Discussing the Heterogeneity of the impact of land transfer on the agricultural green transformation on both sides of the "Hu-Huanyong line" and economically differentiated regions.

The rest of this paper is structured as follows: Section 2 introduces the Theoretical analysis and research hypothesis; Section 3 presents the Empirical strategy and variable selection; Section 4 reports the Empirical results; Section 5 presents the Discussion and Section 6 presents the Conclusions and policy recommendations.

2. Theoretical Analysis and Research Hypothesis

Land transfer plays a crucial role in transitioning from a small-scale peasant economy to large-scale and intensive modern agriculture, directly impacting agricultural green transformation [21]. Firstly, land transfer reduces land fragmentation, leading to improved efficiency by reducing work loss on small plots [33]. This also facilitates mechanization and increases the adoption and energy utilization rates of agricultural machinery [34], reducing energy consumption. Additionally, it enhances factor utilization efficiency through knowledge spillover effects [5] thereby boosting the potential for land carbon sequestration and reducing agricultural carbon emissions. Secondly, land transfer promotes the large-scale management of cultivated land [7], optimizing resource allocation. Specifically, it encourages some farmers to transition away from the agriculture sector, enabling land consolidation and agglomeration management [35]. This enhances the bargaining power and economic efficiency of land operators, leading to decreased pesticide and fertilizer applications and the efficient use of machinery [36].

Finally, land transfer plays a role in lowering the threshold for adopting green production methods [37] It enhances the utilization efficiency of chemical factors and encourages agricultural producers to embrace green production practices, promoting the agricultural green transformation [38]. One reason for this is that continuous planting, resulting from land transfer, reduces the average cost and increases the economic efficiency of farmers adopting green production methods [8]. Moreover, the land market directs land toward larger agricultural workers with comparative advantages in capital, technology, or labor force [39] This, in turn, promotes their medium- and long-term investments in the agricultural sector [40], facilitating the purchase of new agricultural machinery and equipment, which can reduce energy consumption [41]. Furthermore, continuous improvement in land transfer accelerates the development of agricultural infrastructure and boosts farmers' enthusiasm for participating in agricultural technology training [41]. This, in turn, enhances farmers' environmental awareness and encourages the adoption of low-energy and low-emission input elements, further promoting agricultural green transformation. Based on these points, we propose Hypothesis 1:

Hypothesis 1. *Land transfer can improve machinery utilization efficiency and factory utilization efficiency through large-scale production, thus promoting agricultural green transformation.*

Land transfer plays a crucial role in reducing carbon emissions by directly decreasing energy consumption through large-scale agricultural production and the adoption of green production technology [24]. Specifically, land transfer changes the fragmented production and management mode to a large-scale production mode, which is conducive to the efficient utilization of fossil energy. Mechanized production on contiguous plots, for instance, reduces energy consumption per unit of land, which can reduce agricultural carbon emissions [42]. Furthermore, land transfer encourages the adoption of green technology among large-scale farmers [43]. This enables them to utilize new agricultural machinery and equipment with high energy efficiency, thereby lowering energy consumption [44]. The application of green production technology and agricultural green production modes serves as a demonstration effect, further encouraging other farmers to adopt this high-efficiency, high-profit production approach [43], leading to the reduction of carbon emissions [12]. Based on these points, we propose Hypothesis 2:

Hypothesis 2. *Land transfers will reduce agricultural carbon emissions by reducing energy consumption.*

Land transfer can also have an impact on agricultural transformation through agricultural technology progress [45]. When farmers transition to non-agricultural industries and transfer their land, the issue of farmland fragmentation is addressed, which can facilitate large-scale land management [46]. As a result, efficient agricultural mechanized operations and land management become achievable, subsequently promoting agricultural technology progress. Moreover, as land gradually shifts to large-scale agricultural operators, they are more likely to access agricultural loans and government support [47]. This, in turn, reduces financing constraints and loan difficulties for these operators, encouraging them to adopt new production technology and further promoting agricultural technology progress [48].

Agricultural technology progress will promote agricultural green transformation through ecological production methods [49]. Firstly, agricultural technology progress can improve the efficiency of land management and resource utilization, which help farmers reduce energy consumption while maintaining their original output, thereby reducing carbon emissions [50]. Secondly, agricultural technology progress can improve the allocation of elements in the agricultural industry, increasing the marginal productivity of pesticides, fertilizers, and agricultural film, thereby reducing carbon emissions [51]. Finally, agricultural technology progress can also promote the accumulation of farmers' experience and knowledge, which is conducive to reducing agricultural production costs and ultimately realizing the "green transformation" [52]. Thus, we propose Hypothesis 3:

Hypothesis 3. *Land transfer can lower the barriers to using new technology, promote the application of green technology, and promote agricultural green transformation through this agricultural technology progress.*

3. Empirical Strategy and Variable Selection

Considering that the Chinese government issued the Administrative Measures for the Transfer of Contractual Rural Land Management Rights in 2005, it clearly stipulated the basic principles and methods of land transfer. Therefore, this paper uses the panel data of 30 provincial regions in China from 2005 to 2020 and uses a two-way fixed effect model to verify the above three research assumptions: First, test Hypothesis 1, whether the land transfer can promote the agricultural green transformation; second, test Hypothesis 2, that is, whether the land transfer can affect the agricultural carbon emission through energy consumption; and finally, test Hypothesis 3, that is, whether the agricultural technology progress acts as the intermediary variable that the land transfer affects the agricultural green transformation.

3.1. Measurement Model Construction

In order to explore the possible impact of land transfer on the agricultural green transformation, this paper takes agricultural energy consumption and agricultural carbon emission as explained variables and adopts the two-way fixed effect model for analysis. The measurement model is constructed as follows:

$$ei_{it} = \alpha_0 + \alpha_1 ft_{it} + \alpha_2 X_{it} + \mu_i + v_t + \varepsilon_{it} \quad (1)$$

$$ac_{it} = \alpha_0 + \alpha_1 ft_{it} + \alpha_2 X_{it} + \mu_i + v_t + \varepsilon_{it} \quad (2)$$

where *I* stands for province, *t* represents year; ei_{it} represents agricultural energy consumption; ac_{it} represents agricultural carbon emissions; *ei* and *ac* are explained variables in this paper; ft_{it} represents land transfer and is the core explanatory variable, X_{it} indicates a series of control variables, μ_i indicates regional fixed effect, v_t indicates time fixed effect, and ε_{it} indicates random error items. Considering that the use of fossil energy will also produce agricultural carbon emissions, model (3) is constructed to test the carbon emission effect of energy consumption, as follows:

$$ac_{it} = \beta_0 + \beta_1 ei_{it} + \beta_2 ft_{it} + \beta_3 X_{it} + \mu_i + v_t + \varepsilon_{it} \quad (3)$$

where, β_i (i = 1, 2 ... 3) is the parameter to be estimated, and the other variables are the same as above. Theoretical analysis also shows that agricultural technology progress has a significant intermediary effect in the process of land transfer to promote agricultural green transformation; therefore, this paper builds model (4), model (5), and model (6) to investigate the intermediary role of agricultural technology progress:

$$tc_{it} = \gamma_0 + \theta_1 ft_{it} + \theta_2 X_{it} + \mu_i + v_t + \varepsilon_{it} \quad (4)$$

$$ei_{it} = \gamma_0 + \theta_1 tc_{it} + \theta_2 ft_{it} + \theta_3 X_{it} + \mu_i + v_t + \varepsilon_{it} \quad (5)$$

$$ac_{it} = \gamma_0 + \theta_1 tc_{it} + \theta_2 ft_{it} + \theta_3 X_{it} + \mu_i + v_t + \varepsilon_{it} \quad (6)$$

Among them, tc_{it} represents the progress of agricultural technology and is the mediating variable, and the other variables are the same as above.

3.2. Variable Definitions

3.2.1. Explained Variable

Agricultural green transformation: The key to green transformation lies in energy conservation and emission reduction. Therefore, this paper takes two indicators of agricultural energy consumption and agricultural carbon emissions as alternative variables of agricultural green transformation. The specific calculation method is described as follows:

Agricultural energy consumption (ei): This paper, in accordance with the classification of the China energy statistical yearbook (http://www.zgtjnj.org/navibooklist-n3022013309-1.html, accessed on 1 June 2023), selects the annual report of 17 kinds of fossil energy and, using the standard coal conversion coefficient (Table 1), sums the agricultural energy consumption [52]. At the same time, the measured agricultural energy consumption index is reciprocal; the greater the value, the lower the agricultural energy consumption is.

Table 1. Energy type and standard coal conversion coefficient.

Energy Type	Conversion Coefficient	Energy Type	Conversion Coefficient	Energy Type	Conversion Coefficient
Raw coal	0.7143	Other gas	3.5701	Other coking products	1.3000
Cleaned coal	0.9000	Fuel oil	1.4286	Liquefied petroleum gas	1.7143
Briquettes	0.6000	Crude oil	1.4286	Other washed coal	0.2850
Refinery gas	1.5714	Gasoline	1.4714	Other petroleum products	1.2000
Coke	0.9714	Kerosene	1.4714	Natural gas	13.3000
Coke oven gas	6.1430	Diesel oil	1.4571		

Agricultural carbon emissions (ac): Agricultural carbon emissions are mainly carbon emissions caused by agricultural production and operation activities, including the use of chemical factors, as well as carbon emissions caused by production behaviors [53]. These data are extracted from the "China Rural Statistical Yearbook" (China Rural Statistical Yearbook: https://www.yearbookchina.com/naviBooklist-YMCTJ-0.html, accessed on 1 June 2023) from 2006 to 2021. This paper uses the carbon emission coefficient method (Table 2) to calculate agricultural carbon emissions. In order to facilitate the follow-up analysis, the measured agricultural carbon emission index is counted down; the larger the value, the lower the agricultural carbon emission.

Table 2. Types of carbon sources and carbon emission coefficient.

Carbon Source	Carbon Emission Coefficient	Reference Source
Chemical fertilizer	0.8956 kg $C{\cdot}kg^{-1}$	Oak Ridge National Laboratory
Pesticide	4.9341 kg $C{\cdot}kg^{-1}$	Oak Ridge National Laboratory
Agricultural film	5.1800 kg $C{\cdot}kg^{-1}$	Institute of Resource, Ecosystem, and Environment of Agriculture
Diesel	0.5927 kg $C{\cdot}kg^{-1}$	Intergovernmental Panel on Climate Change
Land tilling	312.60 kg $C{\cdot}hm^{-2}$	College of Agronomy and Biotechnology, China Agricultural University
Irrigation	266.48 kg $C{\cdot}hm^{-2}$	He et al. (2022) [54]

3.2.2. Core Explanatory Variable

Land transfer (ft): the ratio of the total area of household contracted land transfer to the area of household contracted land is measured [55]. Land transfer can promote the continuous production of fragmented land and moderate-scale operations, which shows the green transformation reduction effect on agricultural production.

3.2.3. Mediating Variable

Agricultural technology progress (tc): the capital-labor ratio was used to measure agricultural technology progress. According to the above theoretical analysis, agricultural technology progress is an important variable affecting land transfer and agricultural green transformation. Referring to the practice of Xu et al. (2023), this paper adopts the degree of agricultural capital deepening to measure agricultural technology progress [56] and the perpetual inventory method to estimate the capital stock, whose depreciation rate is 5.42%.

3.2.4. Control Variable

According to the existing research [57–59], this paper selects the following control variables. ① Urbanization (urb): Measured by the proportion of urban population to total population. ② Trade dependency (tra): Characterized as the proportion of total imports and exports of agricultural products in the total agricultural product of the region. ③ Educational attainment (edu): Measured by the average years of schooling of the agricultural labor force. ④ Industrial structure adjustment (ins): Measured by the proportion of the sown area of food crops to the total sown area of crops. Descriptive statistics for each variable are shown in Table 3.

Table 3. Descriptive statistics of variables.

Variable Type	Variable Name	Code	N	Mean	Sd	Min	Max
Explained variable	Agricultural energy consumption	ei	480	0.028	0.049	0.003	0.380
	Agricultural carbon emission	ac	480	0.008	0.012	0.001	0.070
Core explanatory variable	Land transfer	ft	480	0.236	0.179	0.014	0.911
Mediating variable	Agricultural technology progress	tc	480	3.938	6.676	0.036	78.068
Control variable	Urbanization	urb	480	0.549	0.141	0.195	0.896
	Trade dependency	tra	480	0.300	0.360	0.016	1.696
	Educational attainment	edu	480	7.678	0.652	5.459	9.838
	Industrial structure adjustment	ins	480	0.654	0.132	0.328	0.971

Simultaneously, to ensure more robust estimation results, the paper addresses the endogeneity among variables, and the findings are presented in Table 4. The correlation coefficient's maximum value among the variables is 0.836, indicating the absence of multicollinearity issues among them.

Table 4. Coefficient of correlation between variables.

	ei	ac	ft	tc	urb	tra	edu	ins
ei	1.000							
ac	0.836	1.000						
ft	0.039	0.213	1.000					
tc	0.349	0.452	0.433	1.000				
urb	0.142	0.393	0.693	0.500	1.000			
tra	0.096	0.357	0.364	0.080	0.673	1.000		
edu	−0.110	0.068	0.506	0.429	0.668	0.375	1.000	
ins	−0.231	−0.261	−0.050	0.099	−0.044	−0.276	0.073	1.000

3.3. Data Sources

This paper examines data from 30 provinces in China (excluding Tibet, Hong Kong, Macao, and Taiwan) spanning from 2005 to 2020. The variables considered in the research include agricultural energy consumption, agricultural carbon emissions, and agricultural technology progress, which were primarily calculated by the author. Data on land transfer is sourced from "National Rural Economic Situation Statistics" spanning from 2006 to 2021, as well as "China's Rural Management Statistical Annual Report" and "2019 Statistical Annual Report on China's Rural Policies and Reforms in 2006."Data on urbanization is obtained from the "China Statistical Yearbook" (China Statistical Yearbook: http://www.stats.gov.cn/sj/ndsj/, accessed on 1 June 2023) from 2006 to 2021. Data on trade dependency is sourced from the "China Agricultural Yearbook" (http://www.zgtjnj.org/navibooklist-n3022050503-1.html, accessed on 1 June 2023) and the "China Agricultural Trade Development Report" from 2006 to 2021. Data on educational attainment is extracted from the "China Population and Employment Statistical Yearbook" (China Population and Employment Statistical Yearbook: https://www.yearbookchina.com/navibooklist-n3022013208-1.html, accessed on 1 June 2023) from 2006 to 2021. Data on industrial structure adjustment is extracted from the "China Rural Statistical Yearbook" (China Rural Statistical Yearbook: https://www.yearbookchina.com/naviBooklist-YMCTJ-0.html, accessed on 1 June 2023) from 2006 to 2021.

4. Empirical Results

4.1. Investigating the Impact of Land Transfer on the Agricultural Green Transformation: An Empirical Test of Hypothesis 1

This paper employs Stata 15 to perform regressions for (1) and (2). Additionally, the Hausman test results show 41.520 (p = 0.000) and 87.200 (p = 0.000), significantly rejecting the null hypothesis. As a result, it is feasible to use a two-way fixed effect model to examine the impact of land transfer on agricultural green transformation, with the outcomes presented in Table 5. The estimated coefficients of land transfer on agricultural green transformation are all significantly positive, confirming the validity of Hypothesis 1. Specifically, the impact coefficient of land transfer on energy consumption is 0.106, which is significantly positive at the 1% level, indicating that every 1% increase in land transfer rate can save energy by 0.106%. This is because land transfer promotes contiguous and large-scale land operations, improves agricultural machinery utilization and energy efficiency, and thus reduces energy consumption per unit of land. Additionally, the continuous implementation of the land transfer policy encourages agricultural entities to adopt clean production technology and advanced machinery, further reducing energy consumption. Likewise, the estimated coefficient of land transfer on agricultural carbon emissions is 0.013, suggesting that every 1% increase in land transfer rate will reduce carbon emissions by 0.103%. Land transfer transfers arable land from low-productivity farmers to high-productivity farmers or agricultural organizations, reducing the cost of production per unit area of land. This prompts agricultural entities to allocate saved costs towards the purchase and application of green production technology, curbing agricultural carbon emissions. Furthermore, this shift enables the transformation from fragmented to large-scale cultivation modes, thereby increasing the adoption rate of agricultural machinery and enhancing land resource allocation efficiency through economies of scale, leading to reduced agricultural carbon emissions.

Table 5. The effect of land transfer on agricultural green transformation.

	ei		ac	
	Coef.	**Std. Err.**	**Coef.**	**Std. Err.**
ft	0.106 ***	0.022	0.013 ***	0.002
urb	−0.026	0.036	−0.009 **	0.004
tra	−0.049 ***	0.012	−0.012 ***	0.001
edu	−0.002	0.008	0.001 *	0.001
ins	0.106 ***	0.066	0.015 **	0.007
_cons	0.193 ***	0.022	0.013 ***	0.002
Time effect	YES		YES	
Regional effect	YES		YES	
N	480		480	
R-sq	0.182		0.385	

Note: * $p < 0.1$, ** $p < 0.05$, *** $p < 0.01$.

4.2. *Investigating the Role of Energy Consumption in the Impact of Land Transfer on Agricultural Carbon Emissions: An Empirical Test of Hypothesis 2*

Theoretical analysis suggests that land transfer can reduce agricultural carbon emissions by curbing agricultural energy consumption. This section aims to empirically test this mechanism. The testing process consists of two steps: first, examining the effect of land transfer on agricultural energy consumption with the regression results presented in Table 5; and second, investigating the combined effect of land transfer and agricultural energy consumption on agricultural carbon emissions. If the coefficient estimates for energy consumption are statistically significant, it indicates that energy consumption acts as an intermediary variable in the relationship between land transfer and agricultural carbon emissions. The results of this test are displayed in Table 6. The estimated impact coefficient of energy consumption on agricultural carbon emissions is 0.005, which is significantly positive at the 1% level. This indicates that energy consumption acts as an intermediary variable in the relationship between land transfer and agricultural carbon emissions. Thus, land transfer can effectively reduce agricultural carbon emissions by suppressing energy consumption, thereby confirming Hypothesis 2.

Table 6. The role of energy consumption in land transfer affecting agricultural carbon emissions.

	Coef.	Std. Err.
ft	0.005 ***	0.001
ei	0.082 ***	0.003
urb	−0.007 ***	0.002
tra	−0.008 ***	0.001
edu	0.002 ***	0.000
ins	0.003 *	0.002
_cons	−0.001	0.004
Time effect	YES	
Regional effect	YES	
N	480	
R-sq	0.806	

Note: * $p < 0.1$, *** $p < 0.01$.

4.3. *Mechanism Test of Agricultural Technology Progress in Land Transfer Affecting Agricultural Green Transformation: An Empirical Test of Hypothesis 3*

To explore the role of agricultural technology progress in the impact of land transfer on agricultural green transformation, this paper considers agricultural technology progress as an intermediary variable. The specific testing process involves two steps: first, examine the effect of land transfer on agricultural technology progress. The significance of the land transfer coefficient indicates its influence on the intermediary variable. Second, assess the joint effect of land transfer and agricultural technology progress on agricultural green transformation. If the estimated coefficients of land transfer and technology progress are significant, it suggests that technological progress acts as an intermediary in the process of agricultural green transformation. The results are presented in Table 7.

Table 7. An examination of the mechanism of agricultural technology progress in the process of land transfer affecting agricultural green transformation.

	Regression1		Regression2		Regression3	
	tc		ei		ac	
	Coef.	Std. Err.	Coef.	Std. Err.	Coef.	Std. Err.
ft	7.234 **	3.444	0.092 ***	0.021	0.011 ***	0.002
tc			0.002 ***	0.000	0.003 ***	0.000
urb	−7.913	5.515	−0.011	0.034	−0.006 **	0.003
tra	−12.869 ***	1.906	−0.025 **	0.012	−0.008 ***	0.001
edu	0.769	1.242	−0.003	0.008	0.001 *	0.001
ins	7.911 *	4.781	−0.203 ***	0.030	−0.015 ***	0.003
_cons	−2.079	10.189	0.197 ***	0.063	0.015 ***	0.006
Time effect	YES		YES		YES	
Regional effect	YES		YES		YES	
N	480		480		480	
R-sq	0.553		0.250		0.545	

Note: * $p < 0.1$, ** $p < 0.05$, *** $p < 0.01$.

Regression 1 reveals that the estimated impact coefficient of land transfer on agricultural technology progress is 7.234, which is significantly positive at the 5% level, indicating that every 1% increase in land transfer rate will increase agricultural technology progress by 7.234%. This is because land transfer facilitates contiguous and large-scale agricultural land operations, which, in turn, enhance farmers' adoption of mechanized operation methods and new green technologies, thereby driving agricultural technology progress. Furthermore, the findings from Regression 2 and Regression 3 show that the estimated impact coefficients of agricultural technology progress on energy consumption and agricultural carbon emissions are 0.092 and 0.011, respectively. This implies that agricultural technology progress serves as an intermediary variable in the relationship between land transfer and agricultural green transformation, verifying Hypothesis 3. The reason behind this is that agricultural technology plays a pivotal role in agricultural transformation. On the one hand, improved agricultural technology directly boosts agricultural output, effectively promoting agricultural green transformation per unit of land. On the other hand, the knowledge spillover effect resulting from advancements in agricultural technology enhances energy use efficiency, promoting agricultural green transformation.

4.4. Robustness Test

To ensure the robustness of the empirical findings regarding the significant role of land transfer in promoting agricultural green transformation, this study employs three methods for robustness testing. Firstly, a shrinking treatment (0.05 tail shrinkage level) approach is applied to mitigate the impact of outliers on the research results by trimming the tails of continuous variables. Secondly, to account for the significant impact of the "Administrative Measures for the Transfer of Rural Land Contractual Management Rights" implemented in 2005, the No. 1 Central document in 2010 requiring the promotion of the land transfer contracts, and the reform of the "separation of three rights" of rural land in 2014, which has greatly accelerated the land transfer and promoted the agricultural green transformation, Sample data from 2005, 2010, and 2014 are eliminated to test the robustness of the previous regression. Additionally, to address any potential endogeneity issues arising from causality problems, where changes in agricultural green transformation may affect the promulgation and implementation of land transfer policies, the explanatory variables are lagged by one period. The results are presented in Table 8. It is evident that, except for some differences in coefficient size, the significance and sign of the core explanatory variables remain consistent with the regression model's results in Table 5. This rigorous confirmation reinforces the previous empirical conclusions that land transfer significantly promotes agricultural green transformation.

Table 8. Robustness test of land transfer affecting agricultural green transformation.

	Winsorize Treatment		**Partial Sample Rejection**		**The Independent Variable Lags One Stage**	
	ei	**ac**	**ei**	**ac**	**ei**	**ac**
ft	0.106 ***	0.013 ***	0.098 ***	0.012 ***	0.112 ***	0.013 ***
	(0.022)	(0.002)	(0.025)	(0.002)	(0.022)	(0.002)
urb	−0.026	−0.009 **	−0.317 ***	−0.057 ***	−0.031	−0.007 **
	(0.036)	(0.004)	(0.075)	(0.007)	(0.035)	(0.003)
tra	−0.049 ***	−0.012 ***	−0.012	−0.007 ***	−0.060 ***	−0.013 ***
	(0.012)	(0.001)	(0.016)	(0.001)	(0.012)	(0.001)
edu	−0.002	0.001 *	0.001	0.002 *	0.002	0.002 **
	(0.008)	(0.001)	(0.009)	(0.001)	(0.008)	(0.001)
ins	−0.188 ***	−0.012 ***	−0.198 ***	−0.013 ***	−0.213 ***	−0.012 ***
	(0.031)	(0.003)	(0.033)	(0.003)	(0.031)	(0.003)
_cons	0.193 ***	0.015 **	0.311 ***	0.035 ***	0.184 ***	0.013 **
	(0.066)	(0.007)	(0.080)	(0.007)	(0.064)	(0.006)
Time effect	YES	YES	YES	YES	YES	YES
Regional effect	YES	YES	YES	YES	YES	YES
N	480	480	390	390	450	450
R-sq	0.182	0.385	0.232	0.495	0.230	0.417

Note: * $p < 0.1$, ** $p < 0.05$, *** $p < 0.01$.

4.5. Regional Heterogeneity Test

The preceding analysis demonstrates that land transfer significantly promotes agricultural green transformation. However, notable disparities exist in land resource endowments and economic development levels in China's rural regions. To delve into this heterogeneity, this paper further divides the research samples into groups based on resource endowments and economic development gaps to explore the varying impact of land transfer on agricultural green transformation.

4.5.1. Effect of Land Transfer on Both Sides of the "Hu-Huanyong Line" on Agricultural Green Transformation

In 1935, geographer Hu Huanyong established an oblique line at a 45-degree angle from Tengchong, Yunnan, to Heihe, Heilongjiang, as a dividing line for population density. This division created two regions: the densely populated southeast, which comprises 43% of the country's land but accommodates 94% of its population, and the sparsely populated northwest. The "Hu-Huanyong line" not only highlights the uneven distribution of China's population but also serves as a significant demarcation for land resource disparities between the southeast and northwest. In the southeast, overpopulation and limited land result in serious land fragmentation, making it an area with relatively strong implementation of land transfer policies. On the other hand, the complex terrain and landforms in the northwest hinder the effective implementation of land transfer policies. Thus, the energy-saving and emission-reduction effects of land transfer may vary across regions.

We employ the "Hu-Huanyong line" as the basis for dividing the research samples and conducting regression analysis. The results are presented in Table 9. In the southeast, the estimated coefficients of land transfer on energy consumption and agricultural carbon emissions are 0.132 and 0.015, respectively. This implies that every 1% increase in the land transfer rate will save energy and reduce emissions by 0.132% and 0.015%, respectively. The high land fragmentation in these areas benefits from an increased land transfer rate, which promotes contiguous and centralized production, thereby improving the utilization efficiency of chemical elements and mechanical equipment, reducing energy consumption and carbon emissions in agricultural production, and ultimately promoting agricultural green transformation. Conversely, the sparsely populated northwest region experiences restrictions in implementing land transfer policies due to its large land area. Additionally, the complex terrain and landforms in this region impede the use of machinery and equipment, limiting the adoption of new green technology. Thus, the impact of land transfer on agricultural green transformation in this region is insignificant.

Table 9. The difference in the effect of land transfer on agricultural green transformation on both sides of the Hu-Huanyong Line.

	Southeast		Northwest	
	ei	**ac**	**ei**	**ac**
ft	0.132 ***	0.015 ***	−0.057	−0.003
	(0.025)	(0.003)	(0.054)	(0.004)
urb	−0.039	−0.011 ***	0.022	0.003
	(0.039)	(0.004)	(0.084)	(0.007)
tra	−0.041 ***	−0.011 ***	0.076	0.006
	(0.012)	(0.001)	(0.079)	(0.006)
edu	0.002	0.002 *	−0.022	−0.001
	(0.009)	(0.001)	(0.016)	(0.001)
ins	−0.257 ***	−0.017 ***	0.084	0.010 *
	(0.035)	(0.004)	(0.065)	(0.005)
_cons	0.200 **	0.016 *	0.136	0.009
	(0.080)	(0.009)	(0.120)	(0.010)
Time effect	YES	YES	YES	YES
Regional effect	YES	YES	YES	YES
N	320	320	160	160
R-sq	0.276	0.463	0.204	0.331

Note: * $p < 0.1$, ** $p < 0.05$, *** $p < 0.01$.

4.5.2. The Impact of Land Transfer on Agricultural Green Transformation in Economically Differentiated Regions

The imbalance of economic structure will aggravate the change in the regional environment [60]. In addition, in more economically developed areas, more capable farmers carry out large-scale land [61] transfers. Therefore, regional economic development differences may have an impact on the effect of land transfer on agricultural green transformation. To explore these differences, this paper divides the research samples into three categories based on economic development: economically developed regions, economically less-developed areas, and economically underdeveloped regions. The goal is to investigate how land transfer impacts agricultural green transformation in regions with varying economic development levels. The results are presented in Table 10. In economically developed areas, the estimated coefficients of land transfer on energy consumption and agricultural carbon emissions are 0.136 and 0.018, respectively. In these regions, fragmented land is prevalent, and land subcontracting, leasing, and swapping with high turnover rates facilitate contiguous and large-scale land planting. This leads to improved utilization efficiency of fossil energy and mechanical equipment, ultimately promoting agricultural "green transformation." Additionally, the high level of economic development in these areas encourages the adoption of new green technology by farmers, further supporting agricultural green transformation.

Table 10. The difference of the effect of land circulation on agricultural green transformation in economic differentiation area.

	Economically Developed Areas		Economically Less-Developed Areas		Economically Underdeveloped Areas	
	ei	**ac**	**ei**	**ac**	**ei**	**ac**
ft	0.136 **	0.018 ***	0.005	0.001 ***	−0.007	0.002
	(0.058)	(0.006)	(0.010)	(0.000)	(0.060)	(0.005)
urb	−0.065	−0.023 **	0.033	0.001	−0.002	−0.001
	(0.095)	(0.010)	(0.022)	(0.001)	(0.059)	(0.005)
tra	−0.000	−0.011 ***	−0.009	−0.000	0.057	0.003
	(0.034)	(0.004)	(0.009)	(0.000)	(0.093)	(0.007)
edu	0.004	0.004	0.001	0.000	−0.012	0.000
	(0.024)	(0.003)	(0.004)	(0.000)	(0.016)	(0.001)
ins	−0.483 ***	−0.026 ***	−0.006	0.001 ***	−0.010	0.002
	(0.082)	(0.009)	(0.015)	(0.000)	(0.069)	(0.005)
_cons	0.291	0.025	0.001	0.002 **	0.150	0.009
	(0.208)	(0.023)	(0.030)	(0.001)	(0.122)	(0.010)
Time effect	YES	YES	YES	YES	YES	YES
Regional effect	YES	YES	YES	YES	YES	YES
N	112	112	208	208	160	160
R-sq	0.443	0.585	0.057	0.583	0.192	0.384

Note: ** $p < 0.05$, *** $p < 0.01$.

Conversely, in economically less developed areas, the estimated coefficient of land transfer on energy consumption is not significant, and the estimated coefficient on agricultural carbon emissions is 0.001. It primarily demonstrates emission reduction effects with limited energy-saving effects. As large producing provinces with extensive land areas, these regions witness improved agricultural production scale through land transfer, promoting mechanized production, and increasing the efficiency of mechanical equipment usage, thereby reducing agricultural carbon emissions. However, overall fossil energy consumption increases due to the expanded production. Economically underdeveloped areas also face challenges in implementing land transfer policies, leading to low land transfer rates. This situation fails to address issues related to fragmented cultivated land and excessive use of chemical elements. Moreover, many of these areas, primarily located in economically backward western regions, employ traditional production methods with limited use of mechanical equipment and new technology. Consequently, the estimated coefficient of land transfer on agricultural green transformation is not significant in these areas.

5. Discussion

Existing studies have focused on the impact of other factors on agricultural green transformation. For instance, Zhang et al. (2022) discovered a nonlinear relationship between agricultural production agglomeration and agricultural green transformation [62]. Wong et al. (2020) found that agricultural insurance significantly inhibits agricultural green transformation [63]. Similarly, Li et al. (2023) identified a nonlinear relationship between urbanization and agricultural green transformation [3]. However, there is a lack of literature investigating the influence of land transfer on agricultural green transformation. In fact, land transfer, as a crucial tool for facilitating large-scale land management [8] is conducive to improving the economic efficiency of land utilization [64] and promoting agricultural green transformation. One of the key contributions of this paper is examining the relationship between land transfer and agricultural green transformation. This further expands the research of Hu et al. (2023) and Wang et al. (2021) [21,36]. This paper not only focuses on the impact of land transfer on agricultural carbon emissions but also comprehensively considers the energy-saving and emission-reduction effects of land transfer. Consequently, it provides a valuable reference for the advancement of agricultural green transformation.

Subsequent studies have demonstrated the close relationship between agricultural energy consumption and agricultural carbon emissions. For instance, Sun et al. (2022) identified fossil energy as a crucial input in large-scale and mechanized agricultural production, directly contributing to carbon emissions [65]. However, limited literature exists on whether land transfer can mitigate agricultural carbon emissions through the reduction of agricultural energy consumption. Thus, the second contribution of this paper is to assess the role of agricultural energy consumption as a mediating variable in the connection between land transfer and agricultural carbon emissions. This finding holds significant value for the government in formulating environmental goals such as "green transformation." The results of this paper show that land transfer can reduce agricultural carbon emissions by reducing agricultural energy consumption, which builds on the research results of Sun et al. (2022) [65].

Additionally, agricultural technology progress is intricately linked to land transfer and agricultural green transformation [66]. The third contribution of this paper is to explore the influence of agricultural technology progress on the relationship between land transfer and agricultural green transformation. This enhances our understanding of the mechanisms underlying land transfer for agricultural green transformation, expanding the study of Hu et al. (2023) and Wang et al. (2021) [21,36]. The results of this paper show that land transfer can promote agricultural green transformation by promoting agricultural technology progress, which expands the research conclusion of Ge et al. (2017) [67]. Moreover, this paper further explores the regional heterogeneity of land transfer in agricultural green transformation. This further expanded the studies of Luo Xuan (2020) [30], Geng, and Luo (2022) [68]; however, they mainly focused on the heterogeneity of land transfer on agricultural carbon emissions in different regions of food function. This paper further expands on this by grouping the study samples according to resource endowment and economic development gap. This segmentation provides vital insights for the government to deepen rural land system reforms and facilitate agricultural green transformation.

Exploring the relationship between land transfer and agricultural green transition is of great significance for agricultural green development, and it is important to consider certain limitations when exploring similar topics. Firstly, this paper solely analyzes land transfers at the provincial level. Enhancing the credibility of research conclusions can be achieved by utilizing county-level data to examine the impact of land transfer on agricultural green transformation. Unfortunately, there is a significant lack of available county-level data currently. Therefore, conducting an analysis with panel data at the prefecture level would be meaningful. Secondly, this study focuses on capital deepening as a measure of agricultural technology progress. However, it is important to acknowledge that mechanical and biological technological progress may also play a role in the relationship between land transfer and agricultural green transformation. Therefore, exploring the effects of other types of technological progress in the land transfer process can yield valuable insights.

6. Conclusions and Policy Recommendations

Based on China's inter-provincial panel data from 2005 to 2020, this paper employs the fixed effect model and the mediation effect model to investigate the impact of land transfer on agricultural green transformation and its internal mechanisms. The findings of the research are as follows: Firstly, land transfer demonstrates an energy-saving effect by reducing energy consumption and an emission reduction effect by lowering agricultural carbon emissions. That is, land transfer plays a significant role in promoting agricultural green transformation. Secondly, land transfer can suppress carbon emissions through the reduction of agricultural energy consumption. Thirdly, agricultural technology progress resulting from land transfer also plays a role in promoting agricultural green transformation. Furthermore, there is heterogeneity in the effect of land transfer on agricultural green transformation. Specifically, land transfer significantly promotes agricultural green transformation on the southeast side of the Hu-Huanyong Line and in economically developed areas. However, on the northwest side of the Hu-Huanyong Line and in economically underdeveloped areas, land transfers do not have a significant impact.

The research conclusions above hold significant policy implications for achieving agricultural green transformation. Firstly, it is essential to optimize the land transfer market. By promoting land transfer, we can achieve intensive and contiguous land operations, thereby boosting the agricultural sector's contribution to China's green transformation efforts. Secondly, when formulating environmental goals, the government should prioritize policy measures aimed at reducing agricultural energy consumption. Thirdly, expediting the agricultural technology's progress is crucial. This can be achieved through the adoption of green production technology, thereby improving energy efficiency and reducing pollution emissions. Lastly, it is imperative to recognize the differentiated layout of land transfers. In economically developed regions, it is essential to encourage farmers to focus on operating efficiency, leading to intensified and conservation-oriented practices to accelerate the agricultural green transformation. On the other hand, in economically underdeveloped areas, it is essential to increase the implementation of land transfers, encouraging farmers to adopt moderate-scale approaches to promote agricultural green transformation.

Author Contributions: G.M. conceived, designed, and conducted this study. D.L. revised the manuscript. T.J. and Y.L. were involved in the analysis and interpretation of the data and funded this study. All authors have read and agreed to the published version of the manuscript.

Funding: This research was funded by the National Social Science Foundation of China's key project (No. 21AJY013) and the Guangxi Philosophy and Social Science Project (No. 22FJY021).

Institutional Review Board Statement: Not applicable.

Informed Consent Statement: Not applicable.

Data Availability Statement: All the data are obtained from the China Statistical Yearbook, the China Rural Statistical Yearbook, the China Rural Economic Situation Statistics, the China's Rural Management Statistical Annual Report, the Statistical Annual Report on China's Rural Policies and Reforms, China Agricultural Yearbook, China Agricultural Trade Development Report, and China Population and Employment Statistical Yearbook (2001–2020). It is available on request from the corresponding author.

Acknowledgments: The authors acknowledge the support provided by their respective institutions.

Conflicts of Interest: The authors declare no conflict of interest.

References

1. Hong, C.; Burney, J.A.; Pongratz, J.; Nabel, J.E.M.S.; Davis, S.J. Global and regional drivers of land-use emissions in 1961–2017. *Nature* **2021**, *589*, 554–561. [CrossRef] [PubMed]
2. Federici, S.; Tubiello, F.N.; Salvatore, M.; Jacobs, H.; Schmidhuber, J. New estimates of CO_2 forest emissions and removals: 1990–2015. *For. Ecol. Manag.* **2015**, *352*, 89–98. [CrossRef]
3. Frank, S.; Beach, R.; Havlík, P.; Valin, H.; Herrero, M.; Mosnier, A.; Hasegawa, T.; Creason, J.; Ragnauth, S.; Obersteiner, M. Structural change as a key component for agricultural non-CO_2 mitigation efforts. *Nat. Commun.* **2018**, *9*, 1060. [CrossRef] [PubMed]
4. Meng, Y.; Liu, L.; Wang, J.; Ran, Q.; Yang, X.; Shen, J. Assessing the impact of the national sustainable development planning of resource-based cities policy on pollution emission intensity: Evidence from 270 prefecture-level cities in China. *Sustainability* **2021**, *13*, 7293. [CrossRef]
5. Wang, J.; Han, P. The impact of industrial agglomeration on urban green land use efficiency in the Yangtze River Economic Belt. *Sci. Rep.* **2023**, *13*, 974. [CrossRef] [PubMed]
6. Vu, T.H.V. Land fragmentation and household income: First evidence from rural Vietnam. *Land Use Policy* **2019**, *89*, 104247.
7. Huang, K.; Deng, X.; Liu, Y.; Yong, Z.; Xu, D. Does off-farm migration of female laborers inhibit land transfer? evidence from sichuan province, China. *Land* **2020**, *9*, 14. [CrossRef]
8. Li, B.; Shen, Y. Effects of land transfer quality on the application of organic fertilizer by large-scale farmers in China. *Land Use Policy* **2021**, *100*, 105124. [CrossRef]
9. Ren, C.; Liu, S.; Grinsven, H.V.; Reis, S.; Gu, B. The impact of farm size on agricultural sustainability. *J. Clean. Prod.* **2019**, *220*, 357–367. [CrossRef]
10. Aha, B.; Ayitey, J.Z. Biofuels and the hazards of land grabbing: Tenure (in)security and indigenous farmers' investment decisions in Ghana. *Land Use Policy* **2017**, *60*, 48–59. [CrossRef]
11. Sheng, Y.; Ding, J.; Huang, J. The relationship between farm size and productivity in agriculture: Evidence from maize production in northern China. *Am. J. Agric. Econ.* **2019**, *101*, 790–806. [CrossRef]
12. Feng, S. Land rental, off-farm employment and technical efficiency of farm households in Jiangxi Province, China. *NJAS Wagening. J. Life Sci.* **2008**, *55*, 363–378. [CrossRef]
13. Xu, D.; Deng, X.; Guo, S.; Liu, S. Labor migration and farmland abandonment in rural China: Empirical results and policy implications. *J. Environ. Manag.* **2018**, *232*, 738–750. [CrossRef] [PubMed]
14. You, H. Impact of urbanization on pollution-related agricultural input intensity in Hubei, China. *Ecol. Indic.* **2016**, *62*, 249–258. [CrossRef] [PubMed]
15. Campi, M.; Dueñas, M.; Fagiolo, G. Specialization in food production affects global food security and food systems sustainability. *World Dev.* **2021**, *141*, 105411. [CrossRef]
16. Luo, J.; Huang, M.; Hu, M.; Bai, Y. How does agricultural production agglomeration affect green total factor productivity? Empirical evidence from China. *Environ. Sci. Pollut. Res.* **2023**, *30*, 67865–67879. [CrossRef]
17. Hou, D.; Wang, X. Inhibition or promotion? The effect of agricultural insurance on agricultural green development. *Front Public Health* **2022**, *10*, 910534. [CrossRef] [PubMed]
18. Tomich, S.T.P. Evolution of land tenure institutions and development of agroforestry: Evidence from customary land areas of sumatra. *Agric. Econ.* **2001**, *25*, 85–101.
19. Xu, X.; Li, C.; Guo, J.; Zhang, L. Land transfer-in scale, land operation scale and carbon emissionsfrom crop planting throughout the life cycle: Evidence from China rural development survey. *Chin. Rural Econ.* **2022**, *11*, 40–58.
20. Yiyun, W.; Xican, X.; Xin, T.; Deming, L.; Baojing, G.; Kee, L.S.; Vitousek, P.M.; Deli, C. Policy distortions, farm size, and the overuse of agricultural chemicals in China. *Proc. Natl. Acad. Sci. USA* **2018**, *115*, 7010–7015.
21. Hu, G.; Wang, J.; Fahad, S.; Li, J. Influencing factors of farmers' land transfer, subjective well-being, and participation in agri-environment schemes in environmentally fragile areas of China. *Environ. Sci. Pollut. Res.* **2023**, *30*, 4448–4461. [CrossRef] [PubMed]
22. Gao, X.; Yasunaga, N.; Inoue, N. Pathways influencing bearers and abandoned farmlands through farmland intermediate management institutions: Using prefectural data in Japan. *Asia-Pac. J. Reg. Sci.* **2023**, 1–27. [CrossRef]
23. Li, L.; Han, J.; Zhu, Y. Does farmland inflow improve the green total factor productivity of farmers in China? an empirical analysis based on a propensity score matching method. *Heliyon* **2023**, *9*, e13750. [CrossRef] [PubMed]
24. Tian, J.; Yu, L.; Xue, R.; Zhuang, S.; Shan, Y. Global low-carbon energy transition in the post-COVID-19 era. *Appl. Energy* **2022**, *307*, 118205. [CrossRef] [PubMed]
25. Rada, N.E.; Fuglie, K.O. New perspectives on farm size and productivity. *Food Policy* **2018**, *84*, 147–152. [CrossRef]
26. Freire-González, J.; Vivanco, D.F.; Puig-Ventosa, I. Economic structure and energy savings from energy efficiency in households. *Ecol. Econ.* **2016**, *131*, 12–20. [CrossRef]
27. Welsch, H.; Ochsen, C. The determinants of aggregate energy use in West Germany: Factor substitution, technological change, and trade. *Energy Econ.* **2005**, *27*, 93–111. [CrossRef]
28. Fei, X. Does where you are from affect how you land? Evidence from land transactions of Chinese manufacturers. *Appl. Econ. Lett.* **2019**, *27*, 525–532. [CrossRef]

29. Yang, L.; Li, Z. Technology advance and the carbon dioxide emission in China—Empirical research based on the rebound effect. *Energy Policy* **2017**, *101*, 150–161. [CrossRef]
30. Luo, S.; He, K.; Zhang, J. The more grain production, the more fertilizers pollution? empirical evidence from major grain-producing areas in China. *China Rural Econ.* **2020**, *1*, 108–131.
31. Ibrahim, R.L.; Al-Mulali, U.; Solarin, S.A.; Ajide, K.B.; Al-Faryan, M.A.S.; Mohammed, A. Probing environmental sustainability pathways in G7 economies: The role of energy transition, technological innovation, and demographic mobility. *Environ. Sci. Pollut. Res. Int.* **2023**, *30*, 75694–75719. [CrossRef]
32. Zhang, Y.; Wu, Q.; Zhao, X.; Hao, Y.; Liu, R.; Yang, Z.; Lu, Z. Study of carbon metabolic processes and their spatial distribution in the Beijing-Tianjin-Hebei urban agglomeration. *Sci. Total Environ.* **2018**, *645*, 1630–1642. [CrossRef] [PubMed]
33. Bradfield, T.; Butler, R.; Dillon, E.; Hennessy, T.; Kilgarriff, P. The effect of land fragmentation on the technical inefficiency of dairy farms. *J. Agric. Econ.* **2021**, *72*, 486–499. [CrossRef]
34. Cong, S. The impact of agricultural land rights policy on the pure technical efficiency of farmers' agricultural production: Evidence from the largest wheat planting environment in China. *J. Environ. Public Health* **2022**, *2022*, 3487014. [CrossRef] [PubMed]
35. Zhou, M.; Zhang, H.; Ke, N. Cultivated Land Transfer, Management Scale, and Cultivated Land Green Utilization Efficiency in China: Based on Intermediary and Threshold Models. *J. Environ. Public Health* **2022**, *19*, 12786. [CrossRef] [PubMed]
36. Wang, H.; Lu, S.; Lu, B.; Nie, X. Overt and covert: The relationship between the transfer of land development rights and carbon emissions. *Land Use Policy* **2021**, *108*, 105665. [CrossRef]
37. Cao, H.; Zhu, X.; Heijman, W.; Zhao, K. The impact of land transfer and farmers' knowledge of farmland protection policy on pro-environmental agricultural practices: The case of straw return to fields in Ningxia, China. *J. Clean. Prod.* **2020**, *277 Pt 1*, 123701. [CrossRef]
38. Cui, N.; Ba, X.; Dong, J.; Fan, X. Does farmland transfer contribute to reduction of chemical fertilizer use? evidence from Heilongjiang Province, China. *Sustainability* **2022**, *14*, 11514. [CrossRef]
39. Jiang, X.; Lu, X.; Liu, Q.; Chang, C.; Qu, L. The effects of land transfer marketization on the urban land use efficiency: An empirical study based on 285 cities in China. *Ecol. Indic.* **2021**, *132*, 108296. [CrossRef]
40. He, K.; Zhang, J.; Zeng, Y. Households' willingness to pay for energy utilization of crop straw in rural China:based on an improved UTAUT model. *Energy Policy* **2020**, *140*, 111373. [CrossRef]
41. Zhang, Y.; Long, H.; Li, Y.; Ge, D.; Tu, S. How does off-farm work affect chemical fertilizer application? evidence from China's mountainous and plain areas. *Land Use Policy* **2020**, *99*, 104848. [CrossRef]
42. Wan, D.; Xue, R.; Linnenluecke, M.; Tian, J.; Shan, Y. The impact of investor attention during COVID-19 on investment in clean energy versus fossil fuel firms. *Financ. Res. Lett.* **2021**, *43*, 101955. [CrossRef] [PubMed]
43. Gao, Y.; Liu, B.; Yu, L.; Yang, H.; Yin, S. Social capital, land tenure and the adoption of green control techniques by family farms: Evidence from Shandong and Henan Provinces of China. *Land Use Policy* **2019**, *89*, 104250. [CrossRef]
44. Xu, Q.; Lu, Y.; Zhang, Y. Subsidy to large-scale farming and food quantity-quality security: Evidence from large-scale farmers. *Econ. Res.* **2022**, *57*, 121–137.
45. Lu, H.; Xie, H.; Yao, G. Impact of land fragmentation on marginal productivity of agricultural labor and non-agricultural labor supply: A case study of Jiangsu, China. *Habitat Int.* **2019**, *83*, 65–72. [CrossRef]
46. Li, Y.; Wu, W.; Liu, Y. Land consolidation for rural sustainability in China: Practical reflections and policy implications. *Land Use Policy* **2018**, *74*, 137–141. [CrossRef]
47. Zhao, H.; Liu, J.; Wu, J. The impact of vertical fiscal asymmetry on carbon emissions in China. *Environ. Sci. Pollut. Res.* **2023**, *30*, 65963–65975. [CrossRef] [PubMed]
48. Yang, Y.; Xue, R.; Zhang, X.; Cheng, Y.; Shan, Y. Can the marketization of urban land transfer improve energy efficiency? *J. Environ. Manag.* **2023**, *329*, 117126. [CrossRef]
49. Miao, Z.; Guo, A.; Chen, X.; Zhu, P. Network technology, whole-process performance, and variable-specific decomposition analysis: Solutions for energy-economy-environment nexus. *IEEE Trans. Eng. Manag.* **2022**, 1–18. [CrossRef]
50. He, P.; Zhang, J.; Li, W. The role of agricultural green production technologies in improving low-carbon efficiency in China: Necessary but not effective. *J. Environ. Manag.* **2021**, *293*, 112837. [CrossRef]
51. Zhang, Y.; Tian, Y.; Wang, Y.; Wang, R.; Peng, Y. Rural human capital, agricultural technology progress and agricultural carbon emissions. *Sci. Technol. Manag. Res.* **2019**, *39*, 266–274.
52. Li, Z.; Li, J. The influence mechanism and spatial effect of carbon emission intensity in the agricultural sustainable supply: Evidence from china's grain production. *Environ. Sci. Pollut. Res.* **2022**, *29*, 44442–44460. [CrossRef] [PubMed]
53. Liu, Y.; Zhang, J.B.; Zhang, L. Analysis of carbon emission efficiency of rice in China under different rice planting patterns based on the DEA-SBM model. *J. China Agric. Univ.* **2018**, *23*, 177–186.
54. He, Y.; Cheng, X.; Wang, F. Study on the regional spillover effects of agricultural carbon emission based on the perspective of agricultural technology diffusion. *J. Agric. Tech. Econ.* **2022**, *4*, 132–144.
55. Kuang, Y.; Yang, J. The effect of total factor productivity growth on rural land transfer. *Economist* **2019**, *3*, 102–112.
56. Xu, B.; Wang, H.; Shen, Z. Impact of structural transformation, technological progress choice on agricultural carbon shadow price: An empirical analysis based on BP technology and a mediating effect model. *Chin. J. Eco-Agric.* **2023**, *31*, 241–252.
57. Wang, S.; Zhang, G. The impact of off-farm employment on the agricultural carbon emission behavior of farmers. *Resour. Sci.* **2013**, *35*, 1855–1862.

58. Fuchs, R.; Brown, C.; Rounsevell, M. Europe's green deal offshores environmental damage to other nations. *Nature* **2020**, *586*, 671–673. [CrossRef] [PubMed]
59. Lei, X.U.; Jie, D.; Jun-Feng, Z.; Lu, L.I. System simulation and policy optimization of agricultural carbon emissions in Hubei Province based on SD model. *Resour. Dev. Mark.* **2017**, *33*, 1031–1035.
60. Cheng, M.; Yao, W. Trend prediction of carbon peak in China's animal husbandry based on the empirical analysis of 31 provinces in China. *Environ. Dev. Sustain.* **2022**, 1–18. [CrossRef]
61. Kuang, Y.; Peng, D. The Mechanism and Empirical Study on the Influence of Farmland Transfer on Factor Market Development—Demonstration Based on Panel Intermediary Effect Model. *Chin. J. Agric. Resour. Reg. Plan.* **2023**, *44*, 250–258.
62. Zhang, H.; Zhang, J.; Song, J. Analysis of the threshold effect of agricultural industrial agglomeration and industrial structure upgrading on sustainable agricultural development in China. *J. Clean. Prod.* **2022**, *341*, 130818. [CrossRef]
63. Wong, H.L.; Wei, X.; Kahsay, H.B.; Gebreegziabher, Z.; Diro, R. Effects of input vouchers and rainfall insurance on agricultural production and household welfare: Experimental evidence from northern Ethiopia. *World Dev.* **2020**, *135*, 105074. [CrossRef]
64. He, D.; Zhang, G.; You, K.; Wu, J.; Guo, S. Property rights and market participation: Evidence from the land titling program in rural China. *J. Chin. Gov.* **2023**, *8*, 110–133. [CrossRef]
65. Sun, D.; Cai, S.; Yuan, X.; Zhao, C.; Gu, J.; Chen, Z.; Sun, H. Decomposition and decoupling analysis of carbon emissions from agricultural economic growth in China's Yangtze River economic belt. *Environ. Geochem. Health* **2022**, *44*, 2987–3006. [CrossRef] [PubMed]
66. Tian, X.; Wu, M.; Ma, L.; Wang, N. Rural finance, scale management and rural industrial integration. *China Agric. Econ. Rev.* **2020**, *12*, 349–365. [CrossRef]
67. Ge, D.; Long, H.; Zhang, Y.; Ma, L.; Li, T. Farmland transition and its influences on grain production in China. *Land Use Policy* **2018**, *70*, 94–105. [CrossRef]
68. Geng, P.; Luo, B. Has the land titling promoted the modernization of rural governance? *Manag. World* **2022**, *38*, 59–76.

sustainability

Article

Impact Mechanism and Effect of Agricultural Land Transfer on Agricultural Carbon Emissions in China: Evidence from Mediating Effect Test and Panel Threshold Regression Model

Ying Tang * and Menghan Chen

School of Public Management, Liaoning University, Shenyang 110036, China
* Correspondence: tangying5440@163.com; Tel.: +86-15004041523

Abstract: In order to identify the mechanism and effect of agricultural land transfer on agricultural carbon emissions, a study was conducted by analyzing the panel data of 30 provincial-level administrative regions from 2005 to 2019. Both the intermediary effect model and panel threshold regression model are applied to test the correlation between agricultural land transfer and agricultural carbon emissions, which provides some clarity on the mechanism of agricultural land transfer affecting agricultural carbon emissions and its future trends. The research results are as follows. Firstly, agricultural land transfer has a positive effect on agricultural carbon emissions, and agricultural factor input plays a mediating role between agricultural land transfer and agricultural carbon emissions. More specifically, the input of agricultural chemical elements has a positive impact on agricultural carbon emissions, while the input of agricultural machinery elements has a negative impact on agricultural carbon emissions. Secondly, under the threshold constraint of the urbanization level, the relationship between agricultural land transfer and agricultural carbon emissions is characterized by an inverted "U" shape, with a threshold value of 0.73. In view of these findings, more attention should be directed to addressing the negative impact of agricultural land transfer on the ecological environment. Furthermore, various targeted measures should be taken to reduce the ecological risk carried by agricultural land transfer, to increase the effort made on achieving the goals of agricultural carbon emission reduction, and to promote the green and sustainable development of the agricultural industry.

Keywords: land use; carbon emissions; intermediary effect model; panel threshold model

Citation: Tang, Y.; Chen, M. Impact Mechanism and Effect of Agricultural Land Transfer on Agricultural Carbon Emissions in China: Evidence from Mediating Effect Test and Panel Threshold Regression Model. *Sustainability* **2022**, *14*, 13014. https://doi.org/10.3390/su142013014

Academic Editors: Lu Zhang, Bing Kuang and Bohan Yang

Received: 1 September 2022
Accepted: 9 October 2022
Published: 11 October 2022

1. Introduction

When it comes to global climate warming, a significant influencing factor for it is the increase in carbon dioxide concentration in the atmosphere due to the social and economic activities of humans [1]. It is a consensus reached among the international community that various measures must be taken possibly soon to reduce carbon emissions in response to the ongoing global climate change. As the world's largest emitter of greenhouse gases, China has committed itself at the 75th United Nations General Assembly to increasing the effort made to cut down on carbon emissions, with effective policies and measures adopted to achieve the "double carbon" goal of carbon peak by 2030 and carbon neutrality by 2060. To achieve this objective, what needs to happen first is to fully understand the overall situation of carbon emissions across China. According to the relevant data, the carbon emissions from agricultural production and land use change account for nearly one fourth of the total [2]. As a large agricultural production country, China contributes about 29% to the total agricultural carbon emissions in Asia and roughly 12% to the total carbon emissions worldwide [3]. Furthermore, it continues to increase at an annual rate of 5% on average [4]. It is estimated that China's agricultural carbon emissions will increase by 30% by 2050 if there are no effective emission reduction measures taken. Obviously, agricultural production contributes significantly to the total carbon emissions in China. Therefore, in order to achieve the "double carbon" objective, it is essential to impose stringent control on the carbon emissions arising from agricultural productions and other relevant activities. At the same time, it is necessary to promote the eco-friendly development of agricultural productions according to the national agricultural green development scheme as part of the 14th five-year plan, which requires the reduction in agricultural carbon emissions. Under this context, there have been many studies conducted by academics on agricultural carbon emissions.

In this respect, the focus of discussion is placed on the factors that affect the scale of carbon emissions. It can be calculated by using the IPCC coefficient method [5], Kaya Porter identity (KPI) method [6], carbon footprint method [7] or others. Having an incremental effect on carbon emission changes, economic scale is the main contributor to increasing carbon emissions [8,9]. Specifically, carbon emissions can be significantly affected by the increase in manufacturing output value and international trade output value in macroeconomic indicators [10]. Furthermore, population size and energy structure are another two important factors in the increase in carbon emissions [11]. The slight changes in the soil carbon cycle may also have a significant impact on the concentration of carbon monoxide in the atmosphere. However, the current technical capacity is insufficient to quantitatively allocate carbon use [12]. The increase in carbon emissions has detrimental effects on the terrestrial climate, as manifested mainly by temperature rise [13]. The utilization intensity of fossil fuels such as coal should be restricted [14], and the carbon emissions from economic activities should be reduced progressively through the popularization of clean energy and technologies, such as solar cells, biomass, hydropower and thermoelectric conversion [15,16]. Apart from that, the scale of carbon emissions should be limited in the form of trading licenses [17]. In China, agricultural carbon emissions are usually characterized by a three-stage change of "up—down—up", and there is a difference between the west and the east [18]. The areas with high total emissions concentrate in those provinces heavily reliant on the agricultural industry [19]. The total carbon emissions are jointly affected by the development of world economy and society and policy intensity [20]. There is an inverted "U" relationship existing between agricultural carbon emissions and economic growth [21], and a "U" relationship existing between environmental regulation and carbon emission efficiency [22]. In addition, the LMDI model [23], Kaya identity [24], STIRPAT model [25], geographical weighted regression model [26] and other methods can be used to conduct quantitative analysis on the influencing factors in agricultural carbon emissions. The results show that agricultural carbon emissions can be significantly reduced by agricultural production efficiency, agricultural structure, agricultural population size, agricultural technology progress and other factors [27,28].

As a market-oriented means to improve the efficiency of rural land resource allocation, rural land transfer relates to society, economy, ecology and more. However, at present, the academic research of agricultural land transfer focuses mainly on its social and economic effects [29–31], and there is little research on the ecological effects of agricultural land transfer. At the same time, to meet the "double carbon" goal and to promote agricultural green development, more attention should be paid to exploring how agricultural land transfer affects agricultural carbon emissions. With the development of agricultural land transfer market and the increase in agricultural land transfer, agricultural land circulation has made significant impact on agricultural ecology [32]. Therefore, it is of much practical significance to analyze how to reduce the ecological risk posed by agricultural land circulation while promoting the moderate-scale practice of agricultural land circulation. Based on the panel data of 30 Chinese provinces from 2005 to 2019, an intermediary effect model and a threshold model are constructed in this study based on theoretical analysis, so as to test the impact path and mechanism of agricultural land transfer on agricultural carbon emissions. Furthermore, the hypothesis is verified, which provides a theoretical reference for effectively promoting agricultural land transfer and reducing agricultural emissions.

The contributions of this study are as follows. Firstly, an intermediary effect model is adopted to test the impact mechanism of agricultural land transfer on agricultural carbon emission in China. Secondly, an analysis is conducted as to the constraints on the relationship between agricultural land transfer and agricultural carbon emissions. Lastly, policy implications are indicated based on the empirical results for the better coordination between agricultural land transfer and agricultural carbon emission.

2. Agricultural Land Transfer and Agricultural Carbon Emission

2.1. Agricultural Land Transfer and Agricultural Production Input

In practice, the specific input mode of production as adopted by the agricultural production subject is affected by the resource endowment of factors, market price and product demand, which leads to a technology selection bias based on labor-saving technology (such as agricultural machinery) or land-saving technology (agricultural chemicals) [33,34]. Under the traditional urban–rural dual registered residence system and the policy that prohibits the circulation of agricultural land, the abundance of rural labor and the scarcity of agricultural land have jointly contributed to the resource endowment characteristics in China. Given a huge national population, land saving technology plays a vital role in improving agricultural production efficiency to make up for the defects of agricultural land resource endowment, which makes China's input of agricultural chemicals far higher than the world average. In recent years, the central government of China has issued a series of policies to promote the orderly circulation of agricultural land, effectively keep the appropriate scale of land resources, and promote the efficiency of agricultural section and increase income of farmers. Under the guidance of the national macro policies, the transfer of agricultural land has developed rapidly. According to the statistics from the Ministry of agriculture and rural sector, there was 35.9 million hm^2 of agricultural land in China at the end of 2018. Agricultural land is transferred among different subjects, accounting for 48.56% of the total. With the development of agricultural land transfer and the breaking of the urban-rural separation pattern in China, the magnitude of rural labor migration and non-agriculturalization continues to improve, which has a significant impact on the factor endowment structure of agricultural production in China [35,36]. For the main body of agricultural land transfer, the increase in agricultural land stock reduces the scarcity and relative price of agricultural land resources, while the continuous outflow of rural populations leads to the relative increase in labor costs. Under this context, the main body of production will adopt labor saving technologies, that is, to increase the input of agricultural machinery and reduce the input of land saving elements. As for the subject who transfers out of agricultural land, agricultural land resources will become scarcer. Therefore, the production subject will adopt land saving technology, that is, to increase the use of agricultural chemicals for the improved output level of agricultural land.

Under the agricultural land transfer policy, the agricultural land transfer in the land market has become increasingly active, thus leading to the optimization and reorganization of agricultural land resources. Through the marginal output equilibrium effect of land market [37], agricultural land will be transferred from the farmers with low production efficiency to major grain growers, professional agricultural enterprises and other modern agricultural production organizations with high production efficiency. In this way, the efficiency of agricultural land utilization can be improved. For the entities who transfer in agricultural land, the expansion of their business may increase the demand for agricultural labor. However, due to the insufficient elasticity of labor supply due to the transfer of agricultural land, it is difficult to meet the demand for agricultural labor after production scale expansion, which will motivate the production entity to invest more in agricultural machinery and equipment for productions, thus further reducing the input of agricultural chemicals [38]. In addition, the transferred entity will concentrate to connect the scattered agricultural land, which is effective in reducing the land fragmentation caused by the decentralized management of farmers. This is conducive to reducing agricultural chemicals input. In addition, since the transferred entity is advantageous in agricultural production capacity and experience, it is easier to reduce the use of traditional agricultural chemicals by applying green and low-risk production technologies [39]. On the contrary, for the entities who transfer out agricultural land, the transfer of agricultural land has reduced the management scale of agricultural land for each entity, which moves the labor force from agricultural production to non-agricultural activities [40]. Therefore, agricultural production has the typical characteristics of concurrent operation. For these farmers, the loss of labor makes it easier to invest more agricultural chemicals for maximum profits. In addition, the stability and duration of agricultural land property rights will have a more significant impact on the investment behavior of farmers, according to the property rights theory. Due to the unstable and short-term agricultural real estate rights, farmers tend to show shortsightedness in their investments. That is to say, farmers, as "economic people", will reject the long-term investment in agricultural land, such as building irrigation and drainage facilities, improving soil quality, etc. Instead, they choose to invest a large amount of agricultural chemicals and make other short-term investments for quick profits [41]. By improving agricultural land circulation policies, the stability of agricultural land property rights can be enhanced, which will motivate farmers to abandon short-term investment for long-term investment [42,43].

2.2. Agricultural Production Input and Agricultural Carbon Emission

Depending on the exact form and function of agricultural input elements, the agricultural element input in agricultural land use activities can be divided into two categories: agricultural chemical element input and agricultural machinery element input. For a long time, the use of chemical fertilizers, pesticides and other agricultural chemical elements in agricultural production activities has played a major role in improving the nutrient content in agricultural soil, reducing the yield loss of crops caused by diseases, insect pests and weeds, improving grain yield and promoting the growth of agricultural economy [44,45]. Given the expanding scale of agricultural land management and the shortage of labor force, the input of agricultural chemistry such as chemical fertilizer provides an effective solution to ensuring grain output [46]. At the same time, the continuous use of agricultural chemicals has also resulted in various issues including excessive carbon dioxide emissions [47], which is more detrimental to the ecological environment. Among them, the contribution of agricultural inputs to agricultural carbon emissions is most significant [48]. The production and utilization of chemical fertilizers are the main factors affecting agricultural carbon emissions [49,50]. Such agricultural chemicals such as chemical fertilizers, pesticides and agricultural film account for about half of the total agricultural carbon emissions [51]. As for the input of agricultural machinery, agricultural machinery technology has a substitution effect on agricultural labor force, which improves the degree of specialization for agricultural productions [52,53]. With the improvement of agricultural mechanization, large-scale agricultural machinery gradually replaces the small, energy-intensive agricultural machinery in the traditional small-scale production, which to some extent curbs agricultural carbon emissions. Meanwhile, the improved level of agricultural machinery utilization significantly promotes the optimization and upgrading of industrial structure and enhances the efficiency of agricultural production, thus reducing agricultural carbon emissions.

Based on the above analysis, the following hypothesis is proposed:

Agricultural land transfer can affect agricultural carbon emissions, with the input of agricultural production materials as an intermediate variable in the impact of agricultural land transfer on agricultural carbon emissions. Among the intermediate variables of agricultural materials input, agricultural chemical factor has a promoting effect on agricultural carbon emissions, while agricultural machinery factor input has an inhibitory effect on agricultural carbon emissions.

3. Materials and Methods

3.1. Analytical Methods

(1) Mediating effect test. In order to verify the research hypothesis proposed in this study, that is, agricultural land transfer affects agricultural carbon emissions by affecting the

input of agricultural chemical elements, the stepwise regression equation is applied to perform a mediating effect test. The design is expressed as follows [54,55]:

$$\ln TC_{it} = \theta_1 + c \ln F_{it} + control_{it} + \varepsilon_{it} \quad (1)$$

$$\ln cp_{it} = \theta_2 + a_1 \ln F_{it} + control_{it} + \varepsilon_{it} \quad (2)$$

$$am_{it} = \theta_2 + a_2 \ln F_{it} + control_{it} + \varepsilon_{it} \quad (3)$$

$$\ln TC_{it} = \theta_3 + c\prime \ln F_{it} + b_1 \ln cp_{it} + b_2 am_{it} + control_{it} + \varepsilon_{it} \quad (4)$$

where $\ln TC_{it}$ represents the interpreted variable of agricultural carbon emissions; $\ln F_{it}$ indicates the explanatory variable of agricultural land transfer; agricultural chemical factor input ($\ln cp$) and agricultural machinery input (am) are intermediate variable; $control_{it}$ refers to the control variable, including agricultural financial level (fsa), agricultural land resource endowment (area), agricultural population scale (popu), agricultural output value structure (pvs), and agricultural planting structure (ps); i,t represent different provinces and time, respectively; ε indicates a random error term.

At the same time, it is considered by some scholars that this method has certain flaws, who suggest using more accurate methods to conduct tests. For example, the bootstrap program developed by Preacher and Hayes [56] not only shows higher test efficiency for mediation effects, but also provides a variety of test program plug-ins for complex models. For researchers, appropriate model plug-ins can be selected to suit their needs. The reported results include the stepwise regression results and the confidence interval of unbiased correction at the 95% significance level. If the confidence interval does not contain 0, it indicates that the intermediary effect exists; otherwise, this effect is non-existent. Therefore, the method as mentioned above is adopted in this study to further verify the robustness of the results about mediating effect.

(2) Panel threshold model. There may be no linearity whether in the relationship between agricultural land transfer and agricultural carbon emissions, or in the relationship between other social and economic factors and agricultural carbon emissions. Therefore, it is necessary to introduce a nonlinear adjustment mechanism to further explore the relationship between agricultural land transfer and agricultural carbon emissions. Herein, the panel threshold regression model proposed by Hansen [57] is adopted to carry out the regression analysis of agricultural land transfer and agricultural carbon emissions, with the urbanization level (the proportion of urban population in the total population) as the threshold dependent variable. The panel threshold model is expressed as follows:

$$\ln TC_{it} = \beta_0 + \alpha \ln TC_{it} + \beta_1 \ln F_{it} \times I(urban_{it} \le \eta) + \beta_2 \ln F_{it} \times I(urban_{it} > \eta) + control_{it} + \varepsilon_{it} \quad (5)$$

where urban represents a threshold dependent variable; η indicates the threshold value; I denotes the indicator function. In two scenarios, one being that the urbanization level falls below the threshold value ($urban_{it} \le \eta$) and the other being that the urbanization level exceeds the threshold value ($urban_{it} > \eta$), the impact of agricultural land transfer on agricultural carbon emissions is β_1 and β_2, respectively. The threshold model can simultaneously estimate the threshold value of the urbanization level and the slope value. The significance of the threshold effect was tested, that is, the original hypothesis H_0; $\beta_1 = \beta_2$. If the original hypothesis is rejected, the alternative hypothesis is accepted, that is, under different urbanization levels, the impact of agricultural land transfer on agricultural carbon emissions varies significantly.

3.2. Variable Definition and Data Source

(1) Explanatory variable: the explanatory variable used in this study is agricultural land transfer, which refers to the transfer of land management rights to other farmers or

organizations by the farmers with land contract management rights in rural areas. According to the existing research results, agricultural land transfer is mostly replaced by cultivated land transfer indicators [58]. Therefore, the transfer area of household contracted farmland in each province is used to represent the transfer of agricultural land in each province as the explanatory variable of this study.

(2) Explained variable: the explained variable used in this study is agricultural carbon emissions, with the narrow sense of agricultural (planting) carbon emissions as the research object. It is defined as the carbon emissions generated during the use of agricultural land, mainly including the carbon emissions generated during the use of chemical fertilizers, pesticides, agricultural films and agricultural diesel, as well as the carbon emissions generated during the irrigation and tillage of agricultural land [59]. The carbon emission accounting formula is expressed as:

$$TC = \sum_{i=1}^{n} O_i = \sum_{i=1}^{n} q_i \times \rho_i \quad (6)$$

where TC represents the total agricultural carbon emission, O_i indicates the carbon emission of each carbon emission form, q_i denotes the quantity of each carbon emission form, and ρ_i refers to the carbon emission coefficient of each form of carbon emissions. The coefficient values of this study are detailed in the research of Ding (2019).

(3) Intermediate variable: agricultural materials input. The input of agricultural materials includes the input of agricultural chemical material and that of agricultural machinery. Among them, the input of agricultural chemical elements includes various agricultural chemicals, such as chemical fertilizers, pesticides and agricultural films, all of which are inputted by the agricultural production entities in the process of crop production. Considering the difficulty in measuring the total input of agricultural chemical material, it can be found out that chemical fertilizer is one of the most important input factors in agricultural production in China, which plays a significant role in promoting grain production [60]. In the meantime, it also contributes significantly to the total agricultural carbon emissions. Therefore, the ratio of fertilizer application to crop planting area in each province is adopted to represent the input of agricultural chemical elements. Referred to as the agricultural machinery and equipment invested by farmers and other production entities in the process of crop production, agricultural machinery input can be used to indicate the level of mechanization in the process of agricultural production. In the existing research results, the total power of agricultural machinery is mostly used to represent the input of agricultural machinery. However, this index is not applicable to accurately indicate the input level of agricultural machinery. This is due to the difficulty in collecting the data on the total power of agricultural machinery at the level of farmers and the fact that the cross regional service of agricultural machinery and the socialized service of agricultural machinery are common in China. Therefore, the total power of regional agricultural machinery is unfit to fully reflect the input of agricultural machinery. Therefore, the comprehensive agricultural machine utilization rate of crop cultivation and harvest as used by the Ministry of Agriculture is adopted in this study to measure the level of agricultural mechanization. This index is the weighted average value of machine cultivation rate, machine sowing rate and machine yield.

(4) Other variables: considering that agricultural carbon emissions may be affected by other factors, other control variables are also introduced into this study, including: ① Agricultural fiscal level: Agricultural finance refers to the government's expenditure on agricultural production activities. The higher the level of expenditure, the more conducive it will be to improving agricultural technology. Furthermore, it has a significant impact on agricultural carbon emissions. In the existing studies, the proportion of fiscal expenditure spent on supporting agriculture to the total agricultural production value is often used to indicate the agricultural financial level. Since the definition of agricultural carbon emissions in this study is specific to planting

carbon emissions, the ration of the total output value of the planting industry to fiscal expenditure on supporting agriculture is used in this study to indicate the agricultural financial level of each province. ② Agricultural land resource endowment: Due to the differences in the amount of agricultural land resources in various regions, there are variations in the status and scale of agricultural production between different regions. Consequently, there are significant differences in agricultural carbon emissions between various regions. Therefore, the per capita cultivated land area of the planting industry in each province is used in this study to indicate the endowment of agricultural land resources in each province. ③ Agricultural population scale: The scale of agricultural population tends to have immediate effects on the regional structure and scale of agricultural production, thus affecting the amount of regional agricultural carbon emissions. Therefore, the number of employees in the planting industry in each province is used in this study to indicate the size of agricultural population. ④ Structure of agricultural output value: It is expressed as the ratio of the output value of planting industry to the total output value of agriculture, forestry, animal husbandry and fishery. ⑤ Agricultural planting structure: It is indicated by the ratio of the sown area of grain crops to the total sown area of crops.

The provincial panel data from 2005 to 2019 are selected for use in this study. Due to the serious lack of data in Tibet, it is excluded from the sample. Finally, 30 provincial administrative regions in mainland China are selected as the research objects. The sample data are sourced from the "China Statistical Yearbook", "China Rural Statistical Yearbook", "China rural operation and management statistical annual report", and "China Agricultural Machinery Industry Yearbook" of the corresponding years. In order to eliminate the impact of variable dimensions and ensure the stability of the data, logarithmic processing is carried out for agricultural land transfer, agricultural carbon emission and agricultural chemical element input. Table 1 lists the descriptive statistics of variables.

Table 1. The descriptive statistics of variables.

Variable Name	Mean	Std. Dev.	Min	Max
Agricultural land transfer (lnF)	12.63	1.38	8.70	15.34
Agricultural carbon emissions (lnTC)	5.27	1.01	2.44	6.77
Agricultural chemical element input (lnCP)	5.82	0.36	4.72	6.68
Agricultural machinery input (am)	0.50	0.24	0.02	1.14
Financial level of agriculture (fsa)	0.39	0.56	5.72	1.74
Agricultural land resource endowment (area)	1.09	0.77	0.30	4.79
Agricultural population size (population)	4.92	3.59	0.15	16.98
Agricultural output value structure (pvs)	0.52	0.09	0.34	0.75
Agricultural planting structure (ps)	0.65	0.13	0.33	0.97

4. Results and Discussion

4.1. Regression Analysis

(1) Benchmark regression. Table 2 shows the baseline regression results obtained for the impact of agricultural land transfer on agricultural carbon emissions. In the absence of control variables, the simple regression of agricultural carbon emissions is performed only on the transfer of agricultural land, with the estimation coefficient being significantly positive at the 1% statistical level. When control variables are introduced and fixed effects are considered for regression estimation, the estimated coefficient of agricultural land transfer remains significantly positive at the 1% statistical level. It is indicated that agricultural land transfer has a significant positive effect on agricultural carbon emissions, as does the endowment of agricultural land resources and the size of agricultural population. Conversely, the level of agricultural finance and agricultural planting structure has a significant negative effect on agricultural carbon emissions.

Table 2. Results of baseline regression.

	lnF	Fsa	Area	Popu	Pvs	Ps	Constant	R^2
Without control variables	0.08 *** (0.01)						4.84 *** (0.13)	0.23
Add control variables	0.10 *** (0.01)	−0.17 *** (0.01)	0.16 *** (0.03)	0.06 *** (0.01)	0.13 (0.20)	−0.79 *** (0.13)	4.77 *** (0.15)	0.51

Note: *** is significant at the level of 1%, and Se values are in brackets.

(2) Intermediary effect test: SPSS 25.0 software and process 4.0 macro program plug-in are applied to conduct regression analysis on the sample data. The results are detailed as follows which are showed in Table 3. In regression equation 1, the impact coefficient of agricultural land transfer on agricultural carbon emissions is 0.29, which passes the test at a significance level of 1%. That is to say, agricultural land transfer has a significant positive impact on agricultural carbon emissions. In the regression equation 2, the influence coefficient of agricultural land transfer on agricultural chemical element input is 0.03, which passes the test at the 5% significance level as well. That is to say, agricultural land transfer has a significant positive impact on agricultural chemical element input. In regression equation 3, the influence coefficient of agricultural land transfer on agricultural machinery factor input is 0.063, which also passes the test at the 1% significance level. That is to say, agricultural land transfer also has a significant positive impact on agricultural machinery factor input. In regression equation 4, the influence coefficient of agricultural land transfer, agricultural chemical element input and agricultural machinery element input on agricultural carbon emissions is 0.30, 0.79 and −0.49, respectively, all of which pass the test at the 1% significance level. That is to say, both agricultural land transfer and agricultural chemical element input have a significant positive impact on agricultural carbon emissions. By contrast, agricultural machinery element input has a significant negative impact on agricultural carbon emissions.

Table 3. Intermediary effect test results of agricultural land transfer on agricultural carbon emissions.

Variables	Regression Equation (1) lnTC		Regression Equation (2) lnap		Regression Equation (3) Am		Regression Equation (4) lnTC	
	β	t	β	t	β	t	β	t
lnF	0.29 (0.020)	15.24 ***	0.03 (0.01)	2.17 **	0.06 (0.01)	8.60 ***	0.30 (0.02)	17.08 ***
lnap							0.79 (0.06)	13.50 ***
am							−0.49 (0.11)	−4.30 ***
fsa	−0.70 (0.04)	−16.681 ***	−0.02 (0.03)	−0.78	−0.05 (0.02)	−3.21 ***	−0.66 (0.04)	−18.25 ***
area	0.15 (0.04)	3.880 ***	−0.17 (0.03)	−5.91 ***	0.11 (0.02)	7.62 ***	0.34 (0.04)	8.99 ***
popu	0.13 (0.01)	15.655 ***	−0.01 (0.01)	−1.96 **	−0.01 (0.00)	−3.01 ***	0.14 (0.01)	19.07 ***
pvs	−0.66 (0.26)	−2.550 **	−0.84 (0.19)	−4.37 ***	0.27 (0.10)	2.67 ***	−0.13 (0.23)	0.58
ps	−0.53 (0.21)	−2.590 ***	−0.16 (0.15)	−1.04	0.28 (0.08)	3.56 ***	−0.27 (0.18)	−1.53
R	0.90		0.40		0.69		0.93	
R^2	0.81		0.16		0.48		0.86	
F	308.59 ***		13.96 ***		67.63 ***		349.05 ***	

Note: **, and *** are significant at the level of 5%, and 1%, respectively, and Se values are in brackets.

From the above results, it can be concluded that agricultural land transfer exerts a partial intermediary effect on agricultural carbon emissions by affecting agricultural material input. Therefore, the first part of the research hypothesis proposed in this study is supported. Moreover, agricultural chemical factor input exerts a positive effect on agricultural carbon emissions, while agricultural machinery factor input has a negative effect on agricultural carbon emissions. Therefore, the second part of the research hypothesis proposed in this study is also supported. In terms of control variables, the impact of agricultural land resource endowment and agricultural population size on agricultural carbon emissions passes the test at the significance level of 1%. Furthermore, the impact coefficient is positive, indicating the promoting effect of agricultural land resource endowment and agricultural population size on agricultural carbon emissions. As for the impact of agricultural financial level on agricultural carbon emissions, it also passes the test at the significance level of 1%. Furthermore, the impact coefficient is negative, which indicates that to a certain extent the target of agricultural carbon emission reduction can be achieved if the local government increases its support for agriculture and promotes the progress in agricultural production technology.

In order to further verify the robustness of the intermediary effect, bootstrap is used to repeatedly extract the sample data for 5000 times and the default 95% unbiased correction interval is used to test the intermediary effect. The results are shown in Table 4. The confidence interval is [0.25, 0.33] and [0.26, 0.33] for the total effect and direct effect, respectively. The confidence interval is [0.01, 0.05] for the intermediary path of "agricultural land transfer → agricultural chemical element input → agricultural carbon emission". The confidence interval is [−0.05, −0.01] for the intermediate path of "agricultural land transfer → input of agricultural machinery factors → agricultural carbon emissions". The confidence interval does not contain 0, which confirms the significance effect propagation paths.

Table 4. Bootstrap test results.

Effect Propagation Path	Coefficient	SE	BootLLCI	BootULCI
Total effect	0.29	0.02	0.25	0.33
Direct effect	0.30	0.02	0.26	0.33
agricultural land transfer → agricultural chemical element input → agricultural carbon emission	0.02	0.01	0.01	0.05
agricultural land transfer → input of agricultural machinery factors → agricultural carbon emissions	−0.03	0.01	−0.05	−0.01

4.2. Threshold Effect Test

Despite agricultural factor input verified as an important medium in the impact of agricultural land transfer on agricultural carbon emissions, the mechanism of this impact may also be affected by other social and economic factors, which leads to a non-linear relationship between them. Therefore, it is necessary to introduce a non-linear mechanism into the model. There are plenty of research results showing an inverted "U" type relationship between urbanization level and environmental pollution [61,62], and agricultural land transfer has a significant impact on urbanization. Therefore, the urbanization level is taken as a threshold dependent variable in this section to analyze the impact of agricultural land transfer on agricultural carbon emissions under the context of different urbanization levels.

(1) Threshold estimate: In this study, Stata 17.0 is applied to repeatedly sample 500 times with the Bootstrap method to test the threshold effect of explanatory variables. The results are shown in Table 5. The urbanization level passes the single threshold test, but the double threshold fails the significance test. At the same time, Figure 1 shows the model likelihood ratio function diagram of the panel threshold model drawn under a single threshold to verify the threshold estimate. The critical value of the LR statistic is 7.35 at the significance level of 5%, and the LR value corresponding to the threshold value of 0.73 falls below 7.35, which is consistent with the reality.

Table 5. Threshold estimation test.

Number of Thresholds	F	p	10%	5%	1%	Threshold	95% Confidence Interval
single	44.85 *	0.082	42.03	50.27	70.61	0.73	[0.72, 0.75]
double	16.23	0.642	42.28	67.54	89.65	0.83	[0.63, 0.85]

Note: * is significant at the level of 10%.

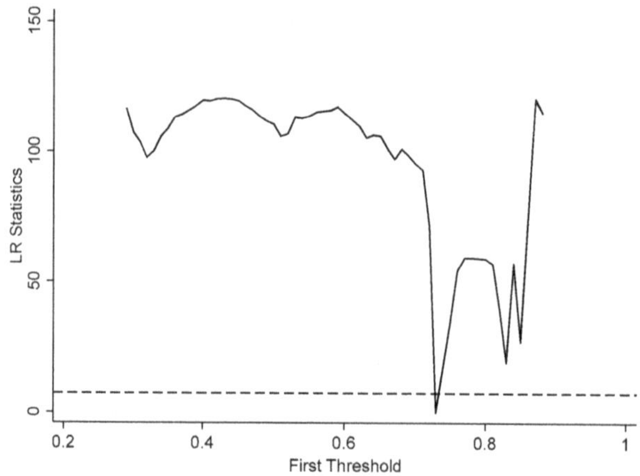

Figure 1. Threshold value and confidence interval of panel threshold model.

(2) Threshold regression results. The panel threshold model is applied to analyze the sample data, with the regression results listed in Table 6. According to the results of panel threshold regression, the impact of agricultural land transfer on agricultural carbon emissions is constrained by the threshold of urbanization level. When urban ≤ 0.73, the impact coefficient of agricultural land transfer on agricultural carbon emissions is 0.06. Agricultural land transfer exerts a positive effect on agricultural carbon emissions. Given the rapid development of rural land transfer, rural labor will concentrate in cities and towns, which improves the urbanization level. At the early stage of urbanization, rural surplus labor definitely increases agricultural capital investment to offset the loss of economic benefits caused by the outflow of agricultural labor, thus increasing agricultural carbon emissions. When urban > 0.73, the impact coefficient of agricultural land transfer on agricultural carbon emissions is -0.06. This is suspected to be due to the fact that the development of urbanization to a certain stage prompts the emergence of "anti-urbanization", as manifested in the flow of labor, capital and other factors back to the countryside, thus improving the conditions of agricultural production and driving the progress in agricultural production technology. In order to mitigate the negative external effects of agricultural production on the ecological environment, the government will also introduce the relevant environmental protection policies and regulations, which can motivate agricultural workers to improve their awareness of green production and increase the use of green and clean energy, thus comprehensively promoting the shift from traditional agricultural production to the green and efficient production characterized by "low input, high output and low pollution". Ultimately, agricultural carbon emissions are reduced. Based on the above research results, the impact of agricultural land transfer on agricultural carbon emissions shows an inverted "U" relationship under the constraint of urbanization level, which rises first and then falls. When the urbanization level exceeds a certain threshold, agricultural land transfer exerts an inhibitory effect on agricultural carbon emissions.

Table 6. Threshold regression results.

Variables	lnTC	Variables	lnTC
lnF(urban ≤ η)	0.06 *** −0.01	pvs	−0.24 ** −0.12
lnF(urban > η)	−0.06 *** −0.02	ps	−0.06 −0.12
fsa	−0.14 *** −0.01	constant	0.69 ** −0.31
area	0.11 *** −0.02	R^2-within	0.69
lnpopu	0.05 *** −0.01	F	89.97

Note: **, and *** are significant at the level of 5%, and 1%.

5. Conclusions and Policy Recommendations

5.1. Conclusions

Based on China's provincial panel data of agricultural land transfer and agricultural carbon emissions from 2005 to 2019, the intermediary effect model is applied in this study to test the impact path and transmission mechanism of agricultural land transfer on agricultural carbon emissions. Furthermore, the panel threshold regression model is used to empirically test the threshold effect of agricultural land transfer on agricultural carbon emissions. On this basis, the following conclusions are drawn:

(1) Agricultural land transfer can affect agricultural carbon emissions through agricultural materials input. Specifically, agricultural chemical factor input has a positive impact on agricultural carbon emissions (0.79), while agricultural machinery factor input has a negative impact on agricultural carbon emissions (−0.49).

(2) The urbanization level exerts a significant single threshold effect on the impact of agricultural land transfer on agricultural carbon emissions. Under the threshold constraint of urbanization level, the relationship between agricultural land transfer and agricultural carbon emissions shows an inverted "U" shape. When the urbanization level falls below 0.73, agricultural land transfer exerts a promoting effect on agricultural carbon emissions. When the urbanization level exceeds 0.73, the transfer of agricultural land has an inhibitory effect on agricultural carbon emissions.

5.2. Policy Recommendations

(1) It is recommended to change the input structure of agricultural elements and reduce the intensity of chemical elements utilization. According to the above research results, the input of agricultural chemical elements can have a promoting effect on agricultural carbon emissions, while the input of agricultural machinery elements can exert an inhibiting effect on agricultural carbon emissions. Different management methods will have an impact on the carbon emissions from agricultural land [63]. Imposing a reasonable control on the input of agricultural chemical elements and improving the level of agricultural mechanization can reduce agricultural carbon emissions. From the perspective of the government, first, it is necessary to effectively regulate the use of agricultural chemicals at the institutional level for ensuring the agricultural ecological safety with institutional strength, including the formulation of relevant laws and regulations to agricultural carbon emissions, the establishment of a monitoring mechanism for the quality of agricultural land ecological environment, the collection of agricultural environmental taxes [64], and the increase in agricultural carbon pollution penalties. Second, the government is supposed to increase the purchase subsidies offered to farmers for using green agricultural chemicals and agricultural machinery as well as include green chemical subsidies and agricultural machinery subsidies in the ecological compensation system. This would encourage farmers to purchase green agricultural chemicals and advanced agricultural machinery [65,66]. Finally, efforts

should be made to improve the awareness of environmental protection among agricultural practitioners. This is essential for environment protection [67,68]. By publicizing the knowledge about ecological and environmental protection through mass media, the internet and other means, agricultural practitioners can better understand that the excessive input of agricultural chemicals is one of the contributors to agricultural carbon emissions. This is conducive to improving the ecological and environmental awareness of agricultural practitioners, which prompts them to reduce agricultural carbon emissions by adopting environmentally friendly technologies. From the perspective of farmers, improving the utilization efficiency of agricultural chemicals is a potential solution to reducing agricultural carbon emission. According to the survey conducted by the Ministry of Agriculture and Rural Affairs of China, the utilization rate of chemical fertilizer for grain crops in China was only 37.8% in 2017, while that of major European countries was about 65% in the same period, which indicates a significant gap. Therefore, it is worth considering the popularization of various efficient fertilization technologies such as soil testing, formulated fertilization, mechanical fertilization, planting and fertilizing, so as to reduce the amount of chemical fertilizer applied while improving the efficiency of chemical fertilizer utilization.

(2) It is suggested that the pace of urbanization can be accelerated to give full play to the inhibitory effect of high urbanization on agricultural carbon emissions. According to the above research, the impact of agricultural land transfer on agricultural carbon emissions is constrained by the threshold of urbanization level. Given the high urbanization level, agricultural land transfer exerts an inhibitory effect on agricultural carbon emissions. As for the potential negative effects of population mobility caused by agricultural land transfer, they include economic and cultural aspects [69]. Therefore, some measures may be suitable for promoting the high-quality improvement of urbanization level through agricultural land transfer. First, the government is supposed to play its role in organization and coordination, with various channels involved in the prompt delivery of employment information to farmers. Meanwhile, it is crucial to increase vocational training for farmers and improve their labor skills and overall quality. This is significant to ensuring that farmers have the ability to perform non-agricultural work and that non-agricultural labor meets market demand. Second, it is necessary to deepen the reform of the registered residence system, accelerate the unified registration and management of urban and rural household registration, promote the synchronous transformation of occupation and identity for non-agricultural employment farmers, reinforce the long-term guarantee mechanism for the citizenization of migrant workers, fully recognize the citizenship of non-agricultural employment farmers, and genuinely integrate non-agricultural employment farmers into the city. Third, the government should put in place the corresponding social security system to reduce potential risks for non-agricultural farmers [70], so as to resolve the problems encountered by the urban farmers in medical care, housing and education received by their children. In the meantime, as a basis for the survival of farmers, agricultural land resources are exposed to certain survival risks for the main body of agricultural land transfer. Therefore, it is essential to improve the effectiveness of rural social security progressively to replace the social security function of rural land, establish the employment security system for those farmers losing their land, help them to find new jobs, and solve their concerns.

Author Contributions: Conceptualization, Y.T.; methodology and software, M.C. All authors have read and agreed to the published version of the manuscript.

Funding: This study was funded by the Social Science Foundation of China (21BGL288).

Institutional Review Board Statement: Not applicable.

Informed Consent Statement: Not applicable.

Data Availability Statement: The sample data are sourced from the corresponding years of "China Statistical Yearbook", "China Rural Statistical Yearbook", "China rural operation and management statistical annual report", and "China Agricultural Machinery Industry Yearbook".

Conflicts of Interest: The authors declare no conflict of interest.

References

1. An, R.; Zhu, G.H. Clustering of economic efficiency of urban energy carbon emissions based on decoupling theory. *Energy Rep.* **2022**, *8*, 9569–9575. [CrossRef]
2. Smith, P.; Bustamante, M.; Ahamad, H. Agriculture, forestry and other land use (AFOLU). In *Climate Change 2014: Mitigation of Climate Change, Contribution of Working Group III to the Fifth Assessment Report of the Intergovernmental Panel on Climate Change*; Cambridge University Press: Cambridge, UK, 2014.
3. Tian, C.S.; Chen, Y. Calculation of China's inter provincial agricultural carbon emissions and evaluation of low-carbon level – based on derived indicators and the application of TOPSIS method. *J. Nat. Resour.* **2021**, *36*, 395–410.
4. Ran, H.G.; Wang, J.H.; Wang, D.X. Study on the changing tendency and counter-measure of carbon emissions produced by agricultural production in China. *Issues Agric. Econ.* **2011**, *32*, 32–38, 110–111.
5. Luo, X.; Ao, X.; Zhang, Z. Spatiotemporal variations of cultivated land use efficiency in the Yangtze River Economic Belt based on carbon emission constraints. *J. Geogr. Sci.* **2020**, *30*, 535–552. [CrossRef]
6. Bennetzen, E.H.; Smith, P.; Porter, J.R. Decoupling of greenhouse gas emissions from global agricultural production: 1970–2050. *Glob. Chang. Biol.* **2016**, *22*, 763–781. [CrossRef]
7. Peter, C.; Helming, K.; Nendel, C. Do greenhouse gas emission calculations from energy crop cultivation reflect actual agricultural management practices?—A review of carbon footprint calculators. *Renew. Sustain. Energy Rev.* **2017**, *67*, 461–476. [CrossRef]
8. Hu, C.Z.; Huang, X.J.; Zhong, T.Y. Analysis of China's Carbon Emission Characteristics and Dynamic Evolution. *China's Popul. Resour. Environ.* **2008**, *3*, 38–42.
9. Li, G.Z.; Li, Z.Z. Empirical analysis on decomposition of carbon emission factors of agricultural energy consumption in China: Based on LMDI model. *Agric. Technol. Econ.* **2010**, *10*, 66–72.
10. Vita, K.S. The Impact of Macroeconomic Indicators on Carbon Emission in Indonesia. *J. Perspekt. Pembiayaan Dan Pembang. Daerah.* **2022**, *10*, 53–62.
11. Shen, L.Y.; Wu, Y.; Lou, Y.L. What Drives the Carbon Emission in the Chinese Cities?—A Case of Pilot Low Carbon City of Beijing. *J. Clean. Prod.* **2018**, *174*, 343–354. [CrossRef]
12. Gougoulias, C.; Clark, J.M.; Shaw, L.Z. Role of Soil Microbes in the Global Carbon Cycle: Tracking the Below-ground Microbial Processing of Plant-derived Carbon for Manipulating Carbon Dynamics in Agricultural Systems. *J. Sci. Food Agric.* **2014**, *94*, 2362–2371. [CrossRef] [PubMed]
13. Arora, V.K.; Boer, G.J.; Tatiana, T.M. Carbon–Concentration and Carbon–Climate Feedbacks in CMIP5 Earth System Models. *J. Clim.* **2013**, *26*, 5289–5314. [CrossRef]
14. Zhao, Q.; Zhou, Y.L.; Fang, Q.S. Analysis of space-time evolution of carbon emissions in central China and its influencing factors. *J. Environ. Sci.* **2022**, *10*, 1–11. [CrossRef]
15. Nyambuu, U.; Semmler, W. Climate Change and the Transition to a Low Carbon Economy—Carbon Targets and the Carbon Budget. *Econ. Model.* **2020**, *84*, 367–376. [CrossRef]
16. Hidalgo, M.C.; Rodríguez, P.; Aumente, R. Energy and Carbon Emission Savings in Spanish Housing Air-conditioning Using Solar Driven Absorption System. *Appl. Therm. Eng.* **2008**, *28*, 1734. [CrossRef]
17. Bertram, G. Tradeable Emmision Permits and the Control of Greenhouse Gases. *J. Dev. Stud.* **1992**, *28*, 423–446. [CrossRef]
18. Tian, Y.; Zhang, J.B.; Li, B. Research on China's agricultural carbon emissions, measurement, time-space comparison and decoupling effect. *Resour. Sci.* **2012**, *34*, 2097–2105.
19. Li, B.; Zhang, J.B.; Li, H.P. Temporal and spatial characteristics of China's agricultural carbon emissions and decomposition of influencing factors. *China's Popul. Resour. Environ.* **2011**, *21*, 80–86.
20. Liu, Z.M.; Huang, X.J.; Lu, X.H. Prediction of China's carbon neutral path under the shared socio-economic path. *J. Geogr.* **2022**, *77*, 2189–2201.
21. Gao, B.; Fang, J.; Lu, X.L. Study on the relationship between regional agricultural carbon emissions and economic growth evolution and their emission reduction potential. *Resour. Environ. Arid Areas* **2017**, *31*, 13–18.
22. Jiang, S.L.; Lu, C.B. Externality and Heterogeneity of Environmental Regulation's Impact on Carbon Emission Efficiency – Analysis Based on Cluster Synergy of Productive Services. *East China Econ. Manag.* **2022**, *9*, 1–14.
23. Zhao, Y.H.; Wang, Z.T.; Zhan, W.X.; Gai, Z.X. Temporal and spatial differentiation characteristics and influencing factors of carbon emissions from farmland use transformation in Heilongjiang Province. *Southwest Agric. J.* **2022**, *10*, 1–11.
24. Dai, X.W.; He, Y.Q.; Zhong, Q.B. A study on the driving factors and contributions of China's agricultural energy consumption and carbon emissions change based on Kaya identity expansion and LMDI index decomposition method. *Chin. J. Ecol. Agric.* **2015**, *23*, 1445–1454.
25. Li, K.Q.; Ma, D.D.; Li, Y.M. Analysis of driving factors and trend prediction of agricultural carbon emissions in Nanjing based on STIRPAT model. *Res. Sci. Technol. Manag.* **2018**, *38*, 238–245.

26. Li, H.; Li, W.; Yao, X.L. Research on spatial-temporal differentiation of agricultural carbon emission influencing factors based on GWR model. *Res. Sci. Technol. Manag.* **2019**, *39*, 238–2451.
27. Xiong, C.H.; Chen, S.; Xu, L.T. Driving factors analysis of agricultural carbon emissions based on extended STIRPAT model of Jiangsu Province, China. *Growth Chang.* **2020**, *51*, 1401–1416. [CrossRef]
28. Han, H.B.; Zhong, Z.Q.; Guo, Y. Coupling and decoupling effects of agricultural carbon emissions in China and their driving factors. *Environ. Sci. Pollut. Res. Int.* **2018**, *25*, 25280–25293. [CrossRef]
29. Hong, M.Y.; Zheng, L.N. Damage to farmers' rights and interests in agricultural land transfer – economic analysis based on bazel's property rights theory. *J. Hebei Univ. Econ. Trade* **2019**, *40*, 14–20.
30. Lan, J.; Li, Q.M. Social multiplier effect of rural land transfer—Analysis of rural governance difficulties and farmers' participation. *J. Cent. China Norm. Univ. Humanit. Soc. Sci. Ed.* **2021**, *60*, 59–68.
31. Zhang, L.; Feng, K.W. A study on the economic performance of the institutional change of rural land transfer—An Empirical Analysis Based on the data of Zaozhuang City from 1958 to 2017. *Price Theory Pract.* **2018**, *04*, 138–142.
32. Cheng, X.Y.; Xin, G.X.; Chen, R.R. Impact of agricultural land transfer on agricultural ecosystem. *Chin. J. Ecol. Agric.* **2016**, *24*, 335–344.
33. Hu, H.; Yang, Y.B. Research on the application of chemical fertilizer by farmers from the perspective of factor substitution—Based on the data of farmers at fixed observation points in rural areas. *Agric. Technol. Econ.* **2015**, *03*, 84–91.
34. Zheng, X.Y.; Xu, Z.G. Resource endowment constraint, factor substitution and induced technological change—Taking the mechanization of grain production in China as an example. *China Econ. Q.* **2016**, *16*, 45–66.
35. Xu, H.X. Urban rural integration development, theoretical basis, realistic motivation and realization conditions. *J. Nanjing Agric. Univ.* **2020**, *20*, 94–101.
36. Liao, L.W.; Long, H.L.; Ma, E.P. Changes in rural labor factors and cultivated land utilization efficiency. *Econ. Geogr.* **2021**, *41*, 148–155.
37. Schultz, T.W. *The Source of Increasing Returns*; Peking University Press: Beijng, China, 2001.
38. Zheng, J.G.; Zhang, R.X.; Zeng, F. Influence of agricultural land transfer on fertilizer input, a case study of Shandong Province. *Resour. Sci.* **2021**, *43*, 921–931.
39. Zhu, P.X.; Su, M.; Yan, J. The influence of the scale and stability of transferred farmland management on Farmers' fertilizer input—A case study of rice production in four counties (cities) of Jiangsu Province. *J. Nanjing Agric. Univ.* **2017**, *17*, 85–94, 158.
40. Han, J.B.; Liu, S.Y.; Zhang, S.F. Rural land ownership confirmation, land transfer and rural labor non-agricultural employment, from the perspective of incomplete contract theory. *Northwest Popul.* **2019**, *40*, 11–22.
41. Besley, T. Property rights and investment incentives, theory and evidence from Ghana. *J. Political Econ.* **1995**, *103*, 903–937. [CrossRef]
42. Qiu, H.G.; Liu, L.; Li, D.W. Business Scale, Land Right Stability and Land Productivity—Empirical Analysis Based on the Survey Data at the Land Plot Level of Four Provinces in China. *China Rural Econ.* **2017**, *06*, 30–43.
43. Ying, R.Y.; He, Z.Z.; Zhou, N. Rural land ownership confirmation, property right status and long-term agricultural investment—A reexamination based on a new round of ownership confirmation reform. *China Rural Obs.* **2018**, *3*, 110–127.
44. Yan, X.; Jin, J.Y.; Liang, M.Z. Yield increasing effect of chemical fertilizer and fertilizer utilization efficiency of major grain crops in China. *Soil* **2017**, *49*, 1067–1077.
45. Yang, J.H. Research on the decoupling relationship between agricultural chemical input and agricultural economic growth—based on the data of 6 provinces and 1 city in East China. *J. Nat. Resour.* **2017**, *32*, 1517–1527.
46. Liu, Q.; Xiao, H.F. Impact of agricultural land operation scale and financial support policy on agricultural carbon emissions. *Resour. Sci.* **2020**, *42*, 1063–1073.
47. Deng, M.J.; Deng, J.J.; Liu, J.Y. Temporal and spatial evolution of carbon emissions and emission reduction potential of fertilizer application for grain crops in China. *Resour. Sci.* **2016**, *38*, 534–544.
48. Liu, Y.; Liu, H.B. Characteristics, influencing factors and peak analysis of agricultural carbon emissions in Shandong Province. *Chin. J. Ecol. Agric.* **2022**, *30*, 558–569.
49. Ding, B.G.; Yang, S.W.; Zhao, Y. Study on the spatio-temporal characteristics and decoupling effect of carbon emissions from the use of cultivated land resources in China. *China Land Sci.* **2019**, *33*, 45–54.
50. Yang, Q.L.; Zhao, R.Q.; Zhao, T. The relationship between agricultural carbon emission efficiency and food security at county level. *China's Agric. Resour. Zoning* **2022**, *5*, 1–17.
51. Zhang, Z.G.; Yuan, Z.; Liu, X. Analysis of the decoupling effect between agricultural carbon emissions and economic growth based on the perspective of input – Taking Henan Province as an example. *Res. Water Soil Conserv.* **2017**, *24*, 272–278.
52. Chen, Y.; Chen, W. Research on the relationship between agricultural mechanization, industrial upgrading and agricultural carbon emissions—Empirical analysis based on dynamic panel data model. *Agric. Technol. Econ.* **2018**, *5*, 122–133.
53. Liu, Q.; Xiao, H.F. What is the logic of agricultural land operation scale affecting agricultural carbon emissions—Mediation of factor input and regulation of cultural quality. *Rural Econ.* **2020**, *5*, 10–17.
54. Wen, Z.G.; Ye, B.J. Intermediary effect analysis—Method and model development. *Prog. Psychol. Sci.* **2014**, *5*, 731–745.
55. Christopher, D.N.; Joshua, P.; James, R. The effects of vocational interests on motivation, satisfaction, and academic performance, Test of a mediated model. *J. Vocat. Behav.* **2021**, *6*, 103583.

56. Kristopher, J.P.; Andrew, F.H. Asymptotic and resampling strategies for assessing and comparing indirect effects in multiple mediator models. *Behav. Res. Methods* **2008**, *40*, 879–891.
57. Hansen, B.E. Threshold effects in non- dynamic panels, Estimation, testing, and inference. *J. Econom.* **1999**, *93*, 345–368. [CrossRef]
58. Zhou, M.; Kuang, B.; Zhou, M. The Spatial and Temporal Evolution of the Coordination Degree in Regard to Farmland Transfer and Cultivated Land Green Utilization Efficiency in China. *Int. J. Environ. Res. Public Health* **2022**, *19*, 10208. [CrossRef] [PubMed]
59. Zhu, Y.Y.; Zhang, Y.; Piao, H.L. Does agricultural mechanization improve agricultural environment efficiency? Evidence from China's planting industry. *Environ. Sci. Pollut. Res. Int.* **2022**, *29*, 53673–53690. [CrossRef] [PubMed]
60. Liao, M.L.; Du, T.T.; Wu, D.D. Factor structure, technical efficiency and Rural Revitalization. *Fujian Forum* **2018**, *4*, 182–187.
61. Deng, S.H.; Guo, L.H. Research on the impact of urbanization process on haze pollution in the Yangtze River Economic Belt—Empirical Analysis Based on spatial panel model. *Res. World* **2019**, *7*, 36–44.
62. Xu, B.; Luo, L.Q.; Lin, B.Q. A dynamic analysis of air pollution emissions in China, Evidence from nonpara-metric additive regression models. *Ecol. Indic.* **2016**, *1*, 346–358. [CrossRef]
63. Yadav, S.; Kumar, R.; Chandra, M.S. Soil Organic Carbon Sequestration and Carbon Pools in Rice Based Cropping Systems in Indo-Gangetic Plains, An Overview. *Int. Res. J. Pure Appl. Chem.* **2020**, *1*, 122–136. [CrossRef]
64. Faisal, M.A.; Jun, L.; Miguel, I.G. A systems approach to carbon policy for fruit supply chains, carbon tax, technology innovation, or land sparing? *Sci. Total Environ.* **2020**, *767*, 144211.
65. Konstantinos, Z.; Konstantinos, P. A Bilevel Linear Programming Model for Developing a Subsidy Policy to Minimize the Environmental Impact of the Agricultural Sector. *Sustainability* **2022**, *14*, 7651.
66. Flavio, G. Rethinking the mineral fertilizer subsidy scheme to promote environmental protection in Italy. *Outlook Agric.* **2021**, *50*, 230–237.
67. Blooshi, L.S.; Ksiksi, T.S.; Gargoum, A.S.; Aboelenein, M. Climate Change and Environmental Awareness, a Study of Energy Consumption among the Residents of Abu Dhabi, UAE. *Perspect. Glob. Dev. Technol.* **2019**, *18*, 582. [CrossRef]
68. Liski, A.H.; Koetse, M.K.; Metzger, M.J. Addressing awareness gaps in environmental valuation, choice experiments with citizens in the Inner Forth, Scotland. *Reg. Environ. Chang.* **2019**, *19*, 2217–2229. [CrossRef]
69. Aragbonfoh, A.F.; Sodiq, S.J. *The Nexus between Nigerian Migrants and the Future of Global Security*; Springer International Publishing: Berlin/Heidelberg, Germany, 2022.
70. Nina, V.U.; Halvard, B.; Neil, A.W. Human security of urban migrant populations affected by length of residence and environmental hazard. *J. Peace Res.* **2021**, *58*, 50–66.

sustainability MDPI

Article

Study on the Spatial-Temporal Evolution of Land Use Ecosystem Service Value and Its Zoning Management and Control in the Typical Alpine Valley Area of Southeast Tibet—Empirical Analysis Based on Panel Data of 97 Villages in Chayu County

Shuping Fan [1], Peng Li [1,*], Qi He [1], Jiaru Cheng [1], Mingfeng Zhang [2], Nan Wu [3], Song Yang [4,5] and Shidong Pan [6]

1 College of Economics and Management, Anhui Agricultural University, Hefei 230036, China
2 Ministry of Natural Resources, Jianghuai Key Laboratory of Arable Land Resources Conservation and Ecological Restoration, Anhui Provincial Institute of Land and Spatial Planning, Hefei 230601, China
3 Anhui Provincial Academy of Eco-Environmental Science Research, Hefei 230071, China
4 University of Chinese Academy of Sciences, Beijing 100049, China
5 Xinjiang Institute of Ecology and Geography, Chinese Academy of Sciences, Wulumuqi 830011, China
6 Anhui Transportation Planning, Design and Research Institute Company Limited, Hefei 230088, China
* Correspondence: 12456007914@stu.ahau.edu.cn

Citation: Fan, S.; Li, P.; He, Q.; Cheng, J.; Zhang, M.; Wu, N.; Yang, S.; Pan, S. Study on the Spatial-Temporal Evolution of Land Use Ecosystem Service Value and Its Zoning Management and Control in the Typical Alpine Valley Area of Southeast Tibet—Empirical Analysis Based on Panel Data of 97 Villages in Chayu County. *Sustainability* **2022**, *14*, 10057. https://doi.org/10.3390/su141610057

Academic Editors: Lu Zhang, Bing Kuang and Bohan Yang

Received: 17 July 2022
Accepted: 8 August 2022
Published: 14 August 2022

Abstract: Under the background of ecological civilization construction and the overall planning of land and space, it is particularly important to explore the land use ecosystem service value and its zoning control. This paper, taking Chayu County, a typical alpine valley area of southeast Tibet as an example and based on the remote sensing interpretation data of three periods in 2000, 2010 and 2020, employs the three-level spatial scale from the village level, the township level to the county level to converge step by step, and uses a series of model algorithms to analyze and calculate the regional ecosystem service value and their dynamic changes, as well as spatial agglomeration and regional type division. The research shows that the land use types mainly consist of forest land, grassland and unused land, whose overall change range is small during the study period. The conversion of land use types is mainly between forest land, grassland and unused land and the land use index generally presents a spatial pattern of "high in the southwest and low in the northeast", showing a decreasing trend to some degree. ESVI generally presents a differentiation pattern of "high in the west and low in the east", with obvious spatial differentiation characteristics of kernel density, significant clustering and distribution characteristics and stable variation range, displaying an overall spatial pattern with characteristics of "dense in the west and sparse in the east, high in the north and low in the south". Based on the administrative village scale, the study area is divided into three different types of land use ecological function areas: habitat maintenance function area, biological protection function area and production support function area. Differentiated approaches to appropriate development and construction and the corresponding optimization paths of ecological protection will be put forward.

Keywords: land use; ecological service value; spatial-temporal evolution; spatial agglomeration; Southeast Tibet

1. Introduction

As a non-renewable resource, land is the most basic material for human production, life and ecology. Land use refers to the management and application of land in a certain period based on the attributes of land itself and the needs of economic and social development [1]. After the modern industrial civilization, the land use structure is out of balance, the environmental quality is deteriorating, and the resource reserves are drying up. Many pollutants caused by this have a serious impact on the balance of the ecological service system, resulting in the continuous decline of its service value. It is necessary to re-examine the balance between land use change and ecological service value so as to provide a scientific basis for the sustainable use of resources [2]. In particular, the rapid progress of urbanization and industrialization has revealed many negative effects related to ecology, climate and human settlements. Land use affects various types, areas and spatial distribution patterns of ecosystems, and also changes the structure, functions and processes of ecosystems, thus affecting the rational allocation of land resources [3]. Therefore, the quantitative research on the impact of land use change on the value of ecological services has become the research frontier and hot topic of many interdisciplinary subjects in the past century.

Ecosystem service function refers to the natural environmental conditions and functions formed and maintained by ecosystems and processes [4]. Ecological service value (ESV), as the core index to measure ecological security, is of great significance to the scientific management of ecosystems and the realization of sustainable development [5,6]. Humans began to study the ecosystem service system in the 1960s, however, due to the limitations of the research environment and technical means, only some research methods were provided, and their value was not quantitatively evaluated [7]. In the 1990s, Costanza et al. defined the research method of ecological service value for the first time, thus laying the research foundation of ESV [8,9]. After the 21st century, many researchers such as Xie Gaodi et al. [10,11] conducted in-depth research on the basis of Costanza's study and formulated the "table of equivalent value of ecological services per unit area of China's terrestrial ecosystem". Since then, they have revised it to varying degrees according to China's land use conditions [12] and ESV has been widely used in the assessment of grassland [13,14], forest [15], farmland [16], cities [17,18], and coastal zones [19]. At present, the methods for estimating ecosystem service value mainly include the functional value method [20,21] and the equivalent factor method [22], but the former involves many parameters and is highly subjective, and the equivalent factor method is widely used [23]. The existing documentary achievements are substantial, which can provide technical ideas, model algorithms and other references for this study.

Throughout the current research, many scholars selected typical representative areas, and took the measurement of land use ecosystem service value as the main body to further explore the laws of space-time evolution and influencing factors. The spatial scale employed by the study is mainly macro and meso, and the micro scale level is scarcely used. It is even more rare to propose zoning differentiation management and control measures from the perspective of spatial zoning [24]. In order to further divide land use ecological function areas and put forward differentiated pattern optimization control measures, this paper, taking Chayu County, a typical alpine valley area in Southeast Tibet as an example and based on the remote sensing interpretation of three periods in 2000, 2010 and 2020 and the formation of 30 m $\times$ 30 m grid data, employs the three-level spatial scale from the village level, the township level to the county level to converge step by step, analyze the changes in the quantity and degree of regional land use, calculate the ecosystem service value and analyze the spatial-temporal evolution characteristics. The research results are expected to provide theoretical basis and technical support for deepening land use ecosystem service value.

2. Overview of the Study Area

Chayu county is a typical area of Southeast Tibet with high mountains and valleys in the western section of Hengduan Mountains (Figure 1). The terrain is high in the northwest and low in the southeast, with a wide vertical height difference. Affected by the Indian Ocean warm current in the south and clamped by derma snow mountain in the north, the high altitude and undisturbed natural environment jointly determine the high sensitivity of the ecosystem. The county covers an area of 31,400 km^2, with forest land taking the absolute advantage, followed by grassland and unused land. The unique topography, climate and hydrothermal conditions make it one of the regions with richest Mountain Biodiversity in Tibet and even China. Having a variety of ecological types and rich biological resources, with forests, wetlands, grasslands, lakes, deserts and other ecosystems distributed, the ecosystem in this county is extremely fragile and has poor anti-interference ability. Once damaged, it is difficult to recover and biodiversity is facing severe challenges. It is extremely important to carry out the research on the value of land-use ecosystem services in this region.

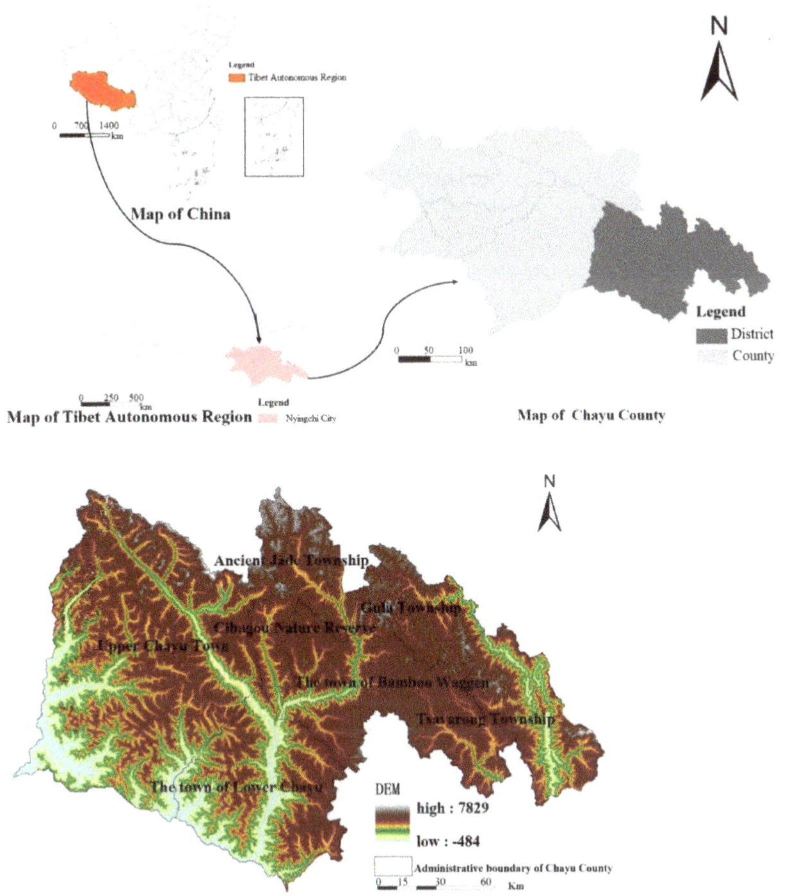

Figure 1. Study area location. (Note: the base map is made based on the standard map (Tibetan s (2020) No. 002) approved by the National Bureau of Surveying and mapping geographic information, and the base map is not modified).

3. Data Sources and Research Methodology

3.1. Sources of Data

Since 2000, China has increased social and economic construction and paid attention to the improvement of the ecological environment. Considering the availability of data and comparative differences, this study, taking 10 years as a period, selects three time spans of the case sites in 2000, 2010 and 2020. The administrative division data of Chayu County are provided by the State Administration of Surveying, Mapping and Geoinformation (https://www.ngcc.cn/ngcc/, accessed on 8 January 2022). The land use data of 30m spatial resolution in 2000, 2010 and 2020 in Chayu County are all from the land use status remote sensing monitoring database downloaded by Globe Land30 (http://www.globallandcover.com/, accessed on 25 December 2021), with 30 m multispectral images as the main data source for production, including (Landsat) TM5, ETM+ and HJ-1 multispectral images. According to the type with the largest area occupied within 30 m, the land use type is reclassified, so that the land use type is divided into 10 types of land cover: arable land, forest, grassland, shrubland, wetland, water body, tundra, artificial bare land and glacier and permanent snow. In view of the needs of this study, forests and shrublands, artificial surfaces, bare land and glaciers and permanent snow are classified as woodland, construction land and unused land, respectively. Data on grain prices, yields and sown areas are from the 2020 China Agricultural Product Price Survey Yearbook and the 2020 Tibet Statistical Yearbook.

3.2. Research Methodology

3.2.1. Land Use Change

(1) Model of land use quantity change

The analysis of the total change of land use type can help to understand the overall situation of regional land change, and the dynamic degree of land use can quantitatively present the speed of regional land use change; the formula is as follows:

$$K = \frac{S_{in} - S_{out}}{S_{i0}} \times \frac{1}{T} \times 100\% \quad (1)$$

where K is the annual change rate of land use type; S_{in} is the inflow area of a certain land type; S_{out} refers to the outflow area of a certain land type; S_{i0} is the area of a certain land type at the initial stage of the stage; T is the span of research years.

(2) Land use degree change model

The comprehensive index of land use degree (L) indicates the degree of human development and utilization of regional land and reflects the two-way impact of land on its natural attributes and human activities, which is an important indicator to measure the depth and breadth of regional land use [25]. The formula is as follows:

$$L = 100 \times \sum_{i=1}^{i=n} A_i \times B_i \quad (2)$$

where L is the index of land use degree; A_i is the classification index of the grade i, referring to the existing studies [26], unused land = 1, forest land, grassland and water area = 2, cultivated land = 3, construction land = 4; B_i is the percentage of land use type area of class i in the total area.

3.2.2. Ecosystem Service Value Assessment

(1) Value revision assessment

Revision based on grain price: according to the research of Xie Gaodi et al. [12], 1/7 of the economic value of the annual natural grain yield of the farmland with an average yield of 1 hm^2 is a standard ecosystem ecological service value equivalent factor. To eliminate the impact of crop price fluctuations in different years on the total value, taking the sown area, the yield and average price of crops of the five major crops (rice, wheat, highland barley, soybean and corn) in Tibet in 2020 as the basic data, the economic value of food crops in the farmland ecosystem per unit area is calculated by the formula as 297.21 yuan/km^2.

$$E_n = \frac{1}{7} \sum_{i=1}^{n} \frac{q_i p_i}{M} \quad (3)$$

where E_n is the economic value (yuan/hm^2) of providing food production services for the farmland ecosystem within the unit area of the study area; n is the main food crops in the study area; q_i is the price of crop i (yuan/kg); p_i is the total yield of crop i (kg); M is the total area of n kinds of food crops (hm^2).

Taking this as a benchmark and taking the spatial-temporal heterogeneity of ecosystems into account, the value coefficient of ecosystem services in the study area needs to be further revised. Referring to the biomass factor table of farmland ecosystems in each region of the country given by Xie Gaodi et al., the biomass factor of farmland ecosystems in the study area is determined to be 0.75, and through revising the biomass factors of various services provided by farmland ecosystems, the value coefficient tables of ecosystem services of different land use types will be generated. The ecosystem services value coefficients of land use type is shown in Table 1.

Table 1. Ecosystem services value coefficients of land use types in the study area. (Unit: RMB /km^2).

Ecosystem Services and Functions	Cultivated Land	Woodland	Grassland	Wetland	Waters	Unused Land
gas exchange	111.45	780.18	178.33	401.23	0.00	0.00
Climate regulation	198.39	601.85	200.62	3811.72	102.54	0.00
water conservation	133.74	713.30	178.33	3455.07	4542.85	6.69
Soil formation and protection	325.44	869.34	434.67	381.17	2.23	4.46
waste disposal	365.57	292.01	292.01	4052.46	4052.46	4.46
Biodiversity conservation	158.26	726.68	242.97	557.27	555.04	75.79
Food production	222.91	22.29	66.87	66.87	22.29	2.23
raw material	22.29	579.56	11.15	15.60	2.23	0.00
Entertainment and leisure	2.23	285.32	8.92	1237.14	967.42	2.23

(2) Value dynamics

The value of ecosystem services of different land use types is equal to the area of different land types multiplied by the corresponding ecosystem service value coefficient, and the change in the value of ecosystem services of land use types in the study area can be expressed by EV (value dynamic degree) of the ecosystem service. The formula is as follows:

$$EV = \frac{EAV_b - EVA_a}{EVA_a} \times \frac{1}{T} \times 100\% \tag{4}$$

where EAV_a and EAV_b are the ecosystem service value of a certain land use type at the initial stage and at the end of the study, respectively; T is the research years.

(3) Sensitivity index

The sensitivity index is employed to analyze the sensitivity of ecosystem services in the study area. The calculation formula is as follows:

$$I = \left| \frac{\frac{ESV_i - ESV_j}{ESV_j}}{\frac{L_i - L_j}{L_j}} \right| \tag{5}$$

where I is the sensitivity index of ecosystem service value; ESV_i is the ecosystem service value in year i; ESV_j is the ecosystem service value in year j; L_i is the index of land use degree in year i; L_j is the index of land use degree in year j.

3.2.3. Spatial Agglomeration of Ecosystem Service Value

(1) Kernel density analysis

Kernel density estimation (KDE), as one of the density estimation methods in the spatial analysis tools built into ArcGIS software, depends on a filter window to define nearby objects.

$$f_n = \frac{1}{nh} \sum_{i=1}^{n} k\left(\frac{x - x_i}{h}\right) \tag{6}$$

where f_n is the estimated value of ESVI distribution kernel density; n is the number of plots; h is the bandwidth or smoothing parameter; k is the kernel density function, and $x - x_i$ is the distance from the measured block x to the sample block x_i.

(2) Spatial autocorrelation analysis

Spatial autocorrelation analysis is an important method and effective means to quantitatively study spatial relationships and analyze spatial patterns. Ecosystem service value is directly related to the distribution of natural geographical elements and the social and economic development of the region. With randomness and structure in space, these factors have geoscience characteristics. Therefore, ecosystem service value, like various geographical entities, has a certain spatial correlation, and geoscience statistical analysis methods such as spatial autocorrelation analysis can be employed [27,28]. The global spatial autocorrelation (GSA) and local spatial autocorrelation (LISA) are used comprehensively to dig into the spatial pattern and evolution characteristics of ESVI, and reveal the correlation between the attribute values of spatial units and other attribute values in adjacent space. The spatial autocorrelation analysis is based on Geodal.18 software to complete.

4. Results and Analysis

4.1. Analysis of Land Use Change

4.1.1. Characteristics of Changes in Land Use Quantity

According to Figure 2, from 2000 to 2020, the county forest land accounted for the largest area, followed by grassland and unused land. The cultivated land distribution transferred gradually from the marginal area in 2000 to the inner river valley while the construction land was mainly distributed in the central area of each township, with the "strip" spatial distribution mainly in the Upper Chayu Town and Lower Chayu Town, and the increase in construction land in the past 20 years was small.

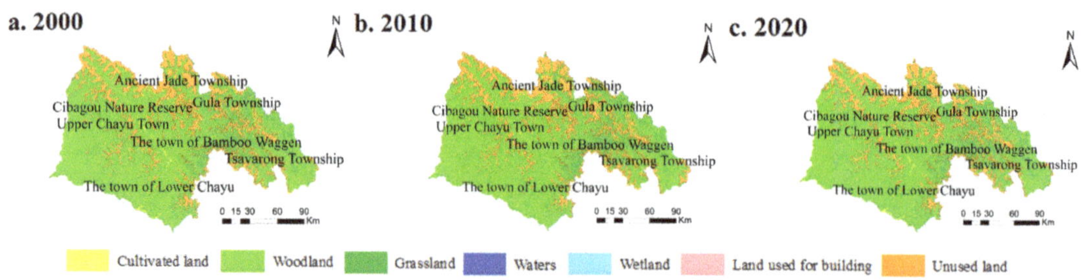

Figure 2. Land use status of the study area from 2000 to 2020.

The proportion of land use types in three different periods of 2000, 2010 and 2020 is shown in Table 2. From the perspective of the overall characteristics of land use types, the changes in the main land use types in the three periods except forest land, grassland and unused land are not obvious due to their small size. The highest proportion of area is forest land, followed by unused land, both of which account for more than 74% of the total area of the study area. From the perspective of the characteristics of land use type change, the area of unused land has increased by 45.44%, with a continuous increase of 3377.26 km^2 in the past 20 years while the grassland area has continued to decrease by 1935.62 km^2 in 20 years, a decrease of 29.29 percentage points. The area of the water area increased first and then decreased and the wetland area remained almost unchanged.

Table 2. Changes in the proportion of different land use types in the study area from 2000 to 2020 (Unit: %).

Land Use Type	2000	2010	2020
Cultivated land	0.27	0.14	0.19
woodland	54.56	53.55	50.04
grassland	21.04	21.76	14.88

Table 2. *Cont.*

Land Use Type	2000	2010	2020
Wetland	0.02	0.01	0.02
waters	0.44	0.37	0.42
land used for building	0.01	0.01	0.03
Unused land	23.66	24.16	34.42

The transfer matrix of different land use types from 2000 to 2020 (Table 3) is obtained by employing the analysis tool of Arc GIS. The main characteristics of land use transfer are as follows: (1) The forest land and grassland have the largest area of transfer-in and transfer-out, 2191.24 km^2 and 2978.08 km^2, respectively, in which the main source of transfer-in is cultivated land and unused land while the main source of transfer-out is unused land and waters; (2) Unused land, as a type of land that has not yet been utilized or is difficult to utilize, has the largest difference between the transfer-in area and the transfer-out area, and its main transfer-in and transfer-out source are cultivated land, forest land and grassland; (3) The transfer-in area and the transfer-out area of cultivated land are 21,934 km^2 and 22,917 km^2, respectively, and the main transfer-in and transfer-out sources are woodland and grassland, including a certain proportion of waters. This shows that the scale replacement between unused land, woodland and grassland in the study area has a great impact on the land use structure. The scale increase in the secondary land types of unused land has mainly been caused by glaciers and permanent snow, and the change of land types has first decreased and then increased in the past 20 years, with the increase from 507,861.61 km^2 to 975,967.01 km^2 in 2020. Affected by the natural environment and geographical location, the high altitude greatly hinders the entry of warm and humid air currents in the southern Indian Ocean, and due to the low temperature, the snow encroaching upon grasslands and woodlands is difficult to melt, making the unused land grow.

Table 3. Land use transfer matrix of the study area for 2000–2020. (Unit: km^2).

2000 \ 2020	Grassland	Cultivated Land	Land Used for Building	Woodland	Wetland	Waters	Unused Land
grassland	3629.95	3.81	1.56	660.35	0.29	5.64	2306.43
Cultivated land	35.35	31.15	3.45	14.08	0.01	0.40	0.97
land used for building	0.19	0.75	1.68	0.10	0.00	0.02	0.03
woodland	824.58	25.13	2.21	14,944.67	0.55	33.83	1304.94
Wetland	0.77	0.03	0.02	0.61	2.27	0.72	1.79
waters	7.09	0.12	0.04	13.64	1.14	81.96	32.98
Unused land	174.48	0.13	0.12	83.56	2.41	9.18	7162.68

4.1.2. Characteristics of Land Use Change

A gradually decreasing distribution pattern of "southwest- northeast" of the land use degree index during the study period can be seen from the spatial distribution map of the land use degree index from 2000 to 2020 (Figure 3). The variable quantity of land use degree from 2000 to 2010 was 3.35, and 3.24 from 2010 to 2020. Since the variable quantity of land use degree in these two periods was greater than 0, the overall study area in these two periods was in development.

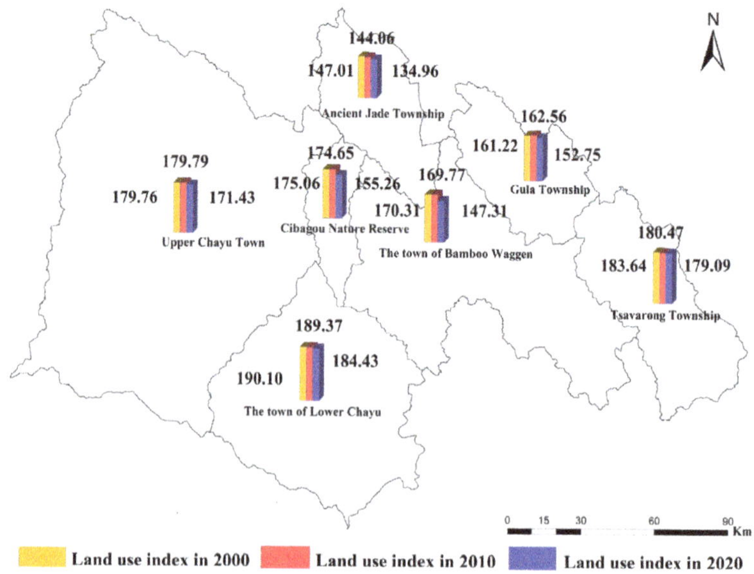

Figure 3. Spatial distribution of land use index in study areas, 2000–2020.

From the specific point of view of each township, the land use degree index showed a decreasing trend of different degrees from 2000 to 2020, of which Zhuwagen Town (23.00) decreased the most, followed by Cibagou Nature Reserve (19.79) and Guyu Township (12.06). The decrease is because of the increase in the scale of unused land. A considerable part of the unused land structure is glaciers and permanent snow, the scale of which has increased during the study period, resulting in a continuous decline in the impact of human activities on land use.

From the specific point of view of each land type, the utilization degree of forest land from 2000 to 2020 is the highest while the utilization degree of construction land is the lowest. The degree of utilization of unused land tended to increase year by year between 2000 and 2020 while the utilization of woodland and grassland showed a downward trend from 2000 to 2020. The degree of utilization of unused land showed an upward trend between 2000 and 2020, and the actual change was more obvious, which was since unused land accounted for a relatively large proportion of the land use structure, with an increase of up to 45.44% during the period. The conversion of land types was mainly based on unused land and grassland, so the variation of land use changed greatly. The actual change in the degree of land use of cultivated land, water areas and construction land between 2000 and 2020 is not obvious mainly due to the small base of the land scale itself.

4.2. *Value Analysis of Ecosystem Services*

4.2.1. Temporal Change Analysis of Ecosystem Services

Overall, ESV showed a decreasing trend between 2000 and 2020, from 964.596 billion yuan in 2000 to 866.642 billion yuan in 2020, a decrease of 16.98%. During the study period, the value of ecosystem services in woodland decreased the most, followed by grassland, with the smallest reduction in cultivated land, of which the reduction in woodland accounted for 70.55% of the total reduction. The value of ecosystem services in unused land increased the most while the wetlands increased the least, of which the increase in unused land accounted for 98.05% of the total increase. During the study period, the proportion of forest land area decreased from 54.56% in 2000 to 50.04% in 2020, but the proportion of ecosystem service value increased from 86.52% to 88.33%, with the net reduction in ESV of 69.108 billion yuan. The proportion of grassland area decreased from 21.04% in 2000 to

14.88% in 2020, and the proportion of ESV decreased from 11.06% to 8.70%, with the net reduction in ESV of 31.238 billion yuan, which shows that the contribution of woodland and grassland to ESV and the regulation of ecological environment are of great significance. The proportion of unused land area increased the most, from 23.66% in 2000 to 34.42% in 2020, and the proportion of ESV also showed an increasing trend, with a net increase in ESV of 3.237 billion yuan. The ratio of EVS from cultivated land to water areas is basically stable, and the ESV of wetlands is relatively small. The ESV changes of different types of land use are shown in Table 4.

Table 4. Changes in ESV of various types of land use in the study area from 2000 to 2020.

Land Use Type		Cultivated Land	Woodland	Grassland	Wetland	Waters	Unused Land
2000	Area (km^2)	85.41	17,135.91	6608.03	6.21	136.97	7432.56
	ESV (RMB 100 million)	13.16	8346.09	1066.44	8.68	140.35	71.24
2010	Area (km^2)	45.24	16,818.29	6833.31	3.54	114.81	7589.31
	ESV (RMB 100 million)	6.97	8191.40	1102.79	4.95	117.65	72.74
2020	Area (km^2)	61.12	15,717.01	4672.41	6.67	131.75	10,809.82
	ESV (RMB 100 million)	9.41	7655.02	754.06	9.32	135.00	103.61
2000–2010	ESV change value	−6.19	−154.70	36.36	−3.73	−22.71	1.50
	ESV change rate	−47.03%	−1.85%	3.41%	−43.00%	−16.18%	2.11%
	Area change value	−40.17	−317.62	225.28	−2.67	−22.16	156.75
	Area change rate	−47.03%	−1.85%	3.41%	−43.00%	−16.18%	2.11%
2010–2020	ESV change value	2.45	−536.38	−348.74	4.38	17.36	30.87
	ESV change rate	35.10%	−6.55%	−31.62%	88.42%	14.75%	42.43%
	Area change value	15.88	−1101.28	−2160.90	3.13	16.94	3220.51
	Area change rate	35.10%	−6.55%	−31.62%	88.42%	14.75%	42.43%
2000–2020	ESV change value	−3.74	−691.08	−312.38	0.64	−5.35	32.37
	ESV change rate	−28.44%	−8.28%	−29.29%	7.41%	−3.81%	45.44%
	Area change value	−24.29	−1418.90	−1935.62	0.46	−5.22	3377.26
	Area change rate	−28.44%	−8.28%	−29.29%	7.41%	−3.81%	45.44%

4.2.2. Spatial Change Analysis of the Value of Ecosystem Services

With the help of ArcGIS spatial analysis technology and the square grid units of 30 m×30 m, the land use data of three periods of the study area were completed. On this basis, the ESVI in each raster cell is measured and analyzed and the spatial interpolation is carried out by kriging. Meanwhile, the natural breakpoint used, and the real situation of the study area fully taken into consideration, the ESVI in each raster is divided into four levels of lower, low, higher, and high according to (1000,2200), (2200), (3400), (3400,4600) and (4600,5800). Then, the spatial pattern distribution map of ESVI in three periods of the study area from 2000 to 2020 is obtained.

It can be seen from Figure 4 that the overall spatial distribution pattern of the study area is "high in the west and low in the east". Specifically, the ESVI in the Middle East region is low while the ESPI in the east-west marginal area is higher. The ESVI in the western parts of the study area, such as Shangcha Town (Buzong Village, Xiba Village, Sports Village), Xiachayu Town (Shama Village, Buba Village, Rima Village) and Cibagou Nature Reserve, are larger while Guyu Township (Boxue Village, Bayi Village, Gujing Village), Gula Township (Shadui Village, Shamei Village, Oyu Village) and the high-altitude area of Ridong (Gada Village, Quwa Village) in the eastern part of Zhuwagen Town have smaller ESVI. ESVI is mainly based on two levels, the high and the low, whose average area accounts for 35% and 30%. The proportion of ESVI low-level area shows an increasing trend while the proportion of area high in ESVI and the ecological service value shows a decreasing trend, and the rate of change of area with low level of ESVI increases first and then decreases.

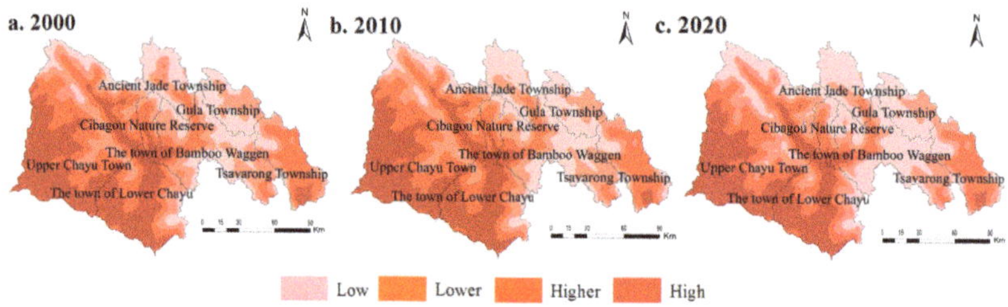

Figure 4. Spatial pattern of ESVI in the study area from 2000 to 2020.

4.2.3. Analysis of Dynamic Change Degree of Ecosystem Service Value

During the study period, the dynamic change degree of the unused land ESV is all positive while that of the other land types is positive or negative. Only the cultivated land's absolute value of ESV dynamic change degree in Xiachayu town is higher than the absolute value of the whole region, and that of the remaining townships is lower than the absolute value of the whole region, among which Zhuwagen Town, Tsavalong Township, Gula Township and Cibagou Nature Reserve have the smallest absolute value, indicating that the cultivated land in the above areas has decreased the most during the study period. The villages and towns where the absolute value of ESV dynamics of forest land is greater than the absolute values of the whole region include Zhuwagen Town, Guyu Township, Gula Township and Cibagou Nature Reserve, indicating that the forest lands in the above areas increased significantly during the study period. Therefore, the ecological environment quality has significantly improved. In terms of grassland, except for Shangchayu Town, Zhuwagen Town and Cibagou Nature Reserve, the absolute value of ESV dynamic change degree in the remaining townships is lower than the absolute value of the whole region, indicating that the large reduction in grassland areas in these places has an impact on the balance of the ecosystem to a certain extent. In regard to unused land, except for Xiachayu Town, Zhuwagen Town and Cibagou Nature Reserve, the absolute value of ESV dynamic change degree of unused land in the other townships is lower than the absolute value of the whole region, which can indicate that the use efficiency of unused land is gradually improving. The ESV dynamics of different land types are shown in Table 5.

Table 5. ESV dynamics of different land types in the study area from 2000 to 2020 (Unit: %).

(Township) Town Name	Cultivated Land	Woodland	Grassland	Wetland	Waters	Unused Land
Entire	−1.42	−0.41	−1.46	0.37	−0.19	2.27
Upper Chayu Town	−0.41	−0.34	−1.53	4.76	0.80	2.05
The town of Lower Chayu	−1.86	−0.15	−0.85	−2.32	−1.07	2.35
The town of Bamboo Waggen	0.00	−0.79	−2.47	1.21	−3.39	3.88
Tsavarong Township	0.00	−0.38	−0.12	1.43	1.16	1.41
Ancient Jade Township	0.64	−1.29	−1.30	−5.00	1.37	1.14
Gula Township	0.00	−0.77	−0.64	0.00	−2.56	1.10
Cibagou Nature Reserve	0.00	−0.83	−3.03	0.00	−4.02	3.97

4.2.4. Sensitivity Analysis of Ecosystem Service Value

Through measuring and analyzing the ecological sensitivity index of land use change in each township from 2000 to 2020, the range of overall ecosystem sensitivity index is [1.4501,4.6137]. With reference to relevant data [29,30], the area of ecosystem sensitivity index less than 1 is a non-sensitive area, and all towns and townships in the county cibagou nature reserves are sensitive areas. According to the numerical size of the sensitivity, the county can be divided into three types of areas: low sensitivity, moderate sensitivity, and high sensitivity. Under this subdivision, from 2000 to 2020, the townships that belong to the low sensitivity areas mainly include Shangchayu, Xiachayu, Zhuwagen, Gula Township and Nature Reserve, where the land use degree index is large while the increase in construction land is also large. However, thanks to the high EVS coefficient of forest land and grassland, the impact on the total amount of ESV is small, indicating that the ecological sensitivity of the above areas is in equilibrium. The moderately sensitive and highly sensitive areas correspond to Tsavalong Township and Gula Township, respectively, mainly located in the lower reaches of the Nu River and the Hengduan Mountains, with an average altitude of more than 2500 m. The types of land use in the region are mainly unused land and grassland, with strong environmental resilience, and the sensitivity index is generally in a good state.

4.3. Spatial Agglomeration Analysis of Ecosystem Service Value

4.3.1. Kernel Density Analysis

The spatial distribution density of ESVI is calculated by using the kernel density function, and the natural breakpoint method is employed to divide the density value into four levels: low density area (0–89), sub-low-density region (89–127), sub-high density (127–166) and high-density region (166–255). The spatial difference in the distribution of ESVI density in the study area is significant (Figure 5), and the fluctuation of kernel density in each year is small.

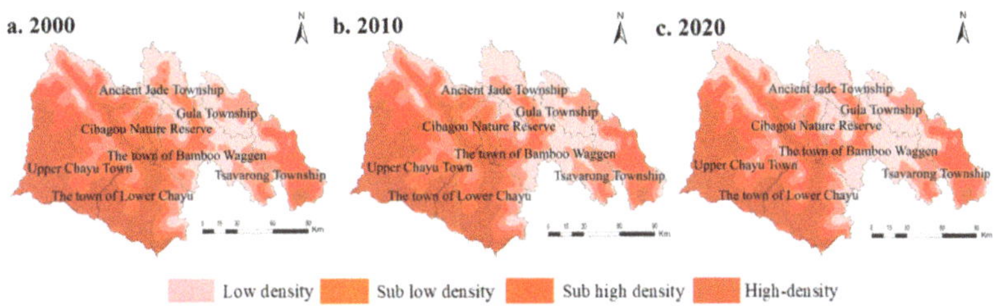

Figure 5. Distribution of kernel density of ecosystem service value per unit area in the study area from 2000 to 2020.

Specifically, in 2000, the ESVI high-density areas were mainly located in Shangchayu Town (Buzong Village, Xiba Village, Sports Village), Lower Chayu Town (Tallinn Village, Shama Village, Xincun), the middle of Cibagou Nature Reserve, the western part of Zhuwagen Town (Baga Village, Xiongjiu Village), and the southern part of Tsavalong Township (Songta Village, Quzhu Village). In 2010, the ESVI high-density areas and the sub-high-density areas showed a contraction trend, with the most obvious contraction in the north of Guyu Township (Boxue Village) and the southeast of Zhuwagen Town (Gada Village). In 2020, the coverage of EVI high-density areas and sub-high-density areas were further reduced, with the scope of ESVI high-density areas in the northeast of Shangchayu Town (Buzong Village) reduced. The changes in 2010 were mainly reflected in the decrease in the sub-high-density and sub-low-density areas of ESVI in the southwest and southeast of the town of Zhuwagen. In summary, the spatial differentiation of ESVI kernel density in the study area during the three study periods is obvious, and the kernel density presents a spatial distribution pattern of "dense in the West and sparse in the East" as a whole.

4.3.2. Spatial Autocorrelation Analysis

Through analyzing the spatial autocorrelation analysis of the ESVI in each grid in the study area, the global Moran'I value in the study area has always been greater than 0.71 in the past 20 years and the p value in most areas has been greater than 0.001, indicating that the ESVI in the townships and towns in the county as a whole has always shown significant positive spatial autocorrelation, some regions displaying obvious spatial aggregation, but the distribution in most regions is relatively random.

Moran'I scatter chart displays the spatial connection pattern between the regional and the surrounding unit attribute. What can be seen from Figure 6 is that the scatter points are mainly distributed in the first quadrant (HH) and the third quadrant (LL) while scatter points distribution in the second quadrant (LH) and the fourth quadrant (HL) is relatively small. Combined with the global Moran'I index, the chart indicates that the spatial distribution intensity of ESVI in different regions of the study area has a high spatial positive correlation and the distribution law is relatively consistent. From 2000 to 2020, the local Moran'I index increased first and then remained unchanged, combined with the situation that the scatters distributed along the trend line increased first and then remained unchanged, which reflected the trend of local spatial autocorrelation in the study area first increased and then remained unchanged.

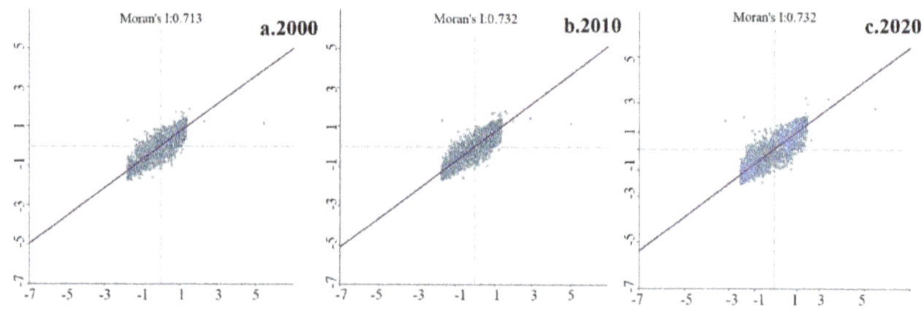

Figure 6. Moran scatter chart of ecosystem service value per unit area in the study area from 2000 to 2020.

Spatial agglomeration and spatial differentiation of ESVI are roughly similar (Figure 7). In 2000, the ESVI high-high agglomeration areas were mainly distributed on the southwest side of Shangchayu Town (Xiba Village, Sports Village), the central area of Lower Chayu Town (Kyoto Village, Tamar Village), a small part of Tsavalong Township (Songta Village, Quzhu Village), several western parts of Zhuwagen Town (Baga Village) and the Cibagou Nature Reserve, which were less affected by human interference and construction land expansion in the spatial area. The low-low agglomeration of ESVI is mainly distributed in Gula Township (Shamei Village, Shadui Village, Longri Village), Guyu Township (Boxue Village, Bayi Village, Gujing Village) and the eastern part of Zhuwagen Town (Gada Village, Jitai Village, Quwa Village), mainly because of the concentrated distribution of unused land in this area and the relatively small distribution of woodland, resulting in low ESVI. ESVI low-high agglomeration areas and high-low agglomeration areas are distributed in a "sporadic" manner within each region. In 2010, Tsavalong Township (Deng Xu Village) and Shangchayu Town (Baya Middle Village) were added to the ESVI high-high agglomeration area. In 2020, the coverage of the high-high agglomeration area in Tsavalong Township was further expanded.

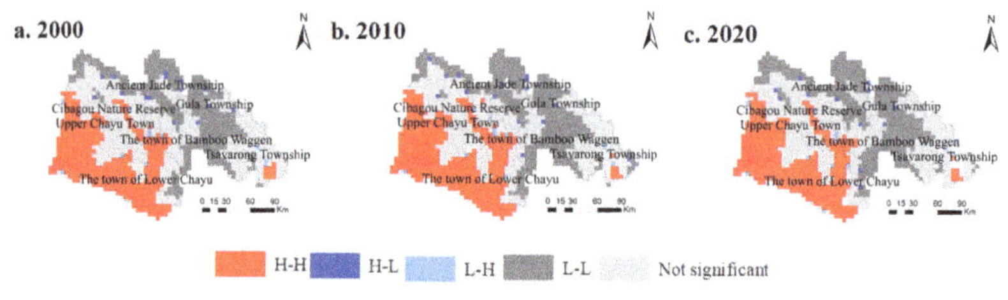

Figure 7. LISA distribution map of ecosystem service value per unit area in the study area from 2000 to 2020.

4.4. Optimization of the Pattern of Ecological Functions of Land Use

Land use ecological function zoning is used to divide the land in an area into different ecological function zones according to the unity of environmental elements such as regional landforms, the similarity between land resources and land use, the current situation of land ecological environment and the future development trend and the relative consistency of governance measures [31]. Based on the administrative village scale and employing the ESVI and K-value clustering method through SPSS software, this paper divides the study area into three different types of land use ecological function types: habitat maintenance function, biological conservation function and production support function. The result of land use ecological function zoning is shown in Figure 8.

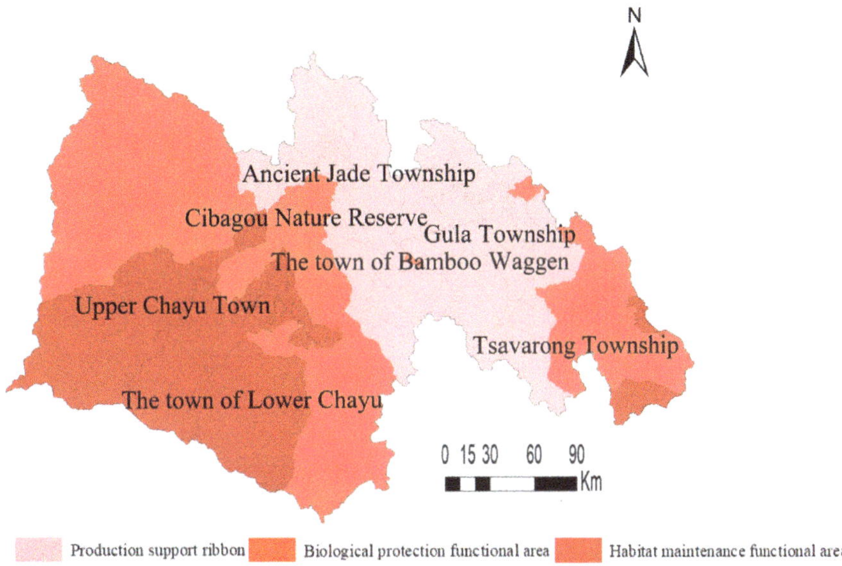

Figure 8. Pattern of land use ecological function zoning in the study area.

4.4.1. Habitat Maintenance Functional Area

This type of area is generally located in the junction of the two towns of Upper and Lower Chayu, the southwestern area of Tsavalong Township and the eastern edge of Gula Township, involving 30 administrative villages. The land use type is mainly woodland and grassland, with strong gas exchange, and regulation, soil formation and protection ability. Therefore, ESVI in this type is at a high level in the county, which plays an important role in improving the climate in the region, purifying the air and improving the quality of the ecological environment. Based on this, this area is identified as habitat maintenance functional area. The reasons are as follows. First, as an area with abundant types of ecological land such as grassland, forest land and waters, measures such as vegetation restoration and habitat restoration should be adopted in accordance with the territorial spatial planning, the comprehensive land improvement and ecological restoration special planning, etc. to strengthen the protection of existing forest land and steadily improve the quality of primitive forest resources. Second, aiming to resolve the problems of weak ecosystem function, disconnected ecological corridors, and fragile ecological barriers, combined with the characteristics of the county ecosystem, the overall planning is referred to forest and grassland resource protection, water and soil conservation, water source conservation and other ecological protection work. The third is to coordinate various ecological elements such as "landscapes, forests, fields, lakes, grass and sand", build a solid ecological security barrier, and form a natural ecological space network protection pattern, making it a veritable "ecological barrier" in the alpine valley area of southeast Tibet.

4.4.2. Biological Protection Functional Area

This type of regional space is distributed in the north-central part of the county in a sheet pattern, involving the largest number of administrative units and the widest coverage, involving 41 administrative villages. The land use types in this area are mainly woodland and grassland. Due to this, raw materials, recreation and leisure capacity in this area is weak, but the biodiversity capacity is strong. Therefore, ESVI in this area is at a relatively high level in the county, especially affected by the radiation of the Cibagou National Nature Reserve, which plays an important role in improving the level of biodiversity in the region and even the entire county, so it is classified as a biological protection functional area. The reasons are as follows. First, biodiversity plays an important role in maintaining and improving regional climatic conditions, which needs to be emphasised. Mainly relying on natural restoration and using artificial treatment as a supplement, focusing on the problems of regional plant damage, biodiversity loss and habitat system destruction, and following the principle of "overall protection, system restoration and comprehensive management" strictly, remediation goals, key areas and major projects will be put forward to protect the biodiversity. Second, to protect and restore biodiversity, more land use space will be released by changing land use patterns and improving land use efficiency, forming a reasonable and compact spatial layout with differentiated development.

4.4.3. Production Support Functional Area

This type of area is mainly scattered in the central river valley adjacent to the two towns of Upper and Lower Chayu and the northeast of Tsavalong Township, besides, there are a small number of areas in the eastern part of Gula Township, involving a total of 26 administrative villages. The land use type is mainly forest land, part of the cultivated land and part of the construction land. With good water conservation capacity and strong food production capacity, it is the main cultivation and production area of the plateau animal husbandry and the plateau forest fruit industry. ESVI within this region is at a relatively low level throughout the county, and therefore, it is classified as production support functional areas. To protect this area, the measures are as follows. First, relying on the unique natural environment and special geographical advantages, emphasis should be laid on cultivating characteristic agriculture and high-efficiency agriculture. Meanwhile, it is necessary to strengthen the prevention and control of non-point source pollution in agriculture and animal husbandry, including the reduction in and control of chemical fertilizers and pesticides, and the prevention and control of livestock and poultry breeding pollution. Second, through multiple methods such as engineering measures, biological measures and agricultural measures, soil pollution control should be strengthened, and measures such as planting green manure and increasing organic fertilizer should be adopted to improve the soil environmental capacity and risk resistance. Third, the strictest system of cultivated land protection system should be implemented to ensure that the quality and function are not reduced, the protection of permanent basic farmland should be strengthened and the illegal occupation of cultivated land should be strictly prohibited, which are in accordance with the basic criterion of "adapting to local conditions, benefiting farmers, and ecological improvement" and focusing on the goal of "harmonious coexistence between man and nature and sustainable development of human society" to strictly adhere to the bottom line of grain production control.

5. Discussion and Conclusions

5.1. Discussion

Chayu County, as a typical high mountain and canyon area in Southeast Tibet, has complex landform and geological structure, but with relatively single natural resource elements, resulting in an extremely fragile ecological environment, especially in the dual interference of natural environment and human factors, and the value of ecosystem services is particularly noteworthy [32]. This study selects typical representative areas, further explores the theory and application of ecosystem service value estimation, and theoretically deepens the scientific connotation and essential requirements, which has important strategic significance for the construction of ecological barrier and ecological civilization in the whole southeast of Tibet from the application of practical achievements.

The typical characteristic of land use in the study area is that the proportion of forest land is the largest, followed by grassland and unused land while other land types of account for a relatively small proportion. In the past 20 years, the fundamental structure of land use has changed relatively little, which is consistent with other research results, further showing that it is difficult for human activities to affect the transformation of land use in this area. It is worth mentioning that the change of cultivated land and construction land which can best represent human development and utilization can basically reflect the degree of development and utilization of land resources and the impact of human activities. It is concluded that the distribution of cultivated land resources in the study area is gradually transferred from the marginal areas to the internal river valley, while the construction land is more concentrated in the central areas of towns and townships, which is also consistent with similar research results [33,34], and synchronized with the population distribution in this region in recent years. It shows that in the areas dominated by natural environment, some local areas are still greatly affected by human effects. At the same time, from the analysis of the land use transfer matrix, the significant feature is that, in the past 20 years, the unused land area has increased by 45.44% and the grassland area has also decreased by nearly 30%, which is mainly affected by natural conditions and the continuously reduced impact of human activities on land use. This is quite different from many other research results, mainly due to the obvious particularity of the study area, which is also the reason for which this case study is chosen.

ESV in this region has decreased by 16.98% in the past 20 years, mainly due to the reduction in ecological service value of forest land and grassland, which leads to a large decline in the region, indicating that the contribution of forest and grassland to regional ESV and the regulation of ecological environment are obvious. More attention should be paid to the protection of biological resources such as forest land and grassland, and human activities should be concentrated as much as possible, and the impact of human intervention should be reduced. The ESVI in the study area presents a distribution pattern of "high in the west and low in the east", specifically, the ESVI in the middle east is relatively low while the ESVI in the eastern and western edge areas is relatively high, mainly in the two levels of relatively high and low, and the area proportion of the relatively low level shows an increasing trend, the key distribution areas of which are analyzed by grid unit and expressed at the village level spatial scale. It will enhance the practical value of this research result and make a breakthrough based on other research results [35,36]. Through using ESDA spatial model to analyze the spatial agglomeration characteristics of regional ESVI, it is concluded that the global Moran 'I value is always greater than 0.71, and the p value of most regions is greater than 0.001, indicating that the regional ESVI as a whole always shows significant positive spatial autocorrelation, and the degree of local spatial autocorrelation first increases and then remains unchanged. This further confirms the scientific value and rationality of using the model algorithm [37]. From the perspective of land use, spatial differentiation management and control, based on the administrative village scale, the whole region is divided into three types of land use ecological function areas. From the goal of how to maintain and improve the ecological service value of the regional land use system, this paper puts forward differentiation management and control measures from a multi-dimensional perspective, so as to improve the service capacity of the entire regional ecosystem, and also provide an optimized path to assisting the land space governance [38].

The alpine valley area of southeast Tibet, whose land use type mainly consists of woodland, grassland and unused land, is regarded as the "top priority" of the ecological civilization construction in the Tibet Autonomous Region. Its ecosystem balance directly affects the water source in the lower reaches of the center, the world's rare animals and plants and the changes in the global climate. During the period from 2000 to 2020, on the one hand, affected by the "returning farmland to forests and grasslands" project and the intensifying phenomenon of non-granitization and non-farming of cultivated land, the area of cultivated land in the alpine valley area of southeast Tibet was greatly reduced. On the other hand, due to population growth and climatic conditions, the area of unused land increased significantly while the area of forest land and grassland decreased. The disturbance of human activities continues to intensify, resulting in gradual changes in the structure of land use types and a continuous decline in the value of ecosystem services. In the past 20 years, the enhancement of land resource development in the alpine valley areas of southeast Tibet has had a negative impact on the value of ecosystem services in the future, while implementing farmland protection and ecological land protection, special attention should be paid to improving land use efficiency, optimizing land use structure, and gradually restoring and improving regional ecosystem service functions [39].

Ecosystem service value is not only affected by the adjustment of land use structure, but also by many social and economic factors, such as climatic conditions, population density, economic level and industrial layout and so on [40]. The calculation of ecosystem service value in this study is based on the equivalent factor method of unit area value. In order to reduce the disconnection between the equivalent table and the current situation of the study area, coefficient correction is made in combination with the actual situation of the region. Despite considering the natural conditions and socio-economic factors affecting ESV, Chayu County, as a typical area of high mountains and valleys in Southeast Tibet, is affected by many factors involving topography, natural disasters and special policies in border areas, greatly different from other general areas. Therefore, it is necessary to further explore a more accurate ESV estimation model algorithm for special areas, focusing on the refinement and specialization of ecosystem classification [41]. With the update and release of high-precision remote sensing data and the continuous enrichment of data collection of positioning observation points, the follow-up will focus on the accounting of ecosystem service value in typical regions, and further explore the basic theory and system method of ecosystem service value accounting under different terrain types and socio-economic models. If the relevant theories and methodologies are further developed, this study will continue to explore in depth, constantly enrich and improve a series of research results, in order to provide an important scientific basis and research foundation for the subsequent study of land use evolution and ecosystem service value. In addition, when carrying out relevant research in the future, special attention should be focused on the spatiotemporal changes of regional ecosystem service value driven by natural factors, socio-economic factors and policy environment, as well as the prediction and simulation research, so as to improve the feasibility and practicality of the current research results.

5.2. Conclusions

The core research task of this study is to explore the characteristics of land use change and the temporal and spatial evolution law of ecosystem service value, to build the zoning pattern of land use ecological functions and put forward differentiated management and control measures. This study, taking Chayu County, a typical alpine valley region in Southeast Tibet, as a typical research object and based on the three periods of remote sensing interpretation data in 2000, 2010 and 2020, employs the three-level spatial scale from the village level, the township level to the county level to converge step by step, to further explore the characteristics of land use evolution, ESV change and space-time response.

From the perspective of the characteristics of land use type change, the area of unused land has increased by 45.44%, with a continuous increase of 3377.26 km^2 in the past 20 years while the grassland area has continued to decrease by 1935.62 km^2 in 20 years, a decrease of 29.29 percentage points. The forest land and grassland have the largest area of transfer-in and transfer-out, 2191.24 km^2 and 2978.08 km^2, respectively, in which the main source of transfer-in is cultivated land and unused land while the main source of transfer-out is unused land and waters. Unused land, as a type of land that has not yet been utilized or is difficult to utilize, has the largest difference between the transfer-in area and the transfer-out area, and its main transfer-in and transfer-out source are cultivated land, forest land and grassland. The transfer-in area and the transfer-out area of cultivated land are 21,934 km^2 and 22,917 km^2, respectively, and the main transfer-in and transfer-out sources are woodland and grassland, including a certain proportion of waters. The scale increase in the secondary land types of unused land has mainly been caused by glaciers and permanent snow, and the change of land types has first decreased and then increased in the past 20 years, with the increase from 507,861.61 km^2 to 975,967.01 km^2 in 2020.

From 2000 to 2020, the land use index of the study area generally presents a spatial pattern of "high in the southwest and low in the northeast". The towns with high index are mainly upper Chayu town and lower Chayu Town, and the towns with low index are mainly Guyu Township and Gula township. During the study period, the land use index of each township showed a decreasing trend to varying degrees. The township with the largest decrease was Zhuwagen Town (23.00), with a decrease of 13.50%. Overall, ESVs showed a decreasing trend between 2000 and 2020, from 964.596 billion yuan in 2000 to 866.642 billion yuan in 2020, a decrease of 16.98%. During the study period, the value of ecosystem services in woodland decreased the most, accounting for 70.55% of the total reduction, followed by grassland, with the smallest reduction in cultivated land. The value of ecosystem services in unused land increased the most while the wetlands the least, of which the increase in unused land accounted for 98.05% of the total increase. From the perspective of the whole region, ESVI in the study area, with obvious spatial differentiation characteristics of kernel density, significant clustering and distribution characteristics and stable variation range, displays an overall spatial pattern with characteristics of "dense in the west and sparse in the east, high in the north and low in the south". Over the past 20 years, the global Moran'I value in the study area has always been greater than 0.71 and the p value in most areas has been greater than 0.001, indicating that the ESVI in the townships and towns in the county has always shown significantly positive spatial autocorrelation, some regions displaying obvious spatial aggregation, but the distribution in most regions is relatively random.

From the perspective of the whole region, the spatial differentiation characteristics of kernel density in ESVI in the study area are obvious, the agglomeration distribution characteristics are significant with stable variation, displaying an overall spatial pattern with characteristics of "dense in the west and sparse in the east, high in the north and low in the south". From the perspective of the agglomeration characteristics of ESVI, the southwest of the study area is dominated by most high ESVI agglomeration characteristics while the central part is dominated by a few high ESVI agglomeration characteristics. The agglomeration characteristics of some areas in the southeastern region are highly concentrated but lacking in contiguity, while the central region maintains the characteristics of low density. In the past 20 years, the ESVI of the townships in the study area has generally shown a differentiation pattern of "high in the west and low in the east" with little change. The high-value areas of ESVI mainly appear in parts of Upper Chayu Town, Lower Chayu Town and Tsavalong Township, which account for a relatively large area of woodland and grassland, and the ESVI low-value areas are mainly distributed in Guyu Township and Gula Township with higher terrain, unused land, woodland and grassland. Specifically, there is a significant positive correlation between ESVI in each township, with high spatial agglomeration, primarily with the high-high aggregation mode and insignificant mode. The high ESVI agglomeration is mainly distributed in a small part of the southwest and southeast of the study area in a sheet pattern, while the insignificant ESVI agglomeration is distributed in the central and eastern marginal areas in a sheet pattern.

Author Contributions: Conceptualization, P.L. and S.F.; Data curation, Q.H., J.C. and M.Z.; Formal analysis, S.F. and P.L.; Investigation, P.L., N.W. and S.Y.; Methodology, S.F. and P.L.; Project administration, M.Z., N.W. and S.Y.; Resources, S.Y. and S.P.; Software, P.L., Q.H. and J.C.; Validation, S.P.; Writing—original draft, P.L.; Writing—review & editing, S.F. All authors have read and agreed to the published version of the manuscript.

Funding: This research is supported by the natural science foundation of Anhui Province, grant number 1908085QG310. Key project of Humanities and Social Sciences Research of Anhui University, grant number SK2021a0155. General project of National Natural Science Foundation of China, grant number 71873003, prosperity and development philosophy and social science fund project of Anhui Agricultural University, grant number 2018zs12zd. Anhui Agricultural University introduces stable talents project, grant number yj2018-60. Anhui College Students' innovation and entrepreneurship training project, grant number S202010364063. National innovation and entrepreneurship training program for college students, grant number 202110364031. Chayu county land and space master plan, grant number 2021–2035 support.

Institutional Review Board Statement: Not applicable.

Informed Consent Statement: Informed consent was obtained from all subjects involved in the study.

Data Availability Statement: All data generated or analyzed during this study are included in this published article.

Conflicts of Interest: The authors declare no conflict of interest.

References

1. Skole, D.L.B., II; Sanderson, S.; Fischer, G.; Leemans, R. Land use and land cover change. *Earth Sci. Front.* **1997**, *4*, 26–33.
2. Costanza, R.; D'Arge, R.; Groot, R.D.; Farber, S.; Grasso, M.; Hannon, B.; Limburg, K.; Naeem, S.; O'Neill, R.V.; Paruelo, J.; et al. The value of the world's ecosystem services and natural capital. *Nature* **1997**, *387*, 253–260. [CrossRef]
3. Daily, G.C.; Soederqvist, T.; Aniyar, S.; Arrow, K.; Dasgupta, P.; Ehrlich, P.R.; Folke, C.; Jansson, A.; Jansson, B.O.; Kautsky, N.; et al. The value of nature and the nature of value. *Science* **2000**, *289*, 395–396. [CrossRef] [PubMed]
4. Etienne, C.; Jessica, P.; Yoann, L.G.; Mathieu, F. Spatial patterns of vineyard abandonment and related land use transitions in Beaujolais (France): A multiscale approach. *Sustainability* **2020**, *12*, 4695.
5. Schneider, A.; Kucharik, L.C.J.; Christopher, J.K. Impacts of urbanization on ecosystem goods and services in the U.S. Corn Belt. *Ecosystems* **2012**, *15*, 519–541. [CrossRef]
6. Huang, M.Y.; Yue, W.Z.; Fang, B.; Feng, S.R. Scale response characteristics and geographic exploration mechanism of spatial differentiation of ecosystem service values in Dabie mountain area. central China from 1970 to 2015. *Acta Geogr. Sin.* **2019**, *74*, 1904–1920.
7. Daily, G.C. *Natures Service: Social Dependence on Nature Ecosystems*; Island Press: Washington, DC, USA, 1997.
8. Robert, C.; Ralph, D.; Rudolf, G. The value of the world's ecosystem services and natural capital. *Nat. Int. Wkly. J. Sci.* **1997**, *387*, 253–260.
9. Liu, F.L.; Yang, R.Y. Evolution of land use patterns in national central cities and its impact on ecosystem service values: A case study of Wuhan city. *Res. Soil Water Conserv.* **2021**, *28*, 177–183+193+2.
10. Xie, G.D.; Zhang, C.X.; Zhang, L.M.; Chen, W.H.; Li, S.M. Improvement of the evaluation method for ecosystem service value based on per unit area. *J. Nat. Resour.* **2015**, *30*, 1243–1254.
11. Xie, G.D.; Zhen, L.; Lu, C.X.; Xiao, Y.; Chen, C. Expert knowledge-based valuation method of ecosystem services in China. *J. Nat. Resour.* **2008**, *23*, 911–919.
12. Xie, G.D.; Zhang, C.X.; Zhang, C.S.; Xiao, Y.; Lu, C.X. The value of ecosystem services in China. *Resour. Sci.* **2015**, *37*, 1740–1746.
13. Sun, H.L.; Chen, Y.N.; Li, W.H.; Li, F.; Mubarek, A. Study on types and ecological services values of the grassland in the ili river Basin. Xinjiang. China. *J. Desert Res.* **2011**, *31*, 1273–1277.
14. Luo, N.N.; Mao, D.H.; Wen, B.L.; Liu, X.T. Climate change affected vegetation dynamics in the Northern Xinjiang of China: Evaluation by SPEI and NDVI. *Land* **2020**, *9*, 90. [CrossRef]
15. Xia, J.Y.; Cao, M.; Xiao, W.; Li, Y.P.; Fu, G.; Wang, W.; Li, J.S. Integrating spatial valuation of ecosystem services into protected area management: A case study of the Cangshan Nature Reserve Dali, China. *Sustainability* **2020**, *12*, 9395. [CrossRef]
16. Liu, M.Z.; Jia, Y.G.; Zhao, J.J.; Shen, Y.J.; Pei, H.W.; Zhang, H.J.; Li, Y.L. Revegetation projects significantly improved ecosystem service values in the agropastoral ecotone of northern China in recent 20 years. *Sci. Total Environ.* **2021**, *788*, 147756. [CrossRef]
17. Tong, S.Q.; Bao, G.; Rong, A.H.; Huang, X.J.; Bao, Y.B.; Bao, Y.H. Comparison of the spatiotemporal dynamics of land use changes in four municipalities of China based on intensity analysis. *Sustainability* **2020**, *12*, 3687. [CrossRef]
18. Chen, J.Y.; Luo, H.X.; Hu, L.L.; Li, Q.H.; Jiang, W.J. Correlation between social economic development and ecosystem service value: A case of Chongqing metropolitan area. *Res. Soil Water Conserv.* **2017**, *24*, 330–336.
19. Yu, L.L.; Zhang, X.X.; Li, Y.F.; Zhao, X.X.; Ma, S.Q.; Cao, H.H.; Cao, J.K. Coastal ecosystem services and their spatial-temporal variation in Haikou. China. *Acta Ecol. Sin.* **2016**, *36*, 2431–2441.
20. Fu, J.; Liu, Y.; Lü, Y. Assessing the soil erosion control service of ecosystems change in the Loess Plateau of China. *Ecol. Complex.* **2011**, *8*, 284–293. [CrossRef]
21. Geng, Q.L.; Ren, Q.F.; Yan, H.M. Target areas for harmonizing the Grain for Green Programmed in China's Loess Plateau. *Land Degrad. Dev.* **2019**, *31*, 325–333. [CrossRef]

22. Costanza, R.; De Groot, R.; Sutton, P.; van der Ploeg, S.; Anderson Sharolyn, J.; Kubiszewski, I.; Farber, S.; Turner, R. Changes in the global value of ecosystem services. *Glob. Environ. Chang.* **2014**, *26*, 152–158. [CrossRef]
23. Wang, Z.Y.; Cao, J.S. Assessing and predicting the impact of multi-scenario land use changes on the ecosystem service value: A case study in the upstream of Xiong' an New Area, China. *Sustainability* **2021**, *13*, 704. [CrossRef]
24. BAI, E.; Bing, X. A review of influences of land use and land cover change on ecosystems. *Chin. J. Plant Ecol.* **2020**, *44*, 543–552. [CrossRef]
25. Li, H.; Zhang, X.Y.; Guo, H.L. Spatial and temporal changes of ecosystem service value in the three gorges reservoir area in the past 30 yeas based on land use. *Res. Soil Water Conserv.* **2021**, *28*, 309–318.
26. Liu, J.Y.; Zhuang, D.F.; Zhang, Z.X.; Gao, Z.Q.; Deng, X.Z. The establishment of land use spatial temporal database and its relative studies in China. *J. Geo-Inf. Sci.* **2002**, *4*, 3–7+4.
27. Daniela, S.; Michelangelo, C.; Annalisa, A.; Donato, M.; Sašo, D. Dealing with spatial autocorrelation when learning predictive clustering trees. *Ecol. Inform.* **2013**, *13*, 22–39.
28. Myong, G.C.; Junko, N. Geographic spatial autocorrelation of morphological characters of the Hemerocallis middendorffii complex (Liliaceae). *Ann. Bot. Fenn.* **1998**, *35*, 183–189.
29. Ma, Y.J.; Huang, X.J.; Xu, M.M.; Zhong, T.Y.; Du, W.X. Sensitivity analysis of ecosystem service value to coastal tideland development in Jiangsu Province. *China Land Sci.* **2006**, *4*, 28–34.
30. Hu, Y.; Qiao, W.F.; He, T.Q. Changes of the land use pattern and ecosystem service value in Yangtze—hwai ecological economic zone. *Resour. Environ. Yangtze Basin* **2020**, *29*, 2450–2461.
31. Sun, B.C.; Ding, Z.L.; Wang, D.Q.; Wang, H.Y. Ecological functional zone of land use of Shuangyashan based on ARCGIS method. *J. Northeast. Agric. Univ.* **2011**, *42*, 109–113.
32. Sun, H.L.; Zheng, D.; Yao, T.D. Protection and construction of the national ecological security shelter zone on Tibetan Plateau. *Acta Geogr. Sin.* **2012**, *67*, 3–12.
33. Alipbeki, O.; Alipbekova, C.; Sterenharz, A.; Toleubekova, Z.; Aliyev, M.; Mineyev, N.; Amangaliyev, K. A spatiotemporal assessment of land use and land cover changes in Peri-Urban areas: A case study of arshaly district, Kazakhstan. *Sustainability* **2020**, *12*, 1556. [CrossRef]
34. Xia, S.F.; Chen, M.Q.; Liu, X.; Cai, H.S. Land resources ecosystem service value accounting of Ganzhou city based on geographic national condition information. *Trans. Chin. Soc. Agric. Mach.* **2019**, *50*, 184–193.
35. Gao, H.J.; Han, H.Q.; Luo, X.Q.; Yu, H.Y.; Han, M.R. Spatial correlation of socioeconomic factors and ecosystem service values n Guizhou province. *Res. Soil Water Conserv.* **2016**, *23*, 262–266.
36. Li, J.C.; Wang, W.L.; Hu, G.Y.; Wei, Z.H. Impacts of land use and cover changes on ecosystem service value in Zoige Plateau. *Acta Ecologica Sinica* **2011**, *31*, 3451–3459.
37. Li, S.; He, F.; Zhang, X.; Zhou, T. Evaluation of global historical land use scenarios based on regional datasets on the Qinghai–Tibet Area. *Sci. Total Environ.* **2019**, *657*, 1615–1628. [CrossRef]
38. Wang, L.Y.; Ke, X.L.; Hatab, A.A. Trade-Offs between Economic Benefits and Ecosystem Services Value under Three Cropland Protection Scenarios for Wuhan City in China. *Land* **2020**, *9*, 117. [CrossRef]
39. Li, T.H.; Li, W.K.; Qian, Z.H. Variations in ecosystem service value in response to land use changes in Shenzhen. *Ecol. Econ.* **2008**, *69*, 1427–1435.
40. Wei, X.D.; Wang, N.; Luo, P.P.; Yang, J.; Zhang, J.; Lin, K.L. Spatiotemporal assessment of land marketization and its driving forces for sustainable urban–rural development in Shaanxi Province in China. *Sustainability* **2021**, *13*, 7755. [CrossRef]
41. Zulpiya, M.M.T.; Hamit, Y.M.T.; Anwar, E.Z.I.Z.; Ajigul, A.B.L.M.T. Response of ecosystem services value to land use/cover change in the Yanqi Basin of Xinjiang, China. *J. Desert Res.* **2014**, *34*, 275–283.

sustainability

MDPI

Article

Digital Financial Inclusion, Cultivated Land Transfer and Cultivated Land Green Utilization Efficiency: An Empirical Study from China

Min Zhou [1], Hua Zhang [2,*], Zixuan Zhang [2] and Hanxiaoxue Sun [1]

1 Faculty of Economics, School of Public Management, Liaoning University, Shenyang 100136, China
2 Faculty of Economics, Sunwah International Business School, Liaoning University, Shenyang 100136, China
* Correspondence: huazhang@lnu.edu.cn; Tel.: +86-187-4155-2777

Abstract: Digital financial inclusion (DFI), characterized by digitalization and inclusiveness, has generally been recognized as a significant promoter of efficiency, effectiveness, and sustainability of agricultural production. Simultaneously, cultivated land green utilization efficiency (CLGUE), which is the significant guarantees of food security, social stability and environmental protection, has attracted increasing attention in the recent decades. In practice, DFI seems to be a vital antecedent of the improvement of CLGUE. However, in the academic field, research on whether and how DFI can affect CLGUE is scarce. In this case, based on triple bottom line theory, this paper theoretically and empirically investigates whether and how DFI can reinforce CLGUE through the mediator of cultivated land transfer (CLT). Using Chinese provincial panel data from 2011 to 2020 and structural equation modelling (SEM) analysis in STATA 16.0, this paper identified the following: (1) DFI can directly facilitate CLGUE; (2) DFI can indirectly improve CLGUE through CLT. (3) DFI has regional heterogeneity in the improvement of CLGUE. Compared to the central and western areas, the positive relationship between DFI and CLGUE in the eastern areas is more obvious; (4) compared with main grain producing and main grain producing and marketing balance areas, the positive relationship in the main grain marketing areas is more obvious. Our research is one of the first to explore the mediating mechanism between DFI and CLGUE from the perspective of CLT.

Keywords: cultivated land green utilization efficiency; digital financial inclusion; cultivated land transfer

Citation: Zhou, M.; Zhang, H.; Zhang, Z.; Sun, H. Digital Financial Inclusion, Cultivated Land Transfer and Cultivated Land Green Utilization Efficiency: An Empirical Study from China. *Sustainability* **2023**, *15*, 1569. https://doi.org/10.3390/su15021569

Academic Editor: Eran Feitelson

Received: 23 November 2022
Revised: 1 January 2023
Accepted: 9 January 2023
Published: 13 January 2023

1. Introduction

Cultivated land occupies 10.20% of the global land surface area, and cultivated land is the main source of grain manufacture and plays a significant role in ensuring ecological security and sustainable development [1,2]. With the rapid growth of the human population, the process of urbanization and industrialization, the shortage of cultivated land and food is steady deterioration in some regions in the world [3–6]. Since the reform and opening-up in 1978, China has undergone rapid urbanization [7]. As the National Bureau of Statistics of China (2021) reported, China's urbanization rate increased from 17.92% in 1978 to 63.89% in 2020. In comparison, from only 2013 to 2015, there was an annual decrease of 354,700 hm^2 of cultivated land due to the construction occupation [8]. Furthermore, one other issue of concern for China is ecological environment issues, resulting in predatory exploitation and irrational utilization of cultivated land. As the National Soil Pollution survey bulletin (2014) reported, the over-standard rate of soil points in China's cultivated land is 19.4%, and sewage irrigation and irrational use of fertilizers, pesticides, etc., are the leading reasons for soil pollution of cultivated land [9]. In the context of ecology civilization construction, China's cultivated land utilization is facing the pressures of transformation from the "extensional" development mode of high-intensity to "connotative" development path of high quality, high efficiency and low pollution.

As a scientific development concept and development method, green development was raised in the Fifth Plenary Session of the 18th CPC Central Committee. Green development's goal is sustainable development, and the basic condition is resource environmental bearing capacity [10]. China's national conditions are large population and less land [11], the per household average cultivated area is merely 0.38 ha, which is lower than the world's average [12]. China's challenge is to feed 20% global population with less than 10% world's cultivated land [13]. Considering both the traditional output of economy and grain, and the positive and negative externalities brought by cultivated land utilization, the transition of cultivated land use to green and efficient is necessary in China [14]. Consequently, the comprehensive analysis on CLGUE and exploring its influence mechanism have some valuable significance in theory and practice for improving the level of ecological civilization construction, and providing more ecological welfare for the people [15,16].

In the present literature, the scientific intension, evaluation index and methods, and the affecting factors of CLGUE have raised attention. Regarding to the concept of CLGUE, there is still no consensus in academia. Scholars have explained it from different perspectives. Lu et al. [14] hold that the goal of cultivated land green utilization was to obtain the maximized economic and social output with the minimized environmental pollution. According to Xie et al. [17], the least costly cost of using cultivated land is combined with the largest economic and ecological impacts by CLGUE. How do we evaluate CLGUE? The existing studies usually measure CLGUE comprehensively; furthermore, the evaluation indexes were selected from "input", "desirable output", and "undesirable outputs" [11,18,19], the methods adopted mostly involve the super-efficient SBM model [11,18,19], non-radial directional distance function (NDDF) approach [17], super-efficiency EBM model [20], etc. Empirical studies have indicated that CLT, cultivated land management scale [11], urbanization rate, GDP per capita, per capita fixed-asset investment in rural areas, the industrial structure [18], agricultural insurance, agricultural subsidies, cultivated land fragmentation [20], farmers' dependence on cultivated land and agricultural added value, farmers' occupational differentiation, agricultural machinery density, and agricultural disaster rate [21] are contributing factors.

Capital is a significance factor of production for farmers' cultivated utilization [22]. The financing problems faced by farmers are crucial during land lease [23]. In order to alleviate the financing constraints of cultivated land operators, a series of promoting DFI policies have been enacted in China. According to the "*China Inclusive Financial Indicators Analysis Report of the PBC* (2018)", the number of mobile banking households in rural areas reached 670 million in 2018. In addition, digitally inclusive financial products and services in rural areas have been continuously enriched, such as "Huinong E Pay", "Nongfa loan", "Yinong loan", "Wing Long loan", etc. The development of DFI would expand the coverage of traditional finance, promote the financial accessibility of remote areas and vulnerable groups effectively, make financial services more geographically penetrating, and alleviate the difficulty and high cost of financing for farmers effectively [24].

The performance of DFI on cultivated land utilization attacked attention of the scholars. The burgeoning trend of DFI in China renders a novel thinking for the upward trajectory of agricultural mechanization to leap out of the foregoing vicious hoop [25]. Digital finance is a significance path to promote agricultural mechanization [26]. Empirical study indicated that that DFI significantly increased farmers' willingness to adopt agricultural technology [27]. Cheng et al. [24] tested the effects of DFI on carbon emissions from cultivated land utilization empirically. The result showed that DFI reduced the intensity of carbon emission [24]. In addition, the study of Zhang [28] showed that DFI could improve the availability of financial credit for rural households, improve the speed and duration of CLT, and accelerate the process of cultivated land utilization to large-scale and intensive.

The change of farmers' willingness in CLT brought by DFI changes resource configuration [28]; however, few studies have paid attention on how this change influences CLGUE. Additionally, CLT can probably solve the land fragmentation problems in China, while this legal arrangement made to prevent land fragmentation has evolved to restrict the use and yield of agricultural lands in some developed countries, for instance, Turkey [29]. Accordingly, whether CLT can further promote CLGUE is uncertain and controversial. Previous studies provide some valuable ideas for the present study, but the mechanism of the influence of DFI on CLGUE has not been revealed. In addition, there are few studies concerning the effects of DFI on CLGUE and heterogeneity. Under the context of rapid development of DFI and large-scale CLT in China, it is of great significance to reveal the influence of DFI on CLGUE and its mechanism. We tried to explore the effects of the development of DFI on CLGUE using the Peking University Digital Financial Inclusion Index and provincial panel data from 2011 to 2020. Furthermore, we intended to demonstrate that the DFI's development could significantly increase CLGUE and that a high level of CLT could significantly improve the positive influence of DFI on CLGUE. The present study elucidates the relationship between DFI and CLGUE and provides new policy references. The structure of this paper is as follows: in the second section, the theoretical basis and proposed hypotheses are introduced. In the third section, methodology is discussed. Then, we report our results and analysis in the fourth section. Finally, we demonstrate our conclusions, contributions, and directions for future research.

2. Theoretical Background and Hypotheses Development

2.1. Triple Bottom Line Theory

This paper regards triple bottom line (TBL) as a convincing framework for integrating distinguishing CLGUE dimensions and identifying the relationships with its antecedents DFI and CLT. The TBL theory can be traced back to accounting and corporate responsibility to orientate firms towards social and environmental protection issues in their operations. The theory consists of three associated components, that is, "profit, planet, people" [30]. The specific contents involved in the theory are presented in Figure 1. The theory points out that the influence operations have on society and the ecological impacts on the environment deserve serious consideration when organizations self-evaluate. With the expansion of the theory, the "sustainability" idea has gradually been introduced in agricultural production [31]. A sustainable agricultural production creates acceptable outputs for its operators but minimizes the environmental damage and adverse impacts on other people [32].

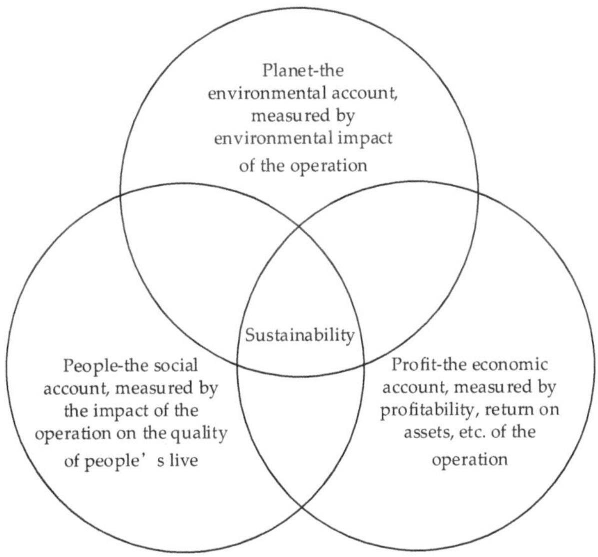

Figure 1. The basic principles of triple bottom line. (Source: Slack et al. [27]).

Accordingly, TBL theory is a significant paradigm for studying the relationship among DFI, CLT, and CLGUE. Specifically, on the one hand, TBL theory is valuable for scholars to measure the expected output parts of CLGUE. Operators should focus on balance of economic, environmental, and societal interests when managing cultivated land. On the other hand, economic, social, and environmental benefits are closely linked [33]. The operators of cultivated land paying attention to social and environmental benefits is conducive to the promotion of economic outputs of cultivated land. The assumption underlying the TBL theory is that a sustainable operation is more likely to stay successful in the long-term than one that focuses on economic goals alone [29]. Simultaneously, economic development can promote the achievement of social and environmental benefits. With the development of DFI, which is the typical product of economic development, operators of cultivated land are more likely to protect the environment and create value for society.

2.2. Hypotheses Development

2.2.1. Digital Financial Inclusion and Cultivated Land Green Utilization Efficiency

On a basis of triple bottom line theory, the rapidly expanding field of DFI, which is the promoter and also the production of economic development, is conductive to the promotion of environmental protection of and value creation for society. To be more specific, in the agricultural production, DFI can efficiently facilitate CLGUE in the following aspects. First, DFI can efficiently control the emission of pollutants (e.g., carbon contamination, pesticide pollution etc.) and the consumption of energy through high-quality financing services, which consequently promote CLGUE. The prior literature has proposed that impediments to technological development caused by high financing costs probably lead to increased energy expenditure and carbon emissions [34,35]. DFI with the lower financing constrains and financing costs can efficiently control the energy expenditure and carbon emissions and subsequently improve CLGUE. Compared with traditional finance, DFI is characterized by digitalization and inclusiveness [36]. In terms of digitalization, scientific analysis of various data generated and processed by digital technology is conducive to achieving green detection [37,38]. For instance, digital technology can be effectively implemented in the field of calculating pesticide and fertilizer application; thus, farmers can accurately calculate the input number of pesticides and fertilizers, so as to avoid pollution caused by excessive input. Consequently, carbon and pollutant emissions can be efficiently controlled and CLGUE will be further improved. [39]. In terms of inclusiveness, convenience of financial service is improved in rural areas, subsequently decreasing the risk and increasing farmers' the quality of investment and credit [40]. Additionally, the rise of green finance from environmental conservation dramatically enhances the green characteristics of finance, which efficiently promotes an increase in energy utilization efficiency and a reduction in carbon emissions [41–43].

Second, DFI can improve the outputs of cultivated land, and subsequently facilitate CLGUE. The extant literature has confirmed that financing constraints have always been obstructive factors that restrict agricultural investments and outputs [44,45]. Thanks to the development of digital technology, DFI provides more efficient financing channels for agricultural production through improving the permeability and enlarging the special scale of financial services [27]. According to the empirical results presented in Zhou et al. (2022) [25], DFI will accelerate farmers' willingness to adopt agricultural technology in China. The more money farmers have at their disposal, the more they are willing to use agricultural technologies to achieve large-scale industrialization [46]; therefore, the overall yield of farmland and green efficiency would increase dramatically [47].

Third, efficient control of inputs is another benefit arising from DFI, which further promotes CLGUE. Changes in factor endowments encourage farmers to choose cheap production factors to replace expensive ones [48]. With the development of DFI, a large number of young workers to nonagricultural sectors and subsequently the supply of rural labor may be insufficient [49]. In this case, farmers adjust the input structure of production factors, using cheap and relatively rich elements, for instance, agricultural machinery to replace labor. Consequently, the cultivated land's green efficiency is efficiently improved though the decrease of inputs.

According to the above analysis, the following hypothesis is proposed:

Hypothesis 1. *Digital financial inclusion is positively correlated with cultivated land green utilization efficiency.*

2.2.2. Digital Financial Inclusion and Cultivated Land Transfer

DFI is a significant antecedent of CLT. First, DFI, as a carrier of information dissemination, can facilitate CLT by reducing transaction costs and information asymmetry [50]. Thanks to the development of digital technologies such as big data technology and cloud computing technology, the speed and efficiency of information dissemination can be improved [25]. Efficient information dissemination and convenient communication can promote cultivated land transfer. Specifically, the essence of CLT is a process of reaching a contract on the cultivated land utilization assets between land transferors and land transferees [51]. There is empirical evidence indicating that low efficiency of information dissemination in rural areas dramatically reduces farmers' cognition of land transfer, increases the transaction costs of land transfer, and consequently restricts the improvement of the CLT system [52]. The development of DFI enables more efficient and accurate access to farmers' property information, land information, and credit records, which relieves the information imbalance between the supply and demand entities of cultivated land [53]. Hence, it reduces the economic costs of land transactions and subsequently facilitates the marketization of CLT.

Second, the development of DFI has brought more nonagricultural entrepreneurial and employment opportunities [54], which further promotes CLT. Specifically, DFI makes it possible to obtain online credit or loans without collateral assets and simultaneously offer financial services with a reasonable interest rate [55]; consequently, it efficiently promotes farmers' entrepreneurial activities. In addition, the expansion of DFI can dramatically accelerate economic growth, especially promoting the development of small and medium firms; thus, small and medium firms can provide more employment opportunities [56,57]. With a large number of farmers employed in nonagricultural sectors, the transfer and contract activities of cultivated land are promoted.

Third, DFI can facilitate CLT through enhancing agricultural mechanization. Traditional finance has great difficulties in supporting the development of agricultural mechanization [58]. When borrowing funds from traditional financial institutions, farmers face a series of challenges such as remote residence, complex terrain, backward transportation, and a lack of collateral and guarantees. DFI greatly expands the scope of financial services, effectively relieves the financial constraints of farmers and accurately identifies the needs of farmers, consequently promoting the application of mechanization in cultivated land [27]. Actively using agricultural machinery to replace labor is conductive to promoting CLT.

Accordingly, we propose the following hypothesis:

Hypothesis 2. *Digital financial inclusion is positively correlated with cultivated land transfer.*

2.2.3. Cultivated Land Transfer and Cultivated Land Green Utilization Efficiency

CLT is related to transferring cultivated land management rights from some individual farmers to professional farmers or economic organizations. It means the transfer of managing rights of cultivated land from low-productivity operators to high-productivity operators, which mitigates cultivated land resource misallocation and effectively promotes the development of CLGUE [59]. First of all, operation entities with higher productivity usually have more technological and cultural advantages. Professional operators improve the efficiency of fertilizer and pesticide utilization, thereby reducing the emissions of carbon and other sources of pollution [60]. Simultaneously, the formal signing of CLT contract is conducive to stabilizing long-term cultivated land management rights, thus helping CLT households to alleviate the concerns of the instability of the cultivated land management right and increase the belief in protecting the cultivated land [61], which contributes to the rational use of chemical fertilizers by CLT households. Therefore, on the whole, farmland transfer can promote CLGUE through transferring management rights to more professional operators.

In addition, CLT policies have a certain impact on grain planting structure. The fertilizer and pesticide usage of food crops is significantly lower than that of other cash crops. Hence, CLT can facilitate CLGUE through adjusting grain planting structure. On the one hand, guaranteeing grain security is an important goal of CLT [62]. It is required to ensure the agricultural use of cultivated land and give priority to grain production, which contributes to increasing the proportion of grain crops, realize the adjustment of planting structure and reduce land pollution [63]. On the other hand, there are significant differences between food crops and nonfood crops in terms of production characteristics, planting management difficulty, and labor demand. Compared with non-food crops, agricultural scale promoted by CLT is more suitable for the production of food crops, which adjusts the planting structure and promotes the sustainable use of cultivated land [64].

In addition, cultivated land transfer can greatly improve cultivated land green utilization through large-scale agricultural modernization. Chen et al. [65] pointed that CLT is an effective path to resolve a contradiction between the farmland fragmentation and the large-scale agricultural modernization. The adoption of agricultural green technology has certain requirements on the scale of operation [66]. For instance, the application of soil testing formula balanced fertilization technology is quite difficult for small-scale farmers for the reason that the technology is time-consuming, high costs, and high technical requirements. Apart from this, the government has strict requirements on the use of chemical fertilizers and pesticides by large-scale farmers. Scale operation can facilitate government and public welfare departments to provide training and guidance on agricultural green technology [67], consequently improving CLGUE.

Accordingly, we assume that:

Hypothesis 3. *Cultivated land transfer is positively correlated with cultivated land green utilization efficiency.*

Hypothesis 4. *Cultivated land transfer mediates the relationship between digital financial inclusion and cultivated land green utilization efficiency.*

3. Materials and Methods

3.1. Model Construction

3.1.1. Measurement of CLGUE

Data envelopment analysis (DEA) is a mathematical programming method for evaluating the relative efficiency of decision-making units (DMUs) with multiple inputs and multiple outputs [68]. The idea of single-input, single-output engineering efficiency was generalized to a multiple-input, multiple-output relative efficiency evaluation [69]. Banker et al. [70] proposed to evaluate the relative efficiency of DMUs by using the variable returns to scale as a criterion, which is the DEA-BCC model. However, neither of the two models could measure the full range of slack variables [71]. To improve the method and eliminate the variation, Tone [72] developed a non-radial and non-angular slack-based measure (SBM) model. The SBM model adds the relaxation variables of the input and output factors to the objective function. Nevertheless, the SBM model cannot measure the efficiency of DMUs with undesirable outputs. Tone [73] took these undesirable outputs into consideration and proposed an SBM model. The SBM-Undesirable-VRS model is set as follows [74]:

$$\rho^* = \min \frac{1-\frac{1}{m}\sum_{i=1}^{m}\frac{s_i^-}{x_{i0}}}{1+\frac{1}{s_1+s_2}\left(\sum_{r=1}^{s_1}\frac{s_r^\delta}{y_{r0}^\delta}+\sum_{r=1}^{s_2}\frac{s_r^b}{y_{r0}^b}\right)}$$
$$s.t.\begin{cases} x_0 = X\lambda + s^- \\ y_0^\delta = Y^\delta\lambda - S^\delta \\ y_0^b = Y^b\lambda + s^b \\ \lambda \geq 0, s^- \geq 0, s^\delta \geq 0, s^b \geq 0 \end{cases} \quad (1)$$

where s^-, s^δ, and s^b correspond to the vectors of relaxation for the inputs, desired, and unwanted, respectively. λ represents the weight vector, and the objective function. ρ^* is the index of CLGUE, which objective value ranges from (0, 1].

3.1.2. Models of Main Effects

Structural equation model (SEM) is significant statistic procedures for testing measurement, functional, predictive, and causal hypotheses. It can not only deal with explicit variables and latent variables, but can also analyze the relationship between multiple explanatory variables, multiple explained variables, and multiple mediation variables [75]. Referring to the relationships between explanatory variable and explained variable, this paper constructed the following path models of the main effects (Formula (2)):

$$clgue_{i,t} = c_1 dfi_{i,t,} + \varepsilon_{i,t} \quad (2)$$

In Formula (2), $clgue_{i,t}$ represents the CLGUE of province i in year t, $dfi_{i,t}$ represents the DFI of province i in year t, c_1 is the path coefficient of DFI influencing CLGUE, $\varepsilon_{i,t}$ is the error term. If the path coefficient c_1 is significantly positive, H1 is verified.

3.1.3. Models of Mediating Effects

According to the relationships among the explanatory variable, mediating variable and explained variable, this paper constructed the following path models of mediating effects (Formula (3)):

$$\begin{cases} clt_{i,t} = a_1 dfi_{i,t} + \varepsilon_{i,t} \\ clgue_{i,t} = b_1 clt_{i,t} + \grave{c}_1 dfi_{i,t} + \varepsilon_{i,t} \end{cases} \quad (3)$$

In Formula (3), $clt_{i,t}$ represents the CLT of province i in year t, a_1 is the path coefficient of DFI influencing CLT, b_1 and $\grave{c}_1$ are the path coefficients of CLT affecting CLGUE and DFI affecting CLGUE, respectively. If the path coefficient a_1 is significantly positive, H2 is verified. If the path coefficient b_1 is significantly positive, H3 is supported. Furthermore, if the mediating path coefficient $a_1 \times b_1$ (*dfi*→*clt*→*clgue*) is significantly positive, H4 is verified.

3.2. Variable Selection and Data Description

(1) Explained Variable: The index of CLGUE was measured by the super-efficient SBM model. In concept of CLGUE and the relevant literature [17,18], twelve variables were selected in the present study to construct the evaluation indicator system of CLGUE, involving input indicators, and desirable and undesirable output indicators (Figure 2).

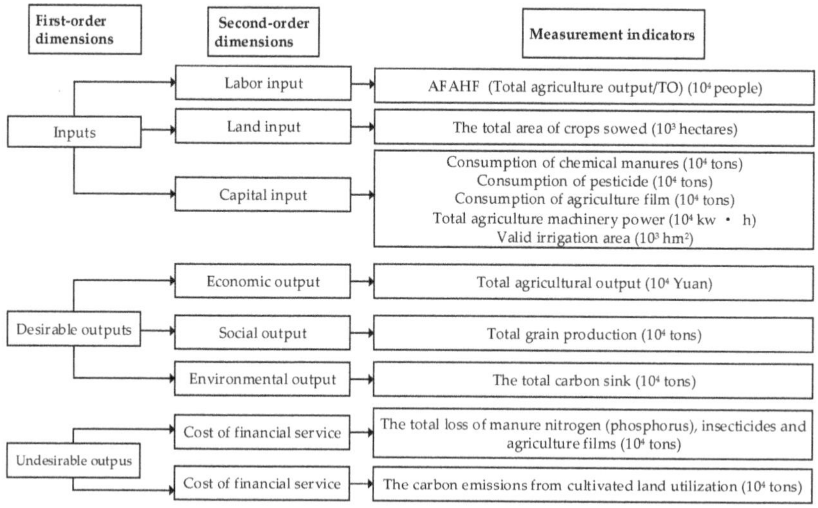

Figure 2. The indicators for measuring CLGUE. Note: AFAHF represents the abbreviation for agricultural, forestry, animal husbandry, and fishery practitioners; TO represents the abbreviation for total output values of agriculture, forestry, animal husbandry, and fishery.

This paper mainly takes carbon emissions and pollution emissions into account as undesired outputs. Total carbon emissions were calculated by multiplying the carbon source by the appropriate carbon emission factors. Based on the literature [18], carbon sources and coefficients include pesticides (4.394 1, kg C/kg), chemical fertilizers (0.895 6, kg C/kg), agriculture films (5.180, kg C/kg), agricultural irrigation (5, kg/hm^2), agricultural machinery (25 kg C/hm^2), total power of agricultural machinery (312.6 kg, C/kW), and agricultural tilling (312.6, kg C/km^2). The calculation formula is set as follows:

$$CECLU_i = \sum C_i = \sum T_i \cdot \delta_i \quad (4)$$

where $CECLU_i$ represents the total carbon emissions from cultivated land utilization, T_i represents the amount of the *i-th* carbon source, and δ_i refers to the *i-th* carbon source's coefficient.

The pollution caused by cultivated land utilization refers mainly to non-point source pollution during cultivated land use. According to previous studies [18,21], nitrogen (phosphorus) fertilizer, pesticide, and agricultural film loss were used to represent pollution emissions. The corresponding loss coefficient refers to the manual of agricultural pollution source coefficient in National Pollution Source Survey. At the same time, the influence of regional differences on the results is considered in the estimation process.

(2) Explanatory Variables: The data source of the DFI index is from the Peking University DFI Index of China. The measurement of DFI is based on the development of innovative digital finance [76]. The index aggregate consists of three dimensions, coverage breath, usage depth, and digitalized level. Figure 3 illustrates the definitions of and relationships between these indicators.

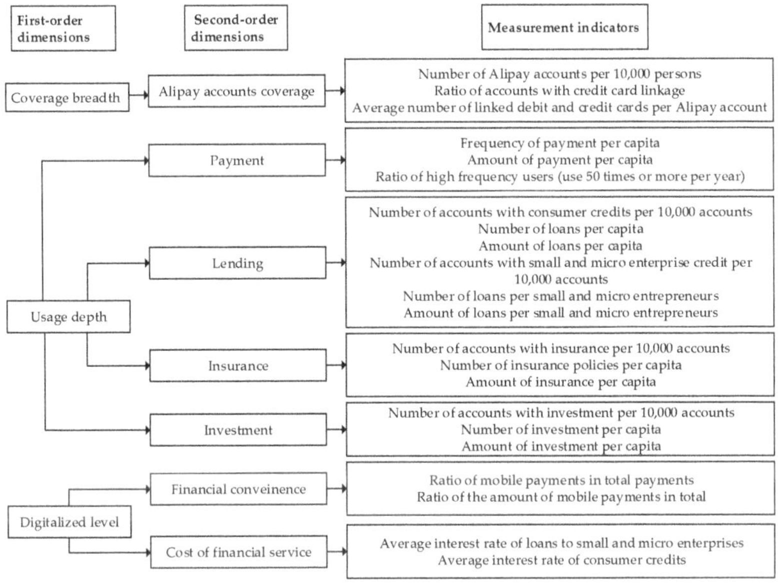

Figure 3. Measurement of DFI Index. (Resource: Guo et al. [76]).

(3) Mediating Variable: The mediating variable was CLT, indicated as the area proportion of CLT to household contracted cultivated land under the household responsibility system in China.

Table 1 illustrates the results of descriptive statistics. First, the average, minimum, and maximum value of DFI is 217.2, 18.33, and 431.9, respectively. It indicates that the levels of DFI of different provinces vary dramatically. Additionally, the levels of DFI of most provinces are at a relatively high level. Second, the average value and standard deviation of CLT are 0.316 and 0.163, respectively, the minimum value is 0.033, and maximum value is 0.911. Accordingly, the ratio of CLT of different provinces varies slightly, and the ratio of CLT of most provinces is at a lower rate. Third, the mean value of CLGUE is 0.704, closer to the maximum value of 1, indicating that the CLGUE of most provinces keeps a higher level. The standard deviation is 0.198, which means that the CLGUE of different provinces varies slightly.

Table 1. Results of descriptive statistics.

Variables	Number	Mean	Std. Dev.	Minimum	Maximum
dfi	300	217.2	96.97	18.33	431.9
clt	300	0.316	0.163	0.033	0.911
clgue	300	0.704	0.198	0.315	1

3.3. Research Region and Data Source

There are 34 provincial-level administrative institutions in China, and large regional differences exist in cultivated land resources, food production, and agriculture development [77]. Hong Kong, Macao, Taiwan, and Tibet were excluded from the empirical research due to lack of available data. Therefore, this present research's subjects consist of 30 provinces or municipalities in mainland China. The CPC Central Committee on the "National Economic and Social Development Seventh Five–Year Plan" (1985) divided the 31 provinces into eastern, central, and western regions. This study also uses this classification (Figure 4).

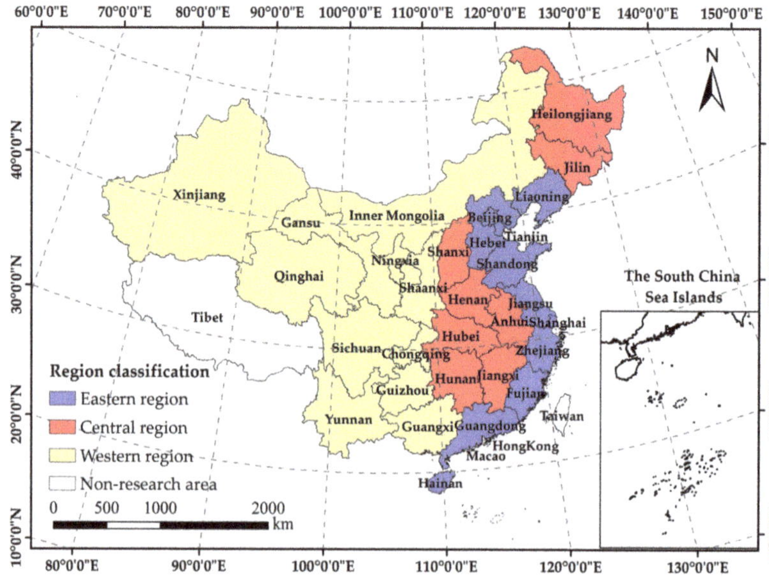

Figure 4. The research area and regional classification.

The information that used to evaluate CLGUE and CLT was gathered from "*China Statistical Yearbook*", "*China Rural Statistical Year-book*", "*China Rural Management Statistical Annual Report*", and China's Rural Policy and Reform Statistical Annual Reports of the resent years, as well as the National Bureau of Statistics of China's website. To fill in the gaps in the individual years' missing data, the interpolation approach was used. In addition, the data source of the DFI index is from Peking University DFI Index of China.

4. Results

4.1. Measurement and Analysis of CLGUE

In this section, Equation (1) was used in this part to compute the CLGUE in China. China's total CLGUE showed a trend toward progressive improvement, from 0.57 in 2011 to 0.92 in 2020, and the average annual growth rate was 5.46% (Figure 5). China has achieved initial success in transformation of cultivated land utilization to being green and efficient. One possible reason is that a number of policies have been formulated to advance the transformation of agricultural production, such as zero growth in fertilizer consumption [78]. In addition, it can be seen that the CLGUE of three regions are characterized by an overall upward trend. Furthermore, large regional differences exist in the average annual growth rates. The average annual growth rates for eastern, central, and western regions, respectively, were 6.24%, 3.70%, and 6.32%. The reason why the annual growth rate of CLGUE in the western region lags behind may be the main grain-producing areas are in central China. The main grain-producing areas play a pivotal role in the process of ensuring national food security. Because of the path dependence, the transformation of cultivated land utilization form "high input and high output" to "green and efficient" is more difficult in main grain-producing areas.

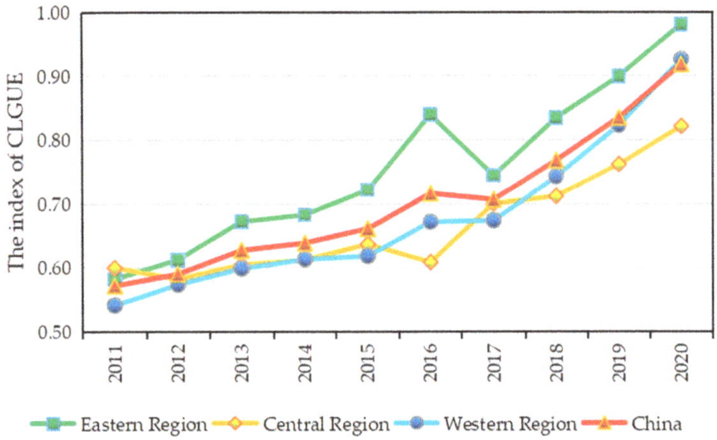

Figure 5. Average value of CLGUE in China, eastern, central, and western regions.

According to Huang and Wang [79], the efficient, relatively high-efficient group, relatively medium-efficient group, and relatively low-efficient groupings, respectively, were assigned to the provinces based on their efficiency values between [1], [0.8, 1), [0.6, 0.8), and [0, 0.6). The spatial–temporal evolution of CLGUE in 30 provinces is shown in Figure 6. In 2011, only Jilin, Heilongjiang, Shanghai, and Qinghai belonged to the efficient group, Beijing, Chongqing, and Ningxia belonged to the relatively medium-efficient group, the other 23 provinces belonged to the relatively low-efficient group. In 2015, Heilongjiang, Shanghai, and Qinghai shifted from the efficient group, while Shandong was moved into the efficient group. The spatial scope of the relatively high-efficiency and medium-efficiency groups emerged as an expanding trend. However, Gansu, Shanxi, Anhui, Yunnan, Inner Mongolia, Zhejiang, Hebei, Guangxi, and Jiangxi still remained in the relatively low-efficient group. In 2020, except Gansu, Shanxi, and Anhui, which still remained in the relatively low-efficient group, CLGUE in other provinces fell into the more efficient group or remained in the efficient group.

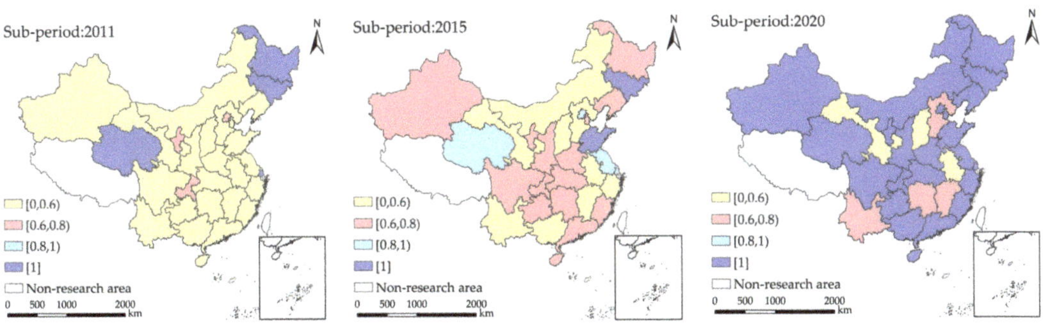

Figure 6. The spatial–temporal evolution of CLGUE.

4.2. Structural Equation Model Results of the Main Effects

The structural equation model results of the main effects are presented in Figure 7 and Table 2. According to the results of model fitting test, the X^2, RMSEA, and SRMR are all less than 0.05. This indicates good goodness of fit of the main effect model [80]. Since the fitting indexes are not used to compare the pros and cons of the models, CFI, AIC, BIC, and other indexes are not reported [81]. Furthermore, the path coefficient of DFI on CLGUE is 0.442, significant at the 1% level. This indicates that DFI can directly improve CLGUE, and hypothesis 1 is supported. Through structural equation model analysis of the main effects, we identified that DFI, characterized by digitalization and inclusiveness, can be a significant promoter of CLGUE. With the development of DFI, CLGUE in China can be dramatically improved.

LR test Pro(>chi) = 0.000 RMSEA = 0.000
SRMR = 0.000

Figure 7. Path diagram and empirical results of main effects. Note: *** represents that it is significant at the 1% level.

Table 2. Results of main effects.

Paths	Coefficients	Standard Error	Z Value	*p* Value	95% Confidence Interval (CI)	
lndfi→clgue1	0.442	0.044	10.020	0.000	0.356	0.529
constant	0.059	0.415	0.140	0.886	−0.754	0.873
variance (e.clgue)	0.805	0.039			0.732	0.885

4.3. Structural Equation Model Results of the Mediating Effects

The structural equation model results of the mediating effects are illustrated in Figure 8 and Table 3. According to the results of model fitting test, the X^2, RMSEA, and SRMR are all less than 0.05. This indicates good goodness of fit of the main effect model [80]. We did not report CFI, AIC, BIC, and other indexes as well. Furthermore, the path coefficient of DFI on CLT is 0.183, passing the test at the 1% significant level. This indicates that DFI is positively correlated with CLT and hypothesis 2 is supported. Then, the path coefficient of CLT on CLGUE is 0.273, significant at the 1% level; hence, CLT is positively correlated with CLGUE and hypothesis 3 is verified. Finally, we investigated the significance of mediating effects. Based on the results of Table 4, the path coefficient of $a_1 \times b_1$ *(dfi→clt→clgue)* is 0.132, significant at the 1% level. It demonstrates that CLT mediates the influencing path of DFI on CLGUE and H4 is verified. Additionally, since the direct path coefficient of DFI on CLGUE is 0.310, also passing the test at the 1% significant level, we identified that CLT has *partial* mediating effects on the relationship between DFI and CLGUE.

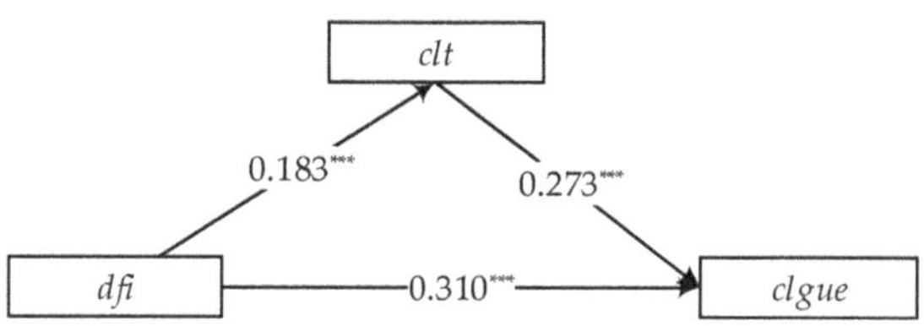

LR test Prob(>chi2) = 0.000 RMSEA = 0.000
SRMR = 0.000

Figure 8. Path diagram and empirical results of mediating effects. Note: *** represents that it is significant at the 1% level.

Table 3. Results of mediating effects.

Paths	Coefficients	Standard Error	Z Value	*p* Value	95% CI	
dfi→clt	0.483	0.042	11.630	0.000	0.402	0.565
constant	−1.894	0.363	−5.220	0.000	−2.605	−1.183
clt→clgue	0.273	0.056	4.910	0.000	0.164	0.382
dfi→clgue	0.310	0.054	5.730	0.000	0.204	0.416
constant	0.576	0.424	1.360	0.174	−0.255	1.407
variance (e.clt)	0.766	0.040			0.691	0.849
variance (e.clgue)	0.748	0.042			0.670	0.834

Table 4. Tests of significance of mediating effects.

Paths	Coefficients	Standard Error	Z Value	*p* Value	95% CI	
dfi→clt→clgue	0.132	0.029	4.470	0.000	0.074	0.190

4.4. Robustness Tests

In this section, we use the method of the substitution of the explained variable to conduct robustness tests. SBM-Undesirable-CRS was selected to appraise the index of CLGUE, SBM-Undesirable-CRS is constructed on the assumption of constant returns to scale. SBM-Undesirable-VRS used in Section 4.1 is constructed on the assumption of variable returns to scale. On a basis of the results illustrated in Table 5, the path coefficient of DFI on CLGUE is 0.497, significant at the 1% level. This indicates that DFI is still positively correlated with CLGUE after the substitution of the explained variable in the main effect analysis. Then, based on the results in Table 6, DFI is positively correlated with CLGUE (0.483, significant at the 1% level), DFI is positively correlated with CLT (0.361, significant at the 1% level), and CLT is positively related to CLGUE (0.282, significant at the 1% level). Meanwhile, in Table 7, the new path coefficient of $a_1 \times b_1$ *(dfi→clt→clgue)* is 0.136, significant at the 1% level. It indicates that CLT still mediates the influencing path of DFI on CLGUE after the substitution of the explained variable in the mediating effect analysis.

Table 5. Results of robustness tests of main effects.

Paths	Coefficients	Standard Error	Z Value	*p* Value	95% CI	
dfi→clgue (new)	0.497	0.041	12.230	0.000	0.418	0.577
constant	−0.650	0.385	−1.690	0.091	−1.403	0.104
variance (e.clgue)	0.753	0.040			0.677	0.836

Table 6. Results of robustness tests of mediating effects.

Paths	Coefficients	Standard Error	Z Value	*p* Value	95% CI	
dfi→clt	0.483	0.042	11.630	0.000	0.402	0.565
constant	−1.894	0.363	−5.220	0.000	−2.605	−1.183
clt→clgue (new)	0.282	0.053	5.280	0.000	0.177	0.386
dfi→clgue (new)	0.361	0.051	7.070	0.000	0.261	0.461
constant	−0.116	0.395	−0.290	0.769	−0.890	0.659
variance (e.clt)	0.766	0.040			0.691	0.849
variance (e.clgue)	0.692	0.042			0.614	0.779

Table 7. Robustness tests of significance of mediating effects.

Paths	Coefficients	Standard Error	Z Value	*p* Value	95% CI	
dfi→clt→clgue (new)	0.136	0.029	4.750	0.000	0.080	0.192

4.5. Heterogeneity Tests of Main Effects

We used the heterogeneity analysis to investigate the different influence of DFI on CLGUE based on different geographical locations. As is illustrated in Table 8, all the path coefficients of DFI on CLGUE in the three areas are significantly positive. The path coefficient of DFI on CLGUE in the eastern areas (0.633) is much higher than those in the central areas (0.228) and western areas (0.415). This indicates that the positive relationship between DFI and CLGUE in the eastern areas is more obvious. Possible reasons for this phenomenon are better facilities for finance and a higher level of financial development in the eastern region [82]. DFI and traditional finance are complementary, provide better financial services for cultivated land operators, and eventually raise CLGUE.

Table 8. Results of heterogeneity tests of main effects (*dfi→clgue*).

	Eastern Areas	Central Areas	Western Areas	MGPAs	MGMAs	GPMBAs
Coefficients	0.633	0.228	0.415	0.408	0.586	0.360
Standard error	0.051	0.105	0.075	0.070	0.071	0.084
Z value	12.370	2.180	5.490	5.840	8.200	4.280
p value	0.000	0.029	0.000	0.000	0.000	0.000
95% CI	0.532	0.024	0.267	0.271	0.446	0.195
	0.733	0.433	0.563	0.546	0.726	0.525

Additionally, heterogeneity analysis on the effects of DFI on CLGUE based on different grain functional was carried out. Referring to Ke et al. [16], 30 provinces can be divided into three categories of main grain-producing areas (MGPAs), main grain-marketing areas (MGMAs), and grain-producing and marketing balance areas (GPMBAs). The results in Table 8 indicate that all the path coefficients of DFI on CLGUE in the three grain-functional areas are significantly positive. The path coefficient of DFI on CLGUE in the MGMAs (0.568) is much higher than those in the MGPAs (0.408) and GPMBAs (0.360). MGMAs are located in the southeast coastal or economically developed provinces, with strict environmental regulation. Empirical analysis showed that environmental regulations had a prominent positive effect on the adoption of green farming practices, such as farmers adopting high efficiency, low toxicity, and low residue pesticides [83].

5. Discussion

This study draws on triple bottom line theory to empirical investigate whether and how DFI can affect CLGUE through CLT. Using a sample of Chinese provincial panel data during the period of 2011–2020 and SEM analyses, this paper draws the following conclusions:

(1) DFI can directly enhance CLGUE. DFI has the characteristics of digitalization and inclusiveness. Scientific analysis of various data generated and processed by digital technology is conducive to achieving green detection. Green finance arising from environmental conservation dramatically enhances the green features of finance, efficiently accelerating the increase of energy utilization efficiency and a reduction in carbon emissions. Apart from the environmental protection effects, DFI can efficiently improve the outputs and control the inputs of cultivated land, which further facilitates cultivated land utilization efficiency.

(2) DFI can indirectly improve CLGUE through cultivated land transfer. CLT means transferring cultivated land management rights from individual farmers to professional

farmers or economic organizations. DFI can facilitate CLT by reducing transaction costs and information asymmetry, providing more nonagricultural entrepreneurial and employment opportunities and enhancing agricultural mechanization. Furthermore, CLT can transfer of managing rights of cultivated land from low-productivity operators to high-productivity operators, subsequently enhancing CLGUE by improving the efficiency of the utilization of fertilizers and pesticides, optimizing grain planting structure and driving large-scale agricultural modernization.

(3) DFI has regional heterogeneity in the improvement of CLGUE. Compared to the central and western areas, the positive relationship between DFI and CLGUE in the eastern areas is more obvious. In addition, compared with major grain producing and main grain producing and marketing balance areas, the positive relationship between DFI and CLGUE in the major grain marketing areas is more obvious.

Our findings make great contributions to the extant literature. In order to guarantee grain security and cultivated land utilization sustainably, the improvement of CLGUE has been more and more widely mentioned in agricultural sustainability in recent years. The extant literature has identified that digital financial inclusion is positively related to the agricultural supply chain [53], the rationalization of rural products' industrial structure and green total factor productivity [82], agricultural production for rural households [84], agricultural high-quality development [85], etc. Nevertheless, studies on the relationship between DFI and low-carbon green utilization of farmland are scarce. In the recent decade, finance characterized by digital and inclusive connotation is developing rapidly in China [34], and seems to be conducive to increasing CLGUE, it is significative to empirical study the influencing mechanism of the emerging financing form on CLGUE. The present paper draws on triple bottom line theory and takes the CLT as the mediating mechanism, revealing how CLT can promote CLGUE in China. Although CLT adversely affects the use and yield of cultivated land in some developed countries [29,86], it has great effects on facilitating CLGUE in China. In China, the ownership rights of cultivated land belong to Chinese government and the operating and managing rights of cultivated land belong to individual farmers. The Chinese cultivated land transfer policy supports the individual farmers in transferring their management rights to large professional households and groups to develop large-scale agricultural operations. The specific forms of transfer include subcontract, transfer, investment, cooperation, leasing, exchange, and other means. Farmers can choose the most suitable way to transfer farmland according to their available funds. The processes of CLT are voluntary, fair, open, and paid. This study theoretically analyzes the impact and mechanism of DFI on CLGUE, constructs a framework mechanism of CLGUE, CLT, and CLGUE, and expands the research's scope on DFI and provides a reference for green agricultural development and digital rural development.

Our findings also provide some practical insights. Firstly, the governments are recommended to increase investments in the research and development of digital financial technologies and applications, so as to continuously extend digital financial inclusion services to wider population. Governments are also suggested to simplify farmland transfer procedures, and widely publicize the subsidy scheme for farmland transfer in order to ensure that the activities of farmland transfer are more transparent, simple, and attractive. With the improvement of digital finance systems and the extension of farmland transfer, cultivated land's green utilization efficiently can be improved. Secondly, since traditional institutions are experience difficulty in offering adequate financial products and services to farmers, financial institutions are recommended to continuously expand the coverage breadth, usage depth, and digitalization level of digital financing services to satisfy farmers' fund demands. Farmers owning sufficient funds will increase their willingness to adopt new technology, introduce large-scale mechanization, and subsequently improve cultivated land green utilization efficiency. Finally, on the one hand, small-scale farmers are suggested to transfer out their land and obtain payments and compensation. They can engage in nonagricultural industry. On the other hand, small-scale farmers are recommended to transfer to other farmers' land and form large-scale a agricultural production, because they have easier access to financial loans and insurance.

Despite these attractive contributions, our research also has limitations. First, our large sample covers Chinese provincial data from 2011 to 2020. Thereby, the generalization of our findings to other countries or regions should be made cautiously. Though our theory is not specific to the China's context, future research may collect data from other countries, especially from developed countries with a maturely developed digital inclusive finance system and different cultivated land transfer policies. Second, owing to time and data constraints, we did not introduce other associated variables in the framework; for instance, the antecedent variables that can affect explanatory variables and the moderating variables that can affect the mechanism are not discussed. We can explore more associated variables in future research to obtain more theoretical and practical inspiration. For instance, we can discuss the antecedent variables of digital financial inclusion (e.g., digital technology, government support), the consequence variables of cultivated land green utilization efficiency (e.g., high-quality agricultural development, sustainably development), and other mediating variables (e.g., level of mechanization, management scale) as well. Last but not least, we measured the intensity of DFI according to the Peking University DFI Index of China. However, due to the rapid development of digital technology, it is difficult for us to cover all of the digital financing channels. In the future, a more scientific measurement method related to DFI can be introduced to reduce measurement errors.

Author Contributions: Conceptualization: M.Z. and H.Z.; methodology, M.Z. and H.Z.; software, H.Z. and H.S.; validation, H.Z. and Z.Z.; data curation, M.Z., H.Z. and H.S.; writing—original draft preparation, M.Z., H.Z. and Z.Z.; writing—review and editing, M.Z. and H.Z.; visualization, M.Z., H.Z., Z.Z. and H.S. All authors have read and agreed to the published version of the manuscript.

Funding: This research was funded by the Social Science Foundation of China (No. 21BGL287) and Education Department of Liaoning Province (No. LJKMR20220424).

Institutional Review Board Statement: Not applicable.

Informed Consent Statement: Not applicable.

Data Availability Statement: Data are available from authors upon reasonable request.

Conflicts of Interest: The authors declare no conflict of interest.

References

1. Lin, X.; Fu, H. Spatial-Temporal Evolution and Driving Forces of Cultivated Land Based on the PLUS Model: A Case Study of Haikou City, 1980. *Sustainability* **2022**, *14*, 14284. [CrossRef]
2. Fu, G.Z.; Bai, W.Q. Advances and prospects of evaluating cultivated land quality. *Resour. Sci.* **2015**, *37*, 0226–0236.
3. Lichtenberg, E.; Ding, C. Assessing farmland protection policy in China. *Land Use Policy* **2008**, *25*, 59–68. [CrossRef]
4. Abass, K.; Adanu, S.K.; Agyemang, S. Peri-urbanisation and loss of arable land in Kumasi Metropolis in three decades: Evidence from remote sensing image analysis. *Land Use Policy* **2018**, *72*, 470–479. [CrossRef]
5. Gorgan, M.; Hartvigsen, M. Development of agricultural land markets in countries in Eastern Europe and Central Asia. *Land Use Policy* **2022**, *120*, 106257. [CrossRef]
6. Berihun, M.L.; Tsunekawa, A.; Haregeweyn, N.; Meshesha, D.T.; Adgo, E.; Tsubo, M.; Masunaga, T.; Fenta, A.A.; Sultan, D.; Yibeltal, M. Exploring land use/land cover changes, drivers and their implications in contrasting agro-ecological environments of Ethiopia. *Land Use Policy* **2019**, *87*, 104052. [CrossRef]
7. Wu, Y.; Shan, L.; Guo, Z.; Peng, Y. Cultivated land protection policies in China facing 2030: Dynamic balance system versus basic farmland zoning. *Habitat Int.* **2017**, *69*, 126–138. [CrossRef]
8. Niu, S.D.; Fang, B. Cultivated land protection system in china from 1949 to 2019: Historical evolution, realistic origin exploration and path optimization. *China Land Sci.* **2019**, *33*, 1–12. [CrossRef]
9. Feng, Y.G.; Peng, J.; Deng, Z.B.; Wang, J. Spatial-temporal variation of cultivated land's utilization efficiency in china based on the dual perspective of non-point source pollution and carbon emission. *China Popul. Resour. Environ.* **2015**, *25*, 18. [CrossRef]
10. Guo, F.Y.; Gao, S.Q.; Tong, L.J.; Ren, J.M. Spatio-temporal evolution track and influencing factors of green development efficiency in Yellow River Basin. *Geogr. Res.* **2022**, *41*, 167–180. [CrossRef]
11. Zhou, M.; Zhang, H.; Ke, N. Cultivated Land Transfer, Management Scale, and Cultivated Land Green Utilization Efficiency in China: Based on Intermediary and Threshold Models. *Int. J. Environ. Res. Public Health* **2022**, *19*, 12786. [CrossRef] [PubMed]

12. FAO. *The State of Food and Agriculture*; Food & Agriculture Organization of the U.N.: Washington, DC, USA, 2016.
13. Lu, X.; Kuang, B.; Li, J.; Han, J.; Zhang, Z. Dynamic Evolution of Regional Discrepancies in Carbon Emissions from Agricultural Land Utilization: Evidence from Chinese Provincial Data. *Sustainability* **2018**, *10*, 552. [CrossRef]
14. Lu, X.; Qu, Y.; Sun, P.; Yu, W.; Peng, W. Green Transition of Cultivated Land Use in the Yellow River Basin: A Perspective of Green Utilization Efficiency Evaluation. *Land* **2020**, *9*, 475. [CrossRef]
15. Zhang, J.F.; He, S.W.; Zhang, G.H.; Zhang, A.L. Ecological Benefit, Spatial Spillover and Fiscal Transfer of Farmland in Watershed: Evidence from the Yangtze River Economic Belt. *Issues Agric. Econ.* **2020**, *12*, 120–132. [CrossRef]
16. Ren, P.; Hong, B.T.; Ma, W.L.; Yuan, Q.Z.; Zhou, J.M. Ecological value estimation of cultivated land based on the IBIS model: A case study of Chong zhou city. *Geogr. Res.* **2016**, *35*, 2395–2406.
17. Xie, H.; Chen, Q.; Wang, W.; He, Y. Analyzing the green efficiency of arable land use in China. *Technol. Forecast. Soc. Chang.* **2018**, *133*, 15–28. [CrossRef]
18. Ke, N.; Lu, X.H.; Kuang, B.; Han, J. Regional differences and influencing factors of green and low-carbon utilization of cultivated land under the carbon neutrality target in China. *China Land Sci.* **2021**, *35*, 67–76. [CrossRef]
19. Zhou, M.; Kuang, B.; Zhou, M.; Ke, N. The Spatial and Temporal Evolution of the Coordination Degree in Regard to Farmland Transfer and Cultivated Land Green Utilization Efficiency in China. *Int. J. Environ. Res. Public Health* **2022**, *19*, 10208. [CrossRef] [PubMed]
20. Qu, Y.; Lyu, X.; Peng, W.; Xin, Z. How to Evaluate the Green Utilization Efficiency of Cultivated Land in a Farming Household? A Case Study of Shandong Province, China. *Land* **2021**, *10*, 789. [CrossRef]
21. Chen, Q.; Xie, H. Temporal-Spatial Differentiation and Optimization Analysis of Cultivated Land Green Utilization Efficiency in China. *Land* **2019**, *8*, 158. [CrossRef]
22. Xu, Y.; Liu, L.-Y.; Pu, L.-J.; Liu, Y.-M.; Xu, F.; Yu, C. Analysis on farmers' behavioral intension of cropland use and its influencing factors in the coastal areas of Northern Jiangsu province. *J. Nat. Resour.* **2022**, *37*, 1643–1653. [CrossRef]
23. Lu, X.M.; Wu, Y.F. Land-renting, Farmers' Agricultural Credit Demands and Credit Constraints: An Analysis of CHFS Data. *Financ. Res.* **2021**, *5*, 40–58.
24. Cheng, Q.W.; Xu, A.X.; Chen, Q. The realization path of agricultural carbon emission reduction under the background of "dual carbon" target. *J. Southwest Minzu Univ. (Humanit. Soc. Sci.)* **2022**, *43*, 115–125.
25. Yan, G.Q.; He, Y.C.; Zhang, X.H. Can the development of digital inclusive finance promote agricultural mechanization? —Based on the perspective of the development of agricultural machinery outsourcing service market. *J. Agrotech. Econ.* **2022**, *1*, 51–64. [CrossRef]
26. Sun, X.T.; Yu, T.; Yu, F.W. The impact of digital finance on agricultural mechanization: Evidence from 1869 counties in China. *Chin. Rural Econ.* **2022**, *2*, 76–93.
27. Zhou, Z.; Zhang, Y.; Yan, Z. Will Digital Financial Inclusion Increase Chinese Farmers' Willingness to Adopt Agricultural Technology? *Agriculture* **2022**, *12*, 1514. [CrossRef]
28. Zhang, Y.Q. Study on the influence and mechanism of digital inclusive finance on rural land transfer: Empirical evidence from CFPS and PKU-DFIIC. *Econ. Manag.* **2022**, *36*, 30–40.
29. Yildiz, O.; Coruhlu, Y.E.; Biyik, C. Registration of agricultural areas towards the development of a future Turkish cadastral system. *Land Use Policy* **2018**, *78*, 207–218. [CrossRef]
30. Slack, N.; Brandon-Jones, A.; Johnston, R. *Essentials of Operations Management*, 7th ed.; Pearson Education Limited: Essex, UK, 2018; pp. 47–49.
31. Smith, G.; Block, L.B.; Ajami, N.; Pombo, A.; Velasco-Aulcy, L. Trade-offs across the water-energy-food nexus: A triple bottom line sustainability assessment of desalination for agriculture in the San quintín Valley, Mexico. *Environ. Sci. Policy* **2020**, *114*, 445–452. [CrossRef]
32. Ellis, M.J.; Sexton, A.; Dunsford, I.; Stephens, N. The triple bottom line framework can connect people, planet and profit in cellular agriculture. *Nat. Food* **2022**, *3*, 804–806. [CrossRef]
33. Norman, W.; Macdonald, C. Getting to the bottom of the triple bottom line. *Bus. Ethics Q* **2004**, *14*, 243–262. [CrossRef]
34. Isaeva, A.; Salahodjaev, R.; Khachaturov, A.; Tosheva, S. The Impact of Tourism and Financial Development on Energy Consumption and Carbon Dioxide Emission: Evidence from Post-communist Countries. *J. Knowl. Econ.* **2021**, *13*, 773–786. [CrossRef]
35. Shi, F.; Ding, R.; Li, H.; Hao, S. Environmental Regulation, Digital Financial Inclusion, and Environmental Pollution: An Empirical Study Based on the Spatial Spillover Effect and Panel Threshold Effect. *Sustainability* **2022**, *14*, 6869. [CrossRef]
36. Lee, C.-C.; Wang, F.; Lou, R. Digital financial inclusion and carbon neutrality: Evidence from non-linear analysis. *Resour. Policy* **2022**, *79*, 102974. [CrossRef]
37. Lee, C.-C.; Yuan, Z.; Wang, Q. How does information and communication technology affect energy security? International evidence. *Energy Econ.* **2022**, *109*, 105969. [CrossRef]
38. Shen, J.H.; Long, Z.; Lee, C.-C.; Zhang, J. Comparative advantage, endowment structure, and trade imbalances. *Struct. Chang. Econ. Dyn.* **2021**, *60*, 365–375. [CrossRef]
39. Dayioğlu, M.A.; Turker, U. Digital Transformation for Sustainable Future—Agriculture 4.0: A review. *J. Agr. Sci. Tarim. Bili.* **2021**, *27*, 373–399. [CrossRef]
40. Wang, Q. Challenges, opportunities and suggestions for the development of China's rural digital financial inclusion. *Front. Econ. Manag.* **2022**, *3*, 0044.

41. Warren, P. The role of climate finance beyond renewables: Demand-side management and carbon capture, usage and storage. *Clim. Policy* **2019**, *19*, 861–877. [CrossRef]
42. Ren, X.; Shao, Q.; Zhong, R. Nexus between green finance, non-fossil energy use, and carbon intensity: Empirical evidence from China based on a vector error correction model. *J. Clean. Prod.* **2020**, *277*, 122844. [CrossRef]
43. Guo, Q.; Su, Z.; Chiao, C. Carbon emissions trading policy, carbon finance, and carbon emissions reduction: Evidence from a quasi-natural experiment in China. *Econ. Chang. Restruct.* **2021**, *55*, 1445–1480. [CrossRef]
44. O'Toole, C.M.; Newman, C.; Hennessy, T. Financing Constraints and Agricultural Investment: Effects of the Irish Financial Crisis. *J. Agric. Econ.* **2013**, *65*, 152–176. [CrossRef]
45. Jing, X.; Jiang, R.; Chen, Z.; Deng, Z. Agricultural output effect of rural finance: An extended regression approach. *E M Ekon Manag.* **2022**, *25*, 4–22. [CrossRef]
46. Wang, Y.Q.; Xu, X.J.; Xiong, H.; Fang, X.F. Research on the influence of inter-linked index insurance and credit contract on the adoption of new technologies in family farms. *Financ. Perspect. J.* **2019**, *33*, 83–91.
47. Lu, F.; Wang, W.; Liu, M.; Liu, M.; Qi, D. The non-linear effect of agricultural insurance on agricultural green competitiveness. *Technol. Anal. Strat. Manag.* **2022**, 1–16. [CrossRef]
48. Skarstein, R. Explaining Productivity Change in Underdeveloped Agriculture. Can the theory of induced innovation do it? *Dev. Stud.* **2013**, *41*, 75–90. [CrossRef]
49. Busato, P.; Berruto, R. Minimising manpower in rice harvesting and transportation operations. *Biosyst. Eng.* **2016**, *151*, 435–445. [CrossRef]
50. Demir, A.; Pesqué-Cela, V.; Altunbas, Y.; Murinde, V. Fintech, financial inclusion and income inequality: A quantile regression approach. *Eur. J. Financ.* **2020**, *28*, 86–107. [CrossRef]
51. Wang, Y.; Li, X.; He, H.; Xin, L.; Tan, M. How reliable are cultivated land assets as social security for Chinese farmers? *Land Use Policy* **2019**, *90*, 104318. [CrossRef]
52. Zhang, F.; Bao, X.; Deng, X.; Xu, D. Rural Land Transfer in the Information Age: Can Internet Use Affect Farmers' Land Transfer-In? *Land* **2022**, *11*, 1761. [CrossRef]
53. Fang, D.; Zhang, X. The Protective Effect of Digital Financial Inclusion on Agricultural Supply Chain during the COVID-19 Pandemic: Evidence from China. *J. Theor. Appl. Electron. Commer. Res.* **2021**, *16*, 3202–3217. [CrossRef]
54. Wang, X.; Fu, Y. Digital financial inclusion and vulnerability to poverty: Evidence from Chinese rural households. *China Agric. Econ. Rev.* **2021**, *14*, 64–83. [CrossRef]
55. Lorente, J.J.C.; Schmukler, S.L. The fintech revolution: A threat to global banking? *Res. Policy Briefs* **2018**, *1*, 105034.
56. Ozturk, I.; Ullah, S. Does digital financial inclusion matter for economic growth and environmental sustainability in OBRI economies? An empirical analysis. *Resour. Conserv. Recy.* **2022**, *185*, 106489. [CrossRef]
57. Khera, P.; Ng, S.; Ogawa, S.; Sahay, R. Measuring Digital Financial Inclusion in Emerging Market and Developing Economies: A New Index. *Asian Econ. Policy Rev.* **2022**, *17*, 213–230. [CrossRef]
58. Belton, B.; Win, M.T.; Zhang, X.; Filipski, M. The rapid rise of agricultural mechanization in Myanmar. *Food Policy* **2021**, *101*, 102095. [CrossRef]
59. Chari, A.; Liu, E.M.; Wang, S.-Y.; Wang, Y. Property Rights, Land Misallocation, and Agricultural Efficiency in China. *Rev. Econ. Stud.* **2020**, *88*, 1831–1862. [CrossRef]
60. Shang, H.; Yi, X.; Yin, C.; Chen, Y.; Zhang, Z. How Does the Stability of Land Management Right (SLMR) Affect Family Farms' Cultivated Land Protection and Quality Improvement Behavior (CLPQIB) in China? *Land* **2021**, *10*, 1052. [CrossRef]
61. Lu, H.; Chen, Y.; Huan, H.; Duan, N. Analyzing Cultivated Land Protection Behavior From the Perspective of Land Fragmentation and Farmland Transfer: Evidence From Farmers in Rural China. *Front. Environ. Sci.* **2022**, *10*, 901097. [CrossRef]
62. Wang, H.; Li, C.; Liu, J.; Zhang, S. Research on Farmers' Willingness of Land Transfer Behavior Based on Food Security. *Sustainability* **2019**, *11*, 2338. [CrossRef]
63. Cao, W.K.; Wu, D.E. Contract system for circulation of the right to rural land contractual management. *Asian Agric. Res.* **2010**, *2*, 53–57.
64. Qiu, T.; Choy, S.B.; Li, S.; He, Q.; Luo, B. Does land renting-in reduce grain production? Evidence from rural China. *Land Use Policy* **2020**, *90*, 104311. [CrossRef]
65. Chen, M.-Q.; Zhong, T.-Y.; Zhou, B.-J.; Huang, H.-S.; He, W.-J. Empirical research on farm households' attitude and behaviour for cultivated land transferring and it's influencing factors in China. *Agric. Econ.* **2010**, *56*, 409–420. [CrossRef]
66. Huang, T.; Xiong, B. Space Comparison of Agricultural Green Growth in Agricultural Modernization: Scale and Quality. *Agriculture* **2022**, *12*, 1067. [CrossRef]
67. Wu, Y.; Xi, X.; Tang, X.; Luo, D.; Gu, B.; Lam, S.K.; Vitousek, P.M.; Chen, D. Policy distortions, farm size, and the overuse of agricultural chemicals in China. *Proc. Natl. Acad. Sci. USA* **2018**, *115*, 7010–7015. [CrossRef]
68. Davtalab-Olyaie, M.; Roshdi, I.; Nia, V.P.; Asgharian, M. On characterizing full dimensional weak facets in DEA with variable returns to scale technology. *Optimization* **2014**, *64*, 2455–2476. [CrossRef]
69. Charnes, A.; Cooper, W.W.; Rhodes, E. Measuring the efficiency of decision making units. *Eur. J. Oper. Res.* **1978**, *2*, 429–444. [CrossRef]
70. Banker, R.D.; Morey, R.C. The Use of Categorical Variables in Data Envelopment Analysis. *Manag. Sci.* **1986**, *32*, 1613–1627. [CrossRef]

71. Niu, H.; Zhang, Z.; Xiao, Y.; Luo, M.; Chen, Y. A Study of Carbon Emission Efficiency in Chinese Provinces Based on a Three-Stage SBM-Undesirable Model and an LSTM Model. *Int. J. Environ. Res. Public Health* **2022**, *19*, 5395. [CrossRef]
72. Tone, K. A slacks-based measure of efficiency in data envelopment analysis. *Eur. J. Oper. Res.* **2001**, *130*, 498–509. [CrossRef]
73. Tone, K. Dealing with undesirable outputs in DEA: A slacks-based measure (SBM) approach. *Present. NAPW III Tor.* **2004**, *2004*, 44–45.
74. Ma, D.; Li, G.; He, F. Exploring PM2.5 Environmental Efficiency and Its Influencing Factors in China. *Int. J. Environ. Res. Public Health* **2021**, *18*, 12218. [CrossRef]
75. Bagozzi, R.P.; Yi, Y. Specification, evaluation, and interpretation of structural equation models. *J. Acad. Mark. Sci.* **2012**, *40*, 8–34. [CrossRef]
76. Guo, F.; Wang, J.; Wang, F.; Kong, T.; Zhang, X.; Cheng, Z. Measuring the development of digital financial inclusion in China: Index compilation and spatial characteristics. *J. Econ.* **2020**, *12*, 1401–1418. [CrossRef]
77. Ke, N.; Zhang, X.; Lu, X.; Kuang, B.; Jiang, B. Regional Disparities and Influencing Factors of Eco-Efficiency of Arable Land Utilization in China. *Land* **2022**, *11*, 257. [CrossRef]
78. Zhou, M.; Hu, B. Decoupling of carbon emissions from agricultural land utilisation from economic growth in China. *Agric. Econ.* **2020**, *66*, 510–518. [CrossRef]
79. Huang, H.P.; Wang, Z.P. Spatial—temporal Differences and Influencing Factors of Agricultural Land Eco-efficiency in Jiangxi Province: Based on the Dual Perspective of Non—point Source Pollution and Carbon Emission. *Resour. Environ. Yangtze Basin* **2022**, *29*, 412–423.
80. Kline, R.B. *Principles and Practice of Structural Equation Modeling*, 4th ed.; The Guilford Press: New York, NY, USA, 2018.
81. Yang, Y. *Structural Equation Model and Stata Application*, 1st ed.; Beijing Peking University Press: Beijing, China, 2021; pp. 42–45.
82. Hong, M.; Tian, M.; Wang, J. Digital Inclusive Finance, Agricultural Industrial Structure Optimization and Agricultural Green Total Factor Productivity. *Sustainability* **2022**, *14*, 11450. [CrossRef]
83. Zhang, Y.; Wang, S.G.; Chen, X. Research on the Influence of Digital Finance on Green Pesticide Application by Farmers —Based on the Information Availability Perspective. *Chin. J. Agric. Resour. Reg. Plan.* 2022. Available online: https://kns.cnki.net/kcms/detail/11.3513.S.20221101.1625.016.html (accessed on 1 August 2022).
84. Liu, Y.; Liu, C.; Zhou, M. Does digital inclusive finance promote agricultural production for rural households in China? Research based on the Chinese family database (CFD). *China Agric. Econ. Rev.* **2021**, *13*, 475–494. [CrossRef]
85. Wang, S.; Chen, Y.B. How can digital financial inclusion promote agricultural high-quality development? The mediating and threshold mechanism. *J. Manag.* **2022**, *35*, 72–87. [CrossRef]
86. Coruhlu, Y.E.; Yildiz, O. Geographical database for object-oriented land division modelling in Turkey. *Land Use Policy* **2017**, *68*, 212–221. [CrossRef]

sustainability MDPI

Article

Landscape Ecological Evaluation of Cultural Patterns for the Istanbul Urban Landscape

Gül Aslı Aksu [1,*], Şermin Tağıl [2], Nebiye Musaoğlu [3], Emel Seyrek Canatanoğlu [4] and Adnan Uzun [5]

1 Department of Landscape Architecture, Kastamonu University, 37150 Kastamonu, Turkey
2 Department of Geography, Izmir Bakircay University, 35665 Izmir, Turkey
3 Department of Geomatics Engineering, Istanbul Technical University, 34469 Istanbul, Turkey
4 Graduate School of Natural and Applied Sciences, Institute of Science, Ege University, 35040 Izmir, Turkey
5 Department of Interior Architecture and Environmental Design, Isik University, 34980 Istanbul, Turkey
* Correspondence: gaaksu@kastamonu.edu.tr or aslibozbay@yahoo.de

Abstract: With the widespread population growth in cities, anthropogenic influences inevitably lead to natural disturbances. The metropolitan area of Istanbul, with its rapid urbanization rate, has faced intense pressure regarding the sustainability of urban habitats. In this context, landscapes comprising patches affected by various disturbances and undergoing temporal changes must be analyzed, in order to assess city-related disturbances. In this study, the main objective was to understand how urbanization changed the function of the spatial distribution of the urban mosaic and, more specifically, its relationship with the size, shape, and connection among land-use classes. For this purpose, we took Besiktas, a district of Istanbul, as the study area. We evaluated the landscape pattern of the urban environment in two stages. First, we used medium-resolution satellite imagery to reveal the general interactions in the urbanization process. Landscape- and class-level landscape metrics were selected to quantify the landscape connectivity, and the distances between classes (green areas and artificial surfaces), patterns, and processes, using five satellite images representing a time span of 51 years (1963, 1984, 1997, 2005, and 2014). The general landscape structure was examined by looking at the temporal–spatial processes of artificial surface and green areas obtained from these medium-resolution satellite images. The trends in selected landscape-level metrics were specified and discussed through the use of a moving window analysis. We then used Pleiades high-resolution satellite imagery (2015) to analyze the landscape structure in more detail. This high-resolution base image allows us to recognize the possibility of classifying basic cultural landscape classes. The findings regarding the spatial arrangement of each class in the areas allocated to 14 cultural landscape classes were interpreted by associating them with the landscape functions. Finally, particulate matter (PM_{10}) concentration data were collected and evaluated as an ecological indicator, in order to reveal the relationships between landscape structure and landscape function. In short, we first evaluated the whole landscape structure using medium-resolution data, followed by the classification of cultural landscapes using high-resolution satellite imagery, providing a time-effective—and, therefore, essential—auxiliary method for landscape evaluation. This two-stage evaluation method enables inferences to be made that can shed light on the landscape functions in an urban environment based on the landscape structure.

Keywords: Pleiades satellite image; landscape pattern; cultural landscapes; landscape function; urban ecosystem; landscape ecology; PM_{10} concentration

Citation: Aksu, G.A.; Tağıl, Ş.; Musaoğlu, N.; Canatanoğlu, E.S.; Uzun, A. Landscape Ecological Evaluation of Cultural Patterns for the Istanbul Urban Landscape. *Sustainability* **2022**, *14*, 16030. https://doi.org/10.3390/su142316030

Academic Editors: Lu Zhang, Bing Kuang and Bohan Yang

Received: 3 November 2022
Accepted: 21 November 2022
Published: 30 November 2022

1. Introduction

Human activities disrupt the balances established in ecosystems, sometimes irreversibly. This deterioration can lead to serious problems, not only aesthetically but also economically and even for human health, in the long term. As environmental problems have reached such levels that they may have a global impact, issues such as the "Environment" and "Sustainability" have become the main topics considered by both the European

Union and United Nations Support Programs. As in Goal 11 "Make cities and human settlements inclusive, safe, resilient and sustainable", described within the UN Sustainable Development Goals (SDGs), such topics are related to our environment and quality of life [1–3]. From this point of view, it is understood that handling the potential of the environment, in line with sustainability principles, has become an international responsibility.

The European Green Deal, the primary goal of which is to regulate the European climate and emissions rates by 2050, has also identified "Preserving and restoring ecosystems and biodiversity" and "Accelerating the shift to sustainable and smart mobility" as main agenda items. The main objective of the strategy determined by the European Commission for the protection of biological diversity until 2020 was to "prevent loss of biodiversity and decline in ecosystem services". The Green Deal aims to improve declining ecosystems by at least 15% by 2030 [4]. The basis of achieving this goal lies in developing an inventory of areas that need to be protected, developed, and improved, in terms of cultural values and biological diversity, based on natural–cultural indicators [5,6].

Human beings consider how the landscape shapes itself to be chaotic or unorganized. As a result, they attempt to control and shape landscape evolution through deliberate and intended actions. However, these actions do not always provide the outcomes they were planned to, as the landscape is composed of many different components, which have their own dynamics and orders of change. This composition can be evaluated as a system that includes its own features and, therefore, we must look at these components in a holistic manner [7–10].

One of the most compelling challenges that humans pose against nature is urbanization. Urbanization combines aspects related to population crowding, thus resulting in denser urban areas and the spread of residents and buildings outside the urban center [11,12]. With the influence of urbanization, the water regime of cities deteriorates, the natural relief changes, natural soil characteristics disappear, and urban heat island effects occur. Changes in the landscape pattern and function can lead to various consequences, starting from the degradation of green systems at the regional scale to the disappearance of biotopes at the local level.

Of all the problems related to urbanization, the most important one is that cities are not sustainable. As cities coexist with human beings and surrounding systems, they are regarded as heterogeneous and complicated, and lack the ability to quickly adapt. Thus, it is tough to predict or monitor the course of events in cities. Nevertheless, in this situation, planning and new designs can be helpful, according to the knowledge of city ecology and sustainability principles [13–15].

Understanding human spatial and material relations and changes in the natural environment is essential to ensuring sustainability. Cities, as energy and material consumption nodes, are not sustainable per se, and they accelerate global ecological degradation. However, cities and their residents play a leading role in ensuring urban sustainability [16,17]. According to the modern understanding of landscape ecology, conceptual models and tools are needed to help analyze and reveal the nature–society interactions at the center of the sustainability debate [15]. With the same awareness, practical tools that refer to protection–utilization balances should be used in planning and management decisions, especially regarding urban landscapes with sensitive balances [18–21].

At the urban scale, habitats are the stepping stones of green network systems, and are essential in urban areas. Urban green areas also have critical functional features for specific purposes, such as filtering the air, balancing noise and the climate, and providing environments for recreational activities. Re-structuring degraded green networks is extremely important for the sustainability of the urban ecosystem. For example, a quality vegetative layer that penetrates structural surfaces has an effect that prevents the formation of heat islands in cities [22,23]. According to Lehmann et al. (2014), urban green areas are essential indicators for ecosystem services in urban habitats, when considered spatially and structurally. In addition, they exhibit properties suitable for the evaluation of microclimatic features [24,25].

Cultural landscapes are areas whose natural features have been changed by human activities, manifesting in the form of layered patterns that leave traces in the landscape. Together with natural features, these layers give a landscape its defining, historical, aesthetic, symbolic, and memorable character. Therefore, it is necessary to define cultural landscapes, develop policies to protect their values, manage the spatial and social changes that occur over time, and enable sustainable uses [26–29]. Due to these high-capacity indicative qualities, many studies have focused on cultural landscapes in an urban environment.

The main concern of natural conservation is protecting species and their habitats. Therefore, identifying and mapping the habitats created by the natural and cultural landscape is essential for the in situ conservation of biological diversity [30].

Cultural landscapes are essential components of the environment shaped by human interventions. They mirror the past and indicate the future while hosting highly diverse anthropogenic uses and natural–cultural heritage. Moreover, landscapes are located between ecosystems and biomes [31]; thus, their quality and diversity constitute a common resource. Therefore, revealing cultural landscapes within urban landscapes and evaluating the relations between them at the regional scale constitute a strong basis for a holistic approach.

Focusing on cultural landscapes in the planning process is of great importance in protecting and promoting biodiversity and supporting sustainable development, increasing the quality of life and comfort of residents [18,19,32–36].

Modern technology has made satellite images with higher spatial resolution available for various applications, such as urban mapping, spatio-temporal change detection, and urban sprawl monitoring [20,21,37].

As manual classification techniques are difficult and time-consuming, it was deemed appropriate to classify satellite images at two different resolutions (medium and high) with the help of remote sensing techniques. Medium-resolution satellite images are divided into two classes using controlled classification techniques, in order to make general evaluations. Then, a high-resolution Pleiades satellite image was divided into 14 cultural landscape classes using the normalized vegetation index (NDVI) (Table 1).

Table 1. Land use/land cover (LU/LC) classes obtained from medium- and high-resolution satellite images.

Satellite	LU/LC Class
Landsat 4, 5, 7 TM, ETM+	Artificial surface, green area
Pleiades (2015)	Garden, openness in garden, grove, openness in grove, cemetery, openness in cemetery, park, openness in park, roadside green area, openness in roadside green area, building, water surface, firm ground, road

The mapping of LU/LC classes provides important outputs for landscape analysis and assessment, and has been widely used in the literature. Evaluating the data obtained as a result of classification using landscape metrics provides access to important findings at the class and landscape level, especially regarding the landscape structure, in a short time. Landscape metrics have been widely used in the literature as an effective tool to reveal the structure and configuration of landscape structure [21,38–49]. In this study, we conduct landscape pattern analysis to evaluate cultural landscapes, and support the findings obtained from this analysis using the PM_{10} concentration, which is an important ecological indicator.

PM has been shown to have a positive relationship with urbanization. Studies examining the relationships between green systems and the concentration of particulate matter in the air have shown that PM density is closely related to the quality and quantity of the green system [50–54]. For this reason, PM_{10} was used as an ecological indicator in this research.

Based on this view, the focal points of this research are as follows: (a) the effect of urbanization on the spatial transformation of cultural landscapes; (b) the interactions between cultural landscape patterns and PM_{10} concentrations; and (c) the relationships between landscape structures and indicative urban habitats.

2. Materials and Methods

2.1. Material

Istanbul is a metropolitan city with an E–W extension, adjacent to the Black Sea and the Marmara Sea, with a characteristic structure similar to the Bosporus, and is in a highly strategic position, in terms of various components. As the research area, the Besiktas district is in the middle of the European Bosporus Side of the city of Istanbul, between 41°02′31″ N. latitudes and 29°00′26″ E. longitudes. Its total surface area, including the buffer zone, is 37.8 km^2 (Figure 1).

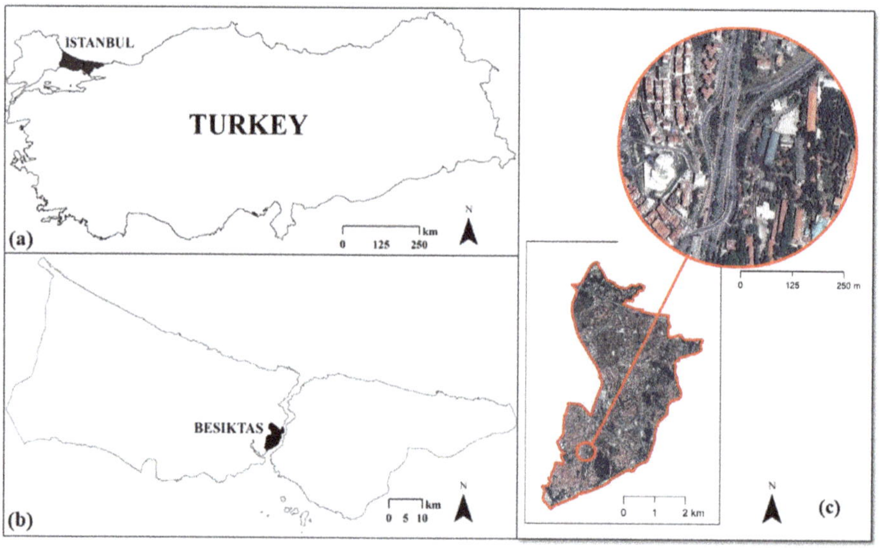

Figure 1. (**a**) Turkey and Istanbul city; (**b**) Istanbul and Besiktas district; (**c**) RGB Pleiades image of Besiktas district.

Besiktas drew our attention, as it is both in the city center and possesses marine and terrestrial transportation opportunities. The heavy pressure of urbanization accelerates landscape changes, bringing the risk of subjecting the green areas of Besiktas to rapid housing transformation. It is noteworthy that the population doubled from 1963 until 1985 (from 107,442 to 204,911 people), then fluctuated slightly in 2000 (190,813 people), 2007 (191,513 people), and 2014 (188,793 people), with relatively small differences (of 1000–3000 individuals) [55].

2.2. Methods

With a holistic perspective, our research aimed to evaluate the landscape pattern and processes at different levels through the use of ecological indicators. Based on the configuration of the landscape pattern, this evaluation revealed the main forces shaping the landscape functions in the urban environment. The flow chart below details the methods applied at different stages of the evaluation (Figure 2).

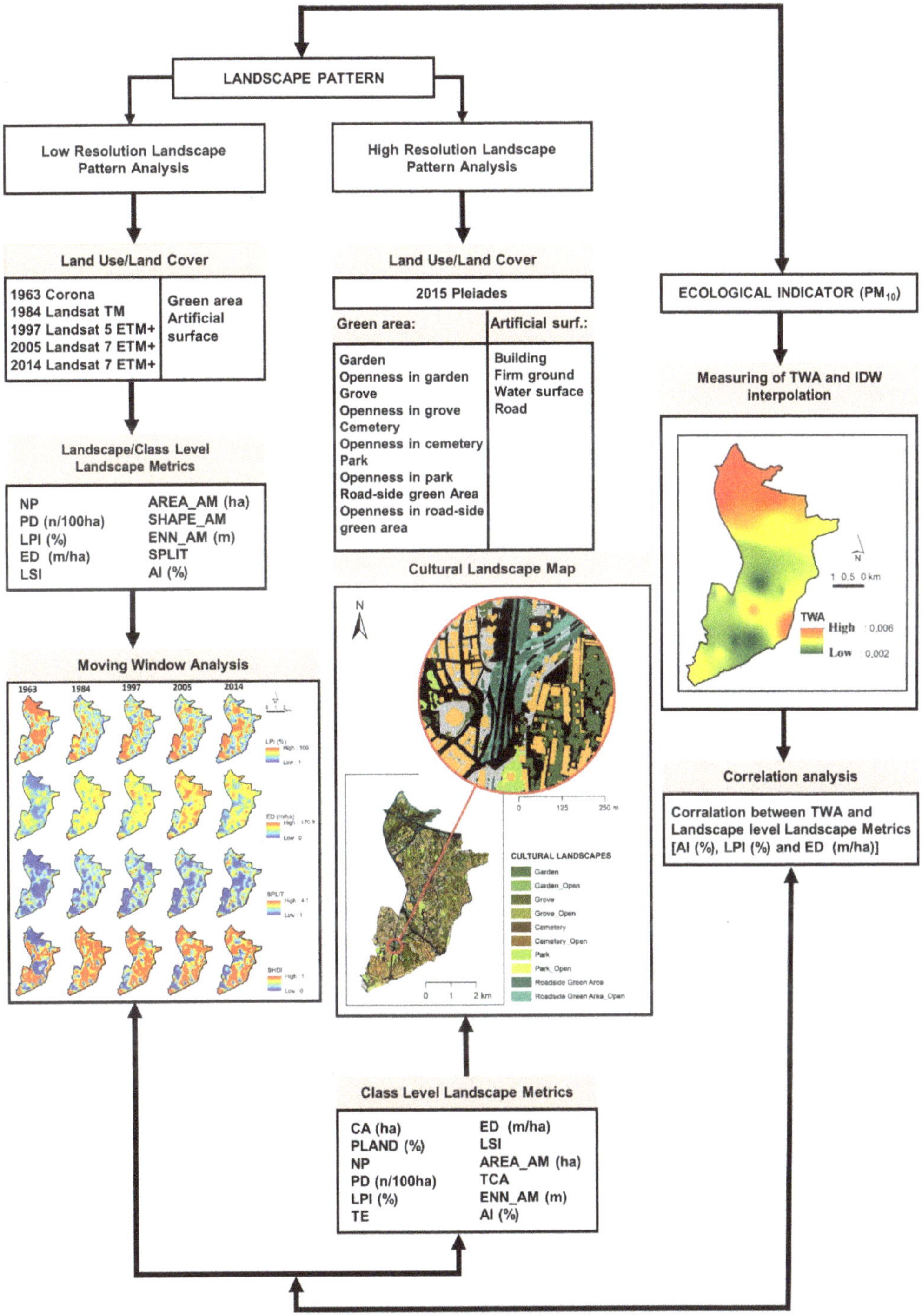

Figure 2. The flow chart of methods applied for the research.

First, we obtained two LU/LC classes (green area, and artificial surface) from medium-resolution images (Landsat) using a pixel-based classification method. Then, we used the normalized difference vegetation index (NDVI) to extract classes from high-resolution satellite imagery. Finally, we obtained a cultural landscape map.

We subjected the classified data to low- and high-resolution landscape pattern analysis, using landscape and class-level landscape metrics for the low-resolution evaluation. We also spatialized landscape-level landscape metrics using the Moving window module of the Fragstats software [44]. Finally, we used class-level landscape metrics for the high-resolution assessment.

We took the PM_{10} concentration as an ecological indicator. For this purpose, we used on-site TWA measurements as a reference and mapped them using the IDW interpolation method. Then, we compared this map with the maps created for landscape-level metrics. For comparison, we evaluated the relationships between PM_{10} concentration and landscape-level metrics using Spearman's coefficient and Pearson's correlation analyses.

As a result, the findings obtained from the two-stage landscape pattern analysis were evaluated with respect to the PM_{10} ecological indicator. Finally, conclusions about the cultural landscape types in the research area could be reached.

2.2.1. Image Processing

The image processing stage was carried out at two levels: medium-resolution images to reveal the general situation of the landscape pattern and determine the transformations exhibited over 51 years, and high-resolution images to reveal cultural landscapes. Satellite images with different resolutions and technical specifications were used for the research (Table 2).

Table 2. Satellite images and features used for the research.

Satellite	Spatial Resolution (m)	Spectral Resolution (μm)	Radiometric Resolution	Temporal Resolution
Landsat 4, 5, 7 TM, ETM+ (1984, 1997, 2005, 2014)	Bants 1, 2, 3, 4, 5, and 7—30 m Bant 6—120 m (for ETM+ Bant 6—60 m, Bant 8—15 m)	B1: 0.441–0.514 B2: 0.519–0.601 B3: 0.631–0.692 B4: 0.772–0.898 B5: 1.547–1.749 B6: 10.31–12.36 B7: 2.064–2.345 B8: 0.515–0.896 (for ETM+)	8 bit	16 days
Pléiades (2015)	2 m multi-bant, 50 cm panchromatic	B1: 0.430–0.550 B2: 0.450–0.620 B3: 0.590–0.710 B4: 0.740–0.940 PAN: 0.470–0.830	12 bit	26 days

LULC changes were examined using five satellite images representing 51 years. For this purpose, Landsat TM and Landsat ETM+ (1984, 1997, 2005, and 2014) images with 30 m × 30 m resolution were utilized. Due to the absence of a Landsat image representing the 1960s, Corona satellite imagery (1963) at 5 m × 5 m resolution was used. These images were recorded on film with cameras, in the form of photographic prints. In addition, scanned and digitized images were used as raster data with 5 m × 5 m spatial resolution. To compare with the data produced from Landsat satellite images, the Corona data were re-sampled to 30 m × 30 m.

The maximum likelihood classification algorithm was used to produce thematic classes [56]. Two different LULC classes were determined: green areas, and artificial surfaces (Figure 3).

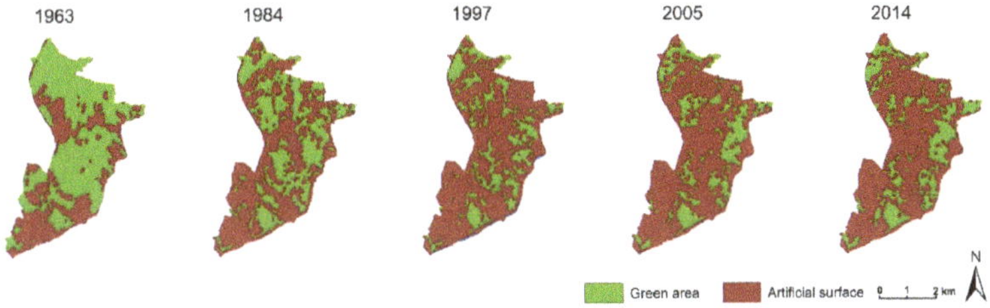

Figure 3. Times series LULC. Spatio-temporal dynamics of landscape structures.

For the study, 100 pixels were selected from the classification results on Landsat (1984, 1997, 2005, and 2014) images, which we compared with old maps and aerial photographs of the area [57]. We used the kappa statistic to test the reliability of comparative data. As general classes were preferred, the accuracy rates were high. Consequently, we reached a kappa accuracy rate of 90–93% (Table 3).

Table 3. Accuracy rates of classified Landsat images.

Classified Image	Overall Accuracy (%)	Kappa Coefficient	Classified Image	Overall Accuracy (%)	Kappa Coefficient
1984	90	0.8607	2005	93	0.8994
1997	90	0.8579	2014	92	0.8772

We obtained above-expected accuracy values for the classes and used them as input in the analysis.

Pleiades satellite images presenting a high spatial resolution were considered appropriate to derive data for distinguishing cultural landscapes. For this reason, we utilized these data for evaluation purposes at this stage [58].

It is possible to differentiate settlement, forest, and agricultural areas with limited values by selecting them according to normalized difference vegetation index (NDVI) values [59,60]. Therefore, we created an NDVI map to distinguish between green and artificial areas in the Pleiades satellite imagery. While sub-classifying artificial surface and green area classes, we used a 1/25,000-scaled base map of the area to mask some layers. In addition, we detailed the sub-classes using the manual digitization method. In total, we obtained 14 cultural landscape classes: garden, openness in garden, grove, openness in grove, cemetery, openness in cemetery, park, openness in park, roadside green area, openness in roadside green area, building, water surface, firm ground, and road.

We used previous maps of the region, satellite images, and Google Earth Pro v.7.3.6., as well as the opinions of experts who know the region, in order to determine the classification of areas and conduct accuracy analyses of the Pleiades satellite imagery results.

2.2.2. Pattern Analysis

As the landscape structure is an essential indicator of the landscape function, it is crucial to obtain information regarding the spatial distribution and arrangement of the LU/LC classes, in terms of perceiving the landscape from a holistic perspective [41,61–63].

The number of patches, the proportion of each patch type, and the spatial arrangement of patches are essential components in determining landscape patterns [64,65]. The landscape composition and configuration affect ecological processes independently and interactively. Therefore, it is vital to understand what component of the landscape pattern is being quantified by a particular metric [41,66]. Some landscape metrics that describe similar landscape characteristics are correlated, but each landscape index reflects a different urban landscape aspect [67]. Regression equations can reveal that the information expressed by landscape metrics is usually not based on a single component, but on the complexity of several components of spatial patterns. Therefore, it is crucial to evaluate the landscape structure using landscape metrics representing a combination of structure, composition, and configuration [65,68,69].

Landscape metrics offer a wide range of options to evaluate the landscape structure, from agricultural areas to mining/quarry sites, from wetlands to forests. The critical issue here is that the expert who makes the evaluation prefers the metric set that will best reveal the spatial arrangement of the landscape structure, depending on the subject investigated [20,67,70–80].

According to several authors, spatial metrics can be used to characterize urban forms. They represent critical determinants such as shape, configuration, and distribution in urban landscape planning, thus providing an opportunity to evaluate the nature of the change in the urban structure [41,46,61,81–88]. Therefore, selecting a complete set of landscape metrics is essential when analyzing the landscape structure [89]. We investigated the landscape composition and configuration of the research area to identify the expanding footprints of habitats using the most appropriate landscape metric combination. In this way, we represent patch complexity, aggregation, and diversity. We used a set of landscape metrics for this research at the class and landscape levels, for all identified time intervals (Table 4). We selected them among the "highly universal and consistent landscape structure components" defined by Cushman et al. (2008) [42]. Previous research focusing on correlational relationships between metrics was also considered [21,47,65,77,90,91]. Topaloğlu et al. (2020) applied principal component analysis (PCA) to summarize the information of a data set containing classes described by several correlated metrics. The findings from this study also helped us to create a complementary but non-repetitive set of metrics [21].

Table 4. Landscape-level metrics used for this research [44].

Metric Name	Abbrev.	Description
Class area (ha)	CA	The total area of the class
Percentage of landscape (%)	PLAND	The percentage of the landscape comprised of a particular patch type
Number of patches	NP	The number of patches of a corresponding patch type (class)
Patch density (n/100 ha)	PD	The number of patches of a corresponding patch type (class) per unit area
Largest patch index (%)	LPI	The area (m^2) of the largest patch in the landscape divided by the total landscape area (m^2)
Total edge (m)	TE	The sum of the lengths (m) of all edge segments in the landscape
Edge density (m/ha)	ED	The sum of the lengths (m) of all edge segments in the landscape, divided by the total landscape area (m^2)
Total core area (ha)	TCA	The sum of the core areas of each patch (m^2)
Landscape shape index	LSI	A standardized measure of patch compactness that adjusts for the size of the patch
Patch area (area-weighted) (ha)	AREA_AM	The area-weight mean patch size
Shape index (area-weighted)	SHAPE_AM	The weighting patches according to their size, on contrary to the LSI in which the total length of edge is compared to a landscape with a standard shape (square) of the same size and without any internal edge
Euclidean nearest-neighbor dist. (A.W.) (m)	ENN_AM	The shortest straight-line distance (m) between a focal patch and its nearest neighbor of the same class
Splitting index	SPLIT	The number of patches obtained by subdividing the landscape into equal-sized patches based on the effective mesh size
Aggregation index (%)	AI	The ratio of the observed number of like adjacencies to the maximum possible number of like adjacencies given the proportion of the landscape comprised of each patch type, given as a percentage
Shannon's diversity index	SHDI	The SHDI equals minus the sum, across all patch types, of the proportional abundance of each patch type multiplied by that proportion
Shannon's evenness index	SHEI	The SHEI equals minus the sum, across all patch types, of the proportional abundance of each patch type multiplied by that proportion, divided by the logarithm of the number of patch types

For low-resolution landscape analysis, the land-use data sets (1963, 1984, 1997, 2005, and 2014) were first converted into grid format (pixel size: 30 m × 30 m). For the high-resolution landscape evaluation, the obtained cultural landscape map (2015) was first converted into grid format (pixel size: 1 m × 1 m), in order to be able to carry out synoptic metric analyses and further analyses using the FRAGSTATS package (v.4) [44]. The 8-cell neighbor rule was applied for standard analyses. We concentrated first on class-level metrics, in order to monitor local impacts and their consequences on region-level changes.

Moving window analysis provides a spatially detailed evaluation of fragmentation indices [44]. Furthermore, moving window analysis allowed us to connect landscape-scale resource utilization to suitability models of setting structure in Besiktas. Wiens (1989) has stated that the moving window size relates measurable patterns to ecological processes [92]. We found that a round-shaped window with a 250 m radius was most effective in generating continuous results with an available cell scale of 30 by 30 m. Moving window analysis was used to output the ED, LPI, SHAPE-MN, and SIDI metrics as maps at the landscape level. Class- and landscape-level metrics were calculated and interpreted for low- and high-resolution data in these stages.

2.2.3. Environmental Indicator (PM_{10})

Particulate matter (PM) was used as an environmental indicator, in order to reveal the effects of landscape changes on the environment. We used PM_{10} values as an indicator for the analysis. Therefore, in situ, we measured PM values using a portable TSI Incorporated DustTrak II Meter. The PM meter obtains a 90° light scattering sensor and a particle volume range of 0.1–10 μm. Considering the 15-minute automatic calculation time of the time-weighted average (TWA) value (8-hour period per day) by the device, we measured PM_{10} for 15 minutes. Measurements were made randomly on days with appropriate weather conditions (no precipitation and wind intensity less than 3 m/h) every three months for one year. The measured values were generalized to the whole area using the inverse distance-weighted (IDW) interpolation method.

We constructed fifty random points considering the existing habitats and extracted dependent and independent values of these points using raster data sets; in particular, we used PM_{10} TWA values as dependent variables and landscape metrics of 2014 obtained from the moving window analysis as independent variables in the Pearson and Spearman correlation analyses. Thus, we obtained the correlation between selected landscape metrics and particulate matter densities. Using this method, we obtained landscape metrics with significant correlation. We tested the individual correlations between independent and dependent variables at significance levels of 0.01 and 0.05 (i.e., $p > 0.01$ or $p > 0.05$).

3. Results and Discussion

The findings of each stage were compared with the findings of the other stages. In this way, a holistic landscape pattern assessment was reached.

3.1. Low-Resolution Landscape Characterization

While the ratio of green areas in Besiktas municipalities was 67.6% in 1963, it decreased to 47.4% in 1984 following the construction of the first Bosporus Bridge in the 1970s. It further decreased to 31.9% in 1997, following the construction of the second Bosporus Bridge in 1988. In this period, especially from the 1980s, new constructions such as hotels, business centers, and shopping malls caused the business area center of Besiktas to develop and rapidly build up in this direction.

Although there were increases in green areas in 2005 and 2014, there was not much recovery (32.3% and 33.8%, respectively). Over the same period, there was a marked increase in artificial surfaces: 32.3% in 1963, 52.6% in 1984, 67.4% in 1997, 67.5% in 2005, and 66.1% in 2014. It is worth noting that there was a rapid acceleration in this increase in 1984, for the abovementioned reasons. The effect of this rapid change between green areas and artificial surfaces on habitat quality and fragmentation was studied, according to the landscape metrics at class and landscape levels. Landscape-level metric assessments allowed changes in the landscape structure to be interpreted and evaluated throughout the research area (Table 5). At the same time, the class-level analyses revealed the changes between habitats. Spatial heterogeneity results are shown at both landscape and class levels (Figures 4 and 5).

Table 5. Landscape-level indexes from 1963 to 2014.

Metrics	Year					Metrics	Year				
	1963	1984	1997	2005	2014		1963	1984	1997	2005	2014
NP	73	118	151	148	134	SHAPE_AM	4.4	7.12	7.6	7.09	6.87
PD (n/100 ha)	4	6.48	8.28	8.11	7.35	ENN_AM (m)	64.5	64.9	64.5	64.5	67.4
LPI (%)	62.8	41.5	65.4	65.2	63.9	SPLIT	2.39	3.66	2.29	2.3	2.38
ED (m/ha)	40.8	76.5	75.5	67.7	69	AI (%)	93.8	88.6	88.7	89.8	89.8
LSI	6.05	9.68	9.64	8.94	8.82	SHDI	0.64	0.7	0.67	0.64	0.65
AREA_AM (ha)	763.2	497.2	794.9	794	764.6	SHEI	0.58	0.63	0.61	0.59	0.59

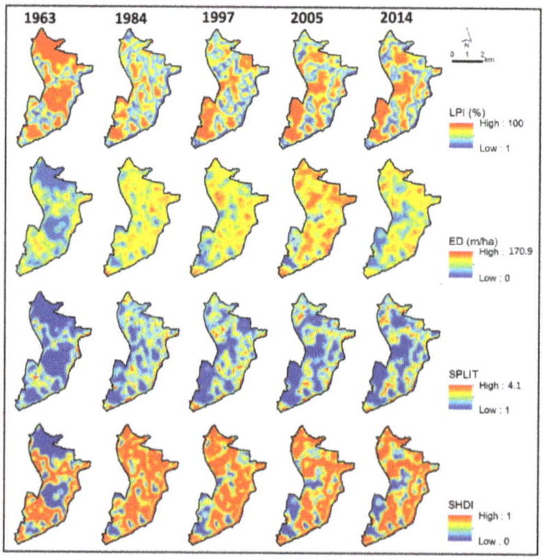

Figure 4. Landscape-level moving window (250 × 250 m) analysis results for LPI, ED, SPLIT, and SHDI indexes.

According to Forman and Godron (1986), the edge density determines the shape of a patch, and can further indicate the distribution of plant and animal species [93]. At the landscape level, especially from 1963 to 1984, the increase in edge metrics indicates that fragmentation became an increasingly dominant factor in the Besiktas landscape. After constructing the first Bosporus Bridge in 1984 and the second Bosporus Bridge in 1997, landscape-level ED reached approximately 76 m/ha. Although this increase was lower in the following period, ED was still higher, compared to 1963. The border between patches is important in forming corridors. Therefore, contrasting patches also indicate connections. According to Ranney et al. (1981), microclimatic changes, wind, and light progression along a high-contrast edge are more likely than on a low-contrast edge in a patch [94]. Patch isolation is also a function of the contrast between the patch and its ecological neighbor. Shape metrics are crucial in revealing the landscape order.

The distance to the nearest neighbor is an indicator defining the distance from a patch to other patches with the same characteristics, which is essential for determining the quality of a habitat. Research has shown that fewer living species in habitats suffer from isolation due to fragmentation; in particular, many studies on birds have discussed this aspect [95,96].

At the landscape level, SHAPE_AM and LSI showed increasing trends, indicating that the landscape pattern became more irregular over time, indicating a disturbance effect due to the presence of people.

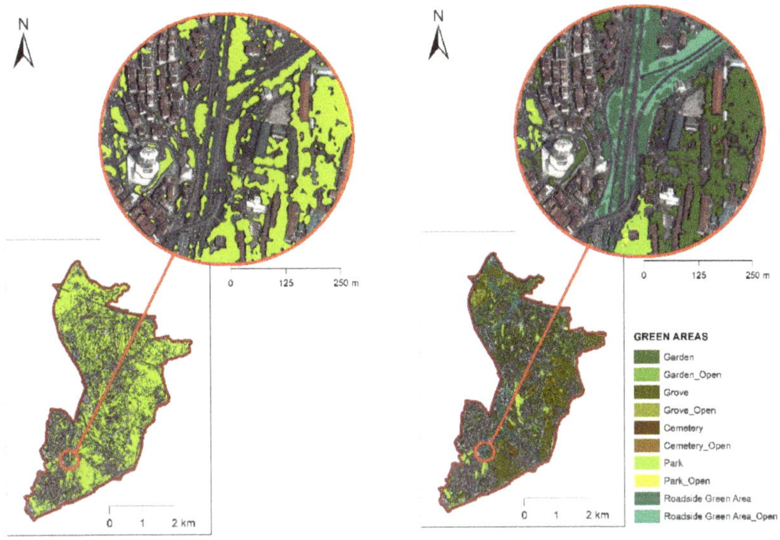

(**a**) (**b**)

Figure 5. The green area layer obtained by NDVI (**a**) was divided into ten cultural landscape classes using a manual classification technique (**b**).

SPLIT is calculated as the number of patches obtained while splitting the entire landscape into equal-sized patches. SPLIT rose from 2.4 to 3.7 during the period 1963–1984, indicating the influence on the landscape and the fragmentation of the natural landscape in this period. The decline and increase in the following period indicate that the landscape is still not stable against this violent change observed in 1984.

From the analysis in Figure 4, it can be observed that the SPLIT value was higher in settlements in 1963, the complication in the entire Besiktas landscape increased in 1984, and it shifted in later periods, with the division of green surfaces being increased.

The SHDI reflects the complexity and heterogeneity in the landscape. The exchange of SHDI in Besiktas was striking from 1963 to 1984. The 6% increase in this index over this period indicates that the heterogeneity of the whole landscape and the number of scattered patches were increasing. During this period, as mentioned above, the construction of the Bosporus Bridge resulted in a rapid change, and scattered landscape patches appeared. In the following period, this change was more stable. The results indicate that the maximum evenness of the area's distribution was 64% in 1963, 70% in 1984, and 65% in 2014. The fact that the index values in the landscape are not too high indicates an irregular distribution of different patch types in the area. Figure 4 shows the spatial variation of the SHDI. In 1963, only heterogeneous structures were observed in places open to settlement. In 1984, all Besiktas landscapes gained heterogeneous structure, which increased until 2014. With the increase in aggregation, patches in the settlement areas reduced in the western part of the research area.

The SHEI value results indicate that large landscape types no longer play a dominant role in the Besiktas district, the average patch area is similar, and the patches tend to show a uniform distribution. Like the SHDI, the SHEI changed remarkably from 1963 to 1984. However, in the following period, it presented an improvement, showing that large landscape types no longer play a dominant role, the average patch area has become more similar, and the patches tend to have a uniform distribution. The indices show that the maximum uniformity of area distribution was 58% in 1963, 63% in 1984, and 59% in

2014. The Besiktas landscapes index values were not very high, indicating an irregular distribution of different patch types in the landscape.

After the holistic evaluation of the landscape pattern, we used class-level landscape metrics to reveal the reasons for the changes in more detail. As the water surfaces in the area obtained from medium-resolution satellite images were distributed in small units (e.g., swimming pools), we did not include them in the class-level metric evaluation. We completed the low-resolution assessment of the research area by calculating the metrics of the green area and artificial surface classes (Table 6).

Table 6. Class-level landscape indexes and the change between each data adapted from 1963, 1984, 1997, 2005, and 2014 classifications show the urban fragmentation process in Besiktas.

CA

Year	1963	1984	Change	1984	1997	Change	1997	2005	Change	2005	2014	Change	Total Change
Green area	19	58	−39	58	102	44	102	88	−14	88	89	1	70
Artificial surface	46	53	−7	53	21	−32	21	42	21	42	37	−5	−9

PD

Year	1963	1984	Change	1984	1997	Change	1997	2005	Change	2005	2014	Change	Total Change
Green area	1.04	3.18	−2.14	3.18	5.6	2.42	5.6	4.82	−0.78	4.82	4.88	0.06	3.84
Artificial surface	2.52	2.91	−0.39	2.91	1.15	−1.76	1.15	2.3	1.15	2.3	2.03	−0.27	−0.49

The results show the increasing fragmentation of green areas and a tendency to transform into small scattered patches in these habitats; however, this increase was not regular. Besiktas district was known as a land of mulberry in the 1960s ("mulberry shake for 2.5 Lira"). When the Bosporus Bridge was on the agenda in the 1970s, Besiktas became a focal point in terms of transportation. As the main arteries—such as Barbaros Boulevard and Buyukdere Avenue—pass through the city's centre, the construction of the Bosporus Bridge already made the central city a knot point. The coastal road, which operated independently from this artery in the past, was thus linked to the interior. This change was the most crucial reason for the changes observed in 1984. The change in the PD peaked in 1997, due to a similar effect in 1988, which brought the second Bosporus Bridge to the square. The ring road of the bridge neighbouring the district from the north—the Trans-European Motorway (TEM)—entered Besiktas with the connection of Levent, serving as an element that increased the demand for new constructions. The renewal of all parks in the Municipality of Besiktas in 2000 helped to decrease the PD in the following period. In this period, the municipality afforested streets and parks, using thousands of tree seedlings, which can be observed as a partial improvement.

LPI

Year	1963	1984	Change	1984	1997	Change	1997	2005	Change	2005	2014	Change	Total Change
Green area	62.81	30.47	32.34	30.47	4.22	−26.25	4.22	6.51	2.29	6.51	7.35	0.84	−55.46
Artificial surface	13.54	41.5	−27.96	41.5	65.44	23.94	65.44	65.19	−0.25	65.19	63.92	−1.27	50.38

The LPI is a highly representative indicator of the proportion of the largest class in the simulated landscape and, at the class level, is considered a parameter reflecting the abundance of classes [65]. Large patches are essential for maintaining more species. In this context, the LPI is one of the most influential metrics of landscape fragmentation. When the LPI was examined in the Besiktas landscape, it did not display a regular change. The LPI index decreased from 1963 to 1984, increased from 1984 to 2005, and then decreased again. Tragically, however, the most significant patch belonged to green areas, and artificial surfaces tended to increase regularly. The increase in aggregation in these areas was expected to have various consequences. Figure 4 indicates that the largest patch in 1963 covered the green areas. In 1984, the patches started to shrink. After 1997, the largest patch was formed of artificial surfaces, with the aggregation of the western settlements. This largest patch appears to be growing in the west–east direction.

ED

Year	1963	1984	Change	1984	1997	Change	1997	2005	Change	2005	2014	Change	Total Change
Green area	39.43	73.3	−33.87	73.3	69.79	−3.51	69.79	64.23	−5.56	64.23	65.71	1.48	26.28
Artificial surface	39.72	75.82	−36.1	75.82	74.89	−0.93	74.89	66.87	−8.02	66.87	67.75	0.88	28.03

Green areas and artificial surfaces showed increases in edge and contrast. In particular, the tendency to increase edge/contrast in green areas may have led to changes in microclimatic conditions, due to the differentiation of wind and light intensity. The ED was low due to the large green surface patch, while the rapid ED increase in 1984 spread to the entire Besiktas landscape. It can be seen, from Figure 4, that it reached its highest value in 2005.

LSI

Year	1963	1984	Change	1984	1997	Change	1997	2005	Change	2005	2014	Change	Total Change
Green area	6.19	11.85	−5.66	11.85	13.64	1.79	13.64	12.47	−1.17	12.47	12.43	−0.04	6.24
Artificial surface	8.82	12.71	−3.89	12.71	10.92	−1.79	10.92	10.37	−0.55	10.37	10.33	−0.04	1.51

The LSI is another important indicator that reflects the heterogeneity of landscape patches [97]. The patch shape quickly became complex in both artificial surfaces and green areas after 1984. This transformation indicates that construction of the Bosporus Bridge formed a breaking point regarding shape irregularity. The LSI values of these two habitats showed an initial upward trend, followed by a decline. Due to the rapid fragmentation, patches with more complicated shapes emerged in both landscapes. As mentioned above, the decrease was related to the aggregation of artificial surfaces and the afforestation of refuges, streets, and parks. However, the changes related to the bridges were focused on artificial surfaces in 1984 and on green areas in 1997. This change indicates that the second bridge had a stronger effect on the geometrical degradation of green areas. According to the settlements, green areas seem to present a more complex shape characteristic due to fragmentation. Buechner (1989) has suggested that the shape of a patch has a particular effect on the mobility of mammals in the patch [98]. In this sense, an increase in shape irregularity in green areas may have led to a significant decrease in the number of mammals, especially in woodlands. On the other hand, the fact that there were many formal irregularities suggests that the core area did not develop in such habitats.

Table 6. *Cont.*

AREA_AM													
Year	1963	1984	Change	1984	1997	Change	1997	2005	Change	2005	2014	Change	Total Change
Green area	1066	375.5	690.5	375.5	44	−331.5	44	56.2	12.2	56.2	58.3	2.1	−1007.7
Artificial surface	131.5	607.4	−475.9	607.4	1159	551.9	1159	1150	−9.3	1150	1127	−23	995.5
The AREA_AM index is essential for representing the degree of aggregation or fragmentation of patches in a spatial manner. According to the simultaneous data, AREA_AM showed the highest index value in artificial landscapes; that is, artificial landscapes had a more scattered distribution. In green areas, the indices were all at low levels, indicating that the patches were of smaller size and presented a scattered distribution. While there was an increasing trend in artificial surfaces, green areas showed a noticeable decline over time. From 1963 to 2014, the AREA_AM values decreased to 58.3 ha in green areas. To the contrary, artificial areas increased to 1127 ha. This change also indicates the dominance of artificial patches, signifying that the artificial landscapes separate green areas and deepen the extent of fragmentation. Therefore, AREA metrics are also important for providing information about the core area. As it protects them from the adverse effects at the edge, the core is an important area for plants and animals [99]. The decrease in AREA_AM at the landscape level indicates that the core area also declines. This situation is an indication of the shrinkage, fragmentation, and even losses of large patches. Rapidly advancing settlements and scattering in settlements can be attributed to the increased core area of artificial surfaces. Accordingly, the loss of or change in species can be hypothesized, especially in the woodland areas of the Istanbul landscape, which has significant ecological importance.													
ENN_AM													
Year	1963	1984	Change	1984	1997	Change	1997	2005	Change	2005	2014	Change	Total Change
Green area	61.51	67.6	−6.09	67.6	72.94	5.34	72.94	71.56	−1.38	71.56	77.73	6.17	16.22
Artificial surface	70.23	62.22	8.01	62.22	60.39	−1.83	60.39	60.79	0.4	60.79	61.77	0.98	−8.46
The difference in ENN_AM between patches was considered together with NP and LPI, providing important information about the urban pattern [76]. At the general landscape level, ENN_AM showed that the distance between similar patches had increased. When examined at the class level, there was an increase in this metric for green areas and a partial decreasing tendency for artificial surfaces, due aggregation.													
SPLIT													
Year	1963	1984	Change	1984	1997	Change	1997	2005	Change	2005	2014	Change	Total Change
Green area	2.53	10.24	−7.71	10.24	129.9	119.7	129.9	100.44	−29.5	100.44	92.46	−7.98	89.93
Artificial surface	42.86	5.7	37.16	5.7	2.33	−3.37	2.33	2.35	0.02	2.35	2.45	0.1	−40.41
For artificial surfaces, SPLIT presented a steady decline; meanwhile, in green areas, it showed a rapid increase until 1997 and a partial decrease afterwards. The SPLIT values provide further proof that the focal patch type in green areas gradually decreased and was divided into smaller patches. In artificial areas, the opposite phenomenon was observed.													
AI													
Year	1963	1984	Change	1984	1997	Change	1997	2005	Change	2005	2014	Change	Total Change
Green area	95.53	88.79	6.74	88.79	84.04	−4.75	84.04	85.66	1.62	85.66	85.99	0.33	−9.54
Artificial surface	90.21	88.5	1.71	88.5	91.42	2.92	91.42	91.92	0.5	91.92	91.85	−0.07	1.64
The AI is an indicator that depicts the degree of aggregation of patches in the landscape [100]. As indicated by the index values examined earlier, it tended to decrease in green areas and increase in artificial surfaces—a sign of loss in green areas and gradual gathering and granular dispersion of artificial surfaces. As mentioned above, these (increasing/decreasing) tendencies were not regular. The spatial variation of the AI was similar to that of the LPI (see Figure 4), as the growth of patches increases the AI.													

By evaluating the metrics at low-resolution level of selected classes within a time-series perspective, we can conclude that the green areas have lost their holistic structure over time, splitting into small units which move away from each other. As a result, they transformed into a complex configuration structure with weakened habitat quality. On the other hand, while the artificial surfaces primarily presented a dispersed and heterogeneous structure, over time, they became closer and formed clusters. This alteration reflects the scattered structure, increased heterogeneity, and disorganized structure of the landscape until 1984. Later, shaping, clustering, and diversity indices reflected a recovery.

The domination of artificial surfaces over the landscape led to the fragmentation of green areas between 1983 and 1997. The construction of Bosporus bridges (one built in 1973 and the other in 1988) and ring roads was the main reason for this disruption. Determining similar spatial transformations with an interim of 14 years is crucial in modelling the impacts of such constructions on the landscape. In particular, through the use of moving window analysis, spatial evaluations were possible, providing us with the means to determine the fragile areas affected by urbanization. These findings comprise essential clues regarding effective landscape analysis method.

3.2. Relationship between Landscape Metrics and PM_{10} Concentartion

Next, we calculated the individual correlations between PM_{10} and landscape-level landscape metrics over the research area. As the PM_{10} measurements were carried out in 2014, the variation between the metric values obtained until 2014 was considered, in order to ensure that the data were comparable. The measured values were generalized and mapped using the inverse distance-weighted (IDW) interpolation method (Table 7).

Table 7. Spearman's coefficient analysis and Pearson correlation analysis results and TWA map created by the IDW interpolation method.

Spearmen's Coefficient Analysis				**Pearson's Correlation Analysis**			
TWA	**AI**	**LPI**	**ED**	**TWA**	**AI**	**LPI**	**ED**
2014	−0.527 (**)	−0.377 (**)	0.385 (**)	2014	−0.359 (*)	−0.351 (*)	0.291 (*)
Diff. 1963–2014	0.380 (**)	0.422 (**)	−0.387 (**)	Diff. 1963–2014	0.397 (**)	0.467 (**)	−0.418 (**)
TWA	**PD**	**SHEI**	**SHDI**	**TWA**	**PD**	**SHEI**	**SHDI**
2014	0.440 (**)	0.317 (*)	0.330 (*)	2014	0.288 (*)	0.317 (*)	0.324 (*)
Diff. 1963–2014	−0.386 (**)	−0.374 (**)	−0.377 (**)	Diff. 1963–2014	−0.403 (**)	−0.462 (**)	−0.466 (**)

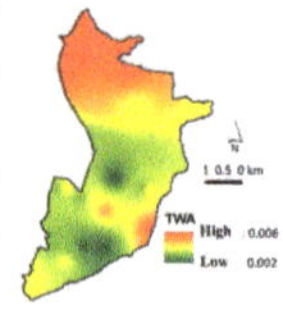

* Correlation is significant at the 0.05 level (2-tailed). ** Correlation is significant at the 0.01 level (2-tailed). N = 50.

Studies have shown that the quality (e.g., biomass, species diversity), size, and shape of green areas affect the PM level. The penetration of vegetation into artificial surfaces in the urban area can facilitate the absorption of particulate matter. The connectivity of the green system also has an important effect on PM concentration [52,101,102].

The landscapes of Besiktas changed rapidly during the period 1963–2014. The associated changes degraded the quality of habitats by causing fragmentation and environmental changes. Therefore, we performed correlation analysis between landscape-level metric values in the relevant metric maps and PM_{10} concentration, measured at test points.

Aggregation significantly affected the PM_{10} level, mainly in residential areas. Considering that the clusters were more abundant in artificial areas, it can be stated that excess clustering on artificial surfaces significantly affects the PM_{10} exposure level. This results in an increase in clusters on artificial surfaces, which may indicate a decrease in ventilation within the city [103]. PM_{10} has been positively correlated with artificial surfaces in previous studies [104,105].

A more heterogeneous and uneven landscape distribution decreases the PM concentration. SHEI—one of the general landscape metrics—presented a significant negative relationship with the PM_{10}. The SHEI reflects the landscape heterogeneity of patches and is sensitive to the distribution of patches. High values of this index indicate dispersed landscapes. When the landscape is better distributed, the relationship between each land-use type and the interaction between the sink and source landscapes will be closer, further reducing PM pollution [106].

The results show that a high PD and a high ED were associated with much higher PM_{10} exposure levels than in less dense and less developed areas. Based on this analysis, we can conclude that landscape metrics are helpful in not only predicting the quality of habitats, but also in estimating the PM_{10} levels and the combination of both parameters, being indicative of urban health.

In the correlation analysis, moderate correlations were observed for the LPI, the AI, and the PD, and near-moderate correlations with the ED, the SHEI, and the SHDI. Based on these significant relationships, in the next step, comments are developed regarding the relationship between the habitat quality and the PM_{10} of cultural landscape classes obtained from high-resolution satellite imagery.

3.3. *High-Resolution Landscape Characterization of Cultural Landscapes*

Through a comprehensive literature review, we compiled species that are likely to live in the research area under normal conditions, included in five fauna groups (birds, small mammals, small butterflies, reptiles, and amphibians), which can indicate the effects on urban habitats [32,33]. In the next step, indicator species of habitats were identified among these listed species. The identified species were observed at 19 test points in field surveys, conducted randomly over two years to obtain representative results for all four seasons, and were associated with habitats in terms of criteria such as shelter, nutrition, and reproduction. Based on this relationship, the green area and artificial surface layers obtained from the high-resolution Pleiades satellite imagery were divided into cultural landscape classes (Figures 5 and 6). The green areas were divided into ten cultural landscapes (garden, openness in garden, grove, openness in grove, cemetery, openness in cemetery, park, openness in park, roadside green area, and openness in roadside green area). These landscapes were manually classified by overlaying the green area layer with the Pleiades satellite imagery (Figure 5).

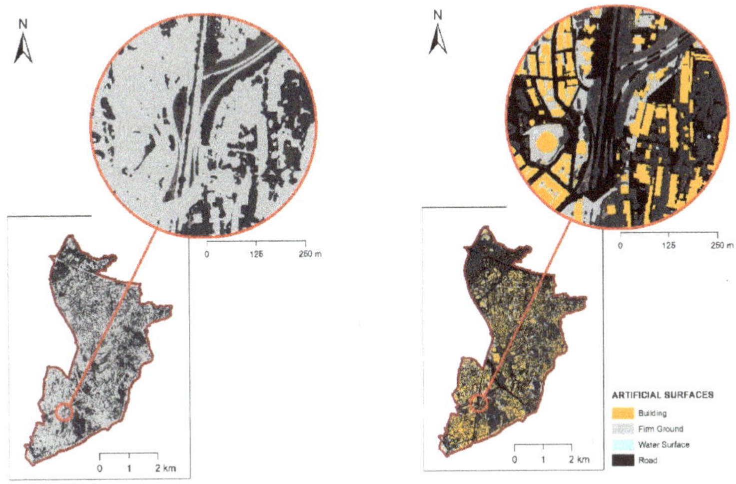

(**a**) (**b**)

Figure 6. The artificial surface layer (**a**) was divided into building, water surface, firm ground, and road landscapes with the help of the base map (**b**).

The artificial surface layer obtained by remote sensing techniques from the Pleiades satellite imagery was divided into four cultural landscapes (building, water surface, firm ground, and road) within the boundaries of the research area. The mentioned cultural landscapes were obtained by cutting the polygons of the current map from the artificial surface layer (Figure 6).

As a result, a cultural landscape map of the research area was obtained by combining the layers obtained for the green areas and artificial surfaces. Furthermore, during the classification process, the habitat requirements of indicator fauna groups were considered (Figure 7).

A unique classification system which focuses on the habitats offered by the urban landscape was chosen. The fact that this unique classification system was taken as a basis while interpreting the unit-corridor matrix relations served as a guide in interpreting the pattern–function relations and priorities.

3.4. Pattern Analysis and Functional Findings of Cultural Landscapes

The 14 cultural landscape classes obtained were subjected to pattern analysis through the use of class-level metrics, and evaluated according to the main landscape functions they reflect (Figure 8).

The PLAND values for determining cultural landscape types indicated that the water surfaces were negligible, the green areas only covered an area of 38.3%, and the artificial surfaces were dominant, with a proportion of 61.65%. These findings demonstrate the impact of urbanization. Among the green areas, gardens had the most significant percentage, while groves had a critical portion. These indicators reveal the classes that should be focused on in landscape planning and management processes. Roads were the class that occupied the most space. The fact that the associated LPI value was also high indicates that this class dominates the landscape in large part.

The cultural landscape classes obtained by considering the indicator fauna groups on the high-resolution satellite image were also evaluated, according to their main landscape functions. We interpreted the findings of the high-resolution analysis regarding the landscape structure in detail, considering the outputs of Aksu and Küçük (2020) on the biotope quality of the research area [19].

Building (1): In urban landscapes, buildings are spatial components that dominate the landscape. In the research area, the spatial distribution of roof surfaces had a ratio of 21.39% (Figure 8). The fact that the NP value was very high, although they covered the surface of the research area at a high rate, shows that small but many units were distributed over the entire area. However, the relatively high TCA and ED values also show that this class is concentrated with holistic structure in certain areas, despite its configuration. The low ENN_AM value also supports this finding. The concentration of buildings which appear as the characteristic structures of many urban landscapes and compete with green spaces reveals the necessity of considering buildings from a different perspective. In the research area, where artificial surfaces dominate the green areas, and in areas with a similar urbanization process, it can be seen how essential the functions of creating habitats and harmonizing the buildings with their environmental potential are. Depending on the floor height, the formation of an artificial topography in the structured urban environment draws attention as the main factor triggering this aspect.

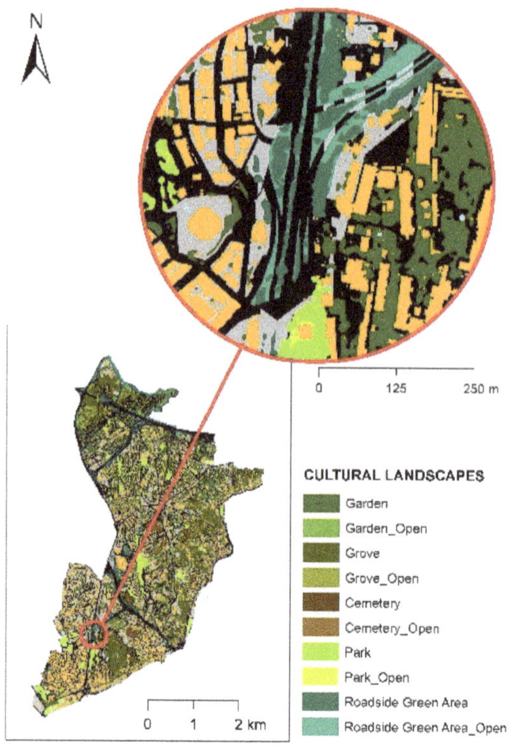

Figure 7. A cultural landscape map was obtained by taking indicator fauna groups as a reference for subcategorizing the green area and artificial surface layers.

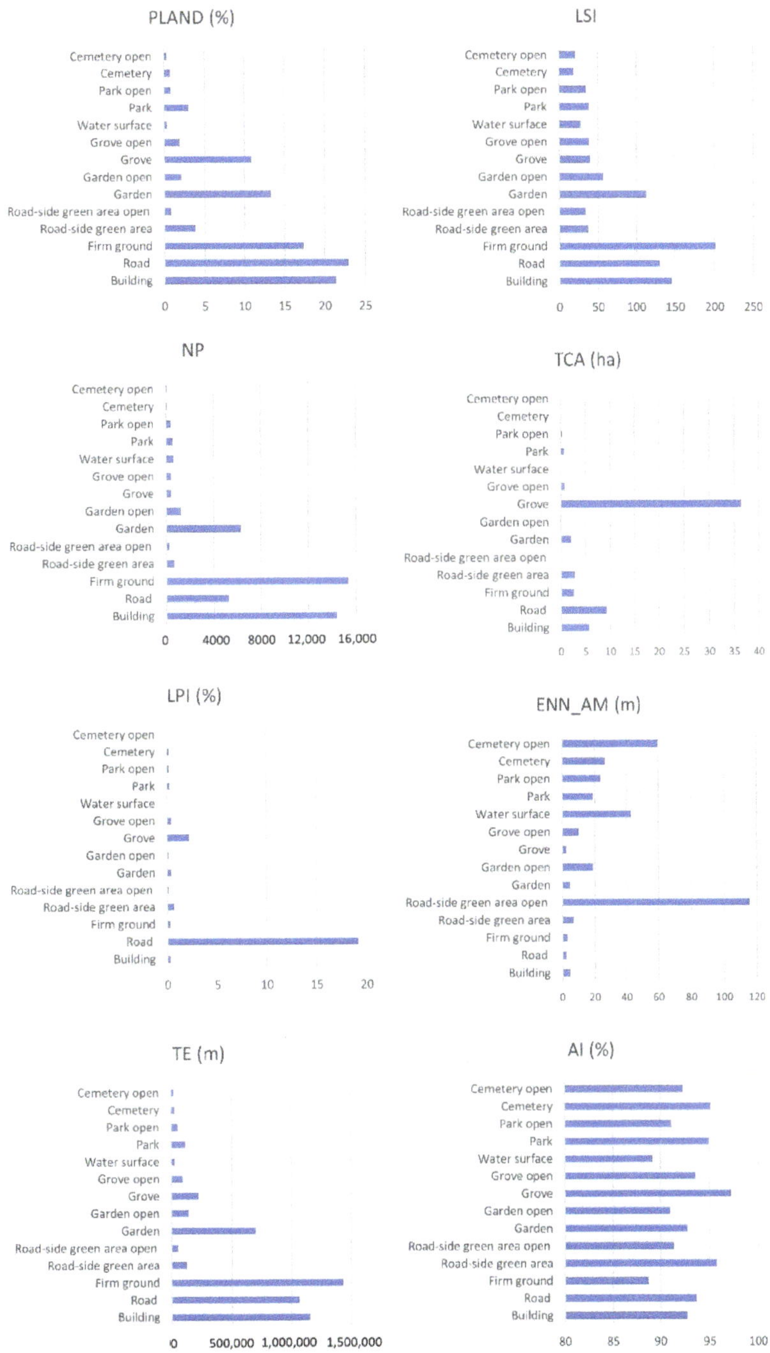

Figure 8. The cultural landscapes' class-level indexes (PLAND, NP, LPI, TE, LSI, TCA, ENN_AM, and AI).

On the other hand, the buildings that dominate the urban ecosystem, as in the research area, should be evaluated in terms of their functional features. For example, they could be used for water regime regulation, eliminating problems related to urban topography, and creating habitats for some living species (e.g., bats, swallows, seagulls, sparrows, crows, reptiles, and so on). Therefore, we can evaluate buildings as components of the urban ecosystem. They may benefit from their existing sustainable potentials (e.g., solar/wind energy harvesting and passive ventilation/lighting systems). Moreover, they may either provide or keep away structures that contribute to climate change (e.g., supporting energy efficiency, preferring green roof-facade systems, and/or smart materials). Although buildings are perceived as disruptive elements of the urban ecosystem, as they create artificial surfaces, they can constitute a shelter and breeding place for many species through simple measures to be taken on the roof and facade surfaces. Considering all of these features, building surfaces should be perceived as important biotopes, and their contribution to the landscape function should be focused on.

Road (2): Road networks are considered a separate class as they constitute a barrier due to their linear structure and have a disintegrating effect on habitats [61]. In the research area, roads were the most dominant class, in terms of the area they covered. Their linear structures play a dominant role in the landscape pattern of the research area. Roads with high ED, TCA, LPI, and LSI values are expected to exhibit a near-geometric character, considering their linear structure. On the other hand, these structures, which may move away from geometry, show how dominant they are in the landscape structure. The low ENN_AM value also supports this finding. Roads close to each other may intersect at many nodules, forming an integrated and complex grid structure.

As the primary factor in the fragmentation of green areas, roads also affect many processes in the urban ecosystem. Due to the insufficient infiltration capacity of the artificial surfaces that dominate the research area, the precipitated water that passes to surface runoff may follow the linear road networks. Again, due to the structure exhibiting continuity along this line, wind flows are artificially directed, thus forming wind corridors. Heavy metals, engine oil, fuel residues, and substances that change pH values (e.g., salting carried out to prevent icing, especially in winter) can accumulate on roads, adversely affecting many landscape functions. Road networks, together with other artificial surfaces, can trigger the formation of urban heat islands. They also carry pollutants, which combine with precipitated water that passes as the surface runoff along the line, thus negatively affecting neighboring habitats. As road networks have a key impact on essential cycles in the urban ecosystem, they should be handled and planned keeping such factors in mind.

Firm Ground (3): Firm ground is typically located as a transition zone between buildings and green areas. For this reason, their spatial and structural features are important. The artificial topography, which is formed depending on the building density, causes the formation of micro-climatic conditions such as wind shadow corridors and increased surface runoff (due to high slope degrees), which are specific to the urban environment [19]. The increase in impermeable firm ground, which generally affects ecological cycles in a negative way, can prevent precipitated water from meeting with the soil, causing many problems due to surface runoff and wasted productive water. Due to these critical properties, hard ground was also included in the classification. It was found that this class—which ranked third in terms of area size—was represented by many units. The fact that the LSI value was the furthest from the geometric indicated that the hard ground presented an organic form. However, it was the class with the highest TE and lowest LPI and TCA values, indicating that the units belonging to the class did not exhibit a holistic character. Therefore, in order to interpret the organic shape structure of the firm ground class areas, it is necessary to focus on the character of the classes to which they are adjacent.

Roadside Green Area (4): In the urban ecosystem, roadside green areas are elements accompanying roads that encourage the fragmentation process, in the form of dissection with their linear structures [61,107]. These features connect green areas in clusters or units. The green texture of roadside green areas differs in terms of the species it contains, dependent on the presence of herbaceous or woody vegetation. For this reason, openness in these areas is considered as a different class. The main factors that negatively affect the vitality and diversity of roadside green areas are gas emissions, wind-shadow canyons, and the selection of wrong plant species for plantations. The woody texture in these areas is vital for flying species such as birds, butterflies, and bats, but can be dangerous for species belonging to other indicator groups, as they accompany the roads. Although they play critical functional roles, when their metric values were examined, it was found that they did not gain an integrated and dominant structure in the manner that roads did. The main reason for this problem was that, compared to the PLAND of roads (22.95%), only 3.85% of the total area consisted of roadside green areas. Together with relatively low TCA, ED, and LPI values, the ENN_AM value was approximately three times that of the road class value, indicating that roadside green areas are insufficient in this urban landscape. Considering that they play an important role as a buffer between green areas and artificial surfaces, in terms of landscape functions and the prevention of deterioration in many functional flows, it is clear that they are essential in the urban ecosystem.

Openings in Roadside Green Areas (5): This biotope is especially important for small butterflies which need openings to live. Many butterfly species could be observed, especially in the roadside green areas where flowering mixed herbaceous vegetation was formed.

Garden (6): Building gardens are important landscapes that act as a buffer between buildings and their environment, ameliorating the disintegrating effect of buildings. This class includes the woody green tissue that forms the immediate surroundings of buildings. This texture is especially suitable for small birds, such as robins and sparrows, and can offer habitat and shelter to reptiles and small mammals. In the research area, garden was the green area class with the highest coverage (13.28%). However, although it covered more area than groves, it was found that this class consisted of many small units with low LPI and TCA and high NP and TE values. In addition, the fact that the LSI value was higher than that of the groves indicated a fragmented structure, rather than exhibiting a more natural structure; the ENN_AM value was also higher than that of the grove units, supporting this fragmented structure.

The plant species preferred in the building gardens determine the animal species that can benefit from that green area. As a result of landscape design implementations, exotic species were commonly encountered along with natural plant species in gardens. Although this situation leads to various problems, it is effective in increasing biological diversity.

Openings in Gardens (7): Herbaceous vegetation and soil areas near buildings are included in this class. This cultural landscape is vital as a home for reptiles such as tortoises, mammals such as rabbits, and small butterflies [19,108].

Grove (8): Groves can host all indicator animal groups, depending on their vegetative diversity, and have high potential for biodiversity [19,109]. They constitute the centers of the green system in urban areas that are in an intensive spatial transformation process. Their protection and development within the urban green system are crucial for the continuity of the whole system. In the research area, groves presented a rate of 10.94%.

The fact that this was the class with the highest TCA value in the research area makes the groves the only alternative for those species that distinguish between edge and core habitats to live in the urban environment. With their qualified core and edge habitats and a wide variety of natural and exotic woody plants, groves are home to many living things in the urban environment. Their holistic nature also enables them to play a dominant and essential role in the green network. The fact that the ENN_AM value was also low is another indicator of the holistic structure of the groves within the research area. Therefore, the development of this class is important to continue the urban ecosystem, in terms of quality and quantity. Regulations effective with respect to the water regime and climatic conditions, improving air quality, and protecting and developing biodiversity, should be included in planning and management processes.

Openings in Groves (9): Although groves generally have tree-dominated dense woody vegetation, there are also open areas, covered with herbaceous vegetation or soil surface. These openings constitute an ideal living environment for creatures that need more light to live and increase the biodiversity of groves. Therefore, it is appropriate to consider them as a separate class as they differ in these features.

Water Surface (10): Water surfaces are vital for all living things in the urban ecosystem and indispensable for many species as a habitat. However, it was determined that the water surfaces in the research area were very few and insufficient, in terms of quality. Although the PLAND ratio of water surfaces was the lowest, the high NP and ENN_AM values and low AI values indicate that the water surfaces in the study area were typically represented as small disconnected units. According to the experience gained from field studies, most of these small water surfaces are swimming pools that are cleaned with chemicals. Therefore, the water surfaces, which were already insufficient in terms of area, are also weak in terms of quality. This situation constitutes a problem that disrupts the continuity of the urban ecosystem and, in this respect, urgently needs to be addressed.

Park (11): Park areas are the class representing woody vegetation in public areas under the responsibility of the metropolitan municipality or district municipality. In these areas, where intensive use is generally seen, species that have adapted to human life attract attention. Considering the PLAND value, when the edge–core area relations (TE and TCA) and LPI values (third place) of the parks were examined, we found that they can constitute a stepping stone between the block units formed by groves and gardens. For this reason, it is essential to manage the design and arrangement processes of parks in the research area with this awareness. The connector positions of parks in the green network should be considered both in the selection of plant species and in the design of artificial surfaces.

Openings in Parks (12): This class includes openings within park areas covered with herbaceous vegetation or soil cover. Although these openings are not expected to serve timid species in park areas where human utilization is intense, they are considered a separate class, allowing specific species that have adapted to human activities and which need openings to live.

Cemetery (13): Cemeteries are areas where physical interventions such as pesticides and pruning are made at a minimum level. In addition, as they are not exposed to intense human use, they constitute a quiet environment. In this respect, they are important habitats for relatively timid species that cannot find shelter in other urban biotopes. The age of a cemetery is essential, in terms of the vegetation quality. While old cemeteries host old trees, the vegetation of new cemeteries consists mainly of bushes or young trees with lean structures. As such, no species that need tree hollows to shelter in were observed in new cemeteries.

However, ancient cemeteries may serve as a stepping stone for many species, especially between groves and other biotopes. All the cemeteries in the research area were areas with a certain age of tree texture. Although they are ecologically precious areas, they constitute a small percentage in the research area (0.76%). Therefore, when examined in terms of PLAND, NP, and ENN_AM values, cemeteries should be considered in terms of providing shelter to different species with the integrated units they form, even though they cannot be considered in a connective position within the landscape structure.

Openings in Cemeteries (14): Openings with herbaceous vegetation or soil-covered surfaces in cemeteries are considered a separate class, as they have different characteristics appealing to different species.

3.5. PM_{10} Concentration and Habitat Relations

The fact that the NP-dependent PD value, which is positively correlated with PM_{10}, was high for the building and firm ground classes indicates that hardscapes play an essential role in controlling the particulate matter density in the urban ecosystem. The choice of green systems or smart materials with the ability to absorb pollutants, especially as roof and facade materials, can significantly contribute to balancing the PM concentration in the entire urban landscape.

Considering the negative correlation between the AI and PM_{10}—which we interpret as decreasing PM while clustering increases—the high AI values in groves and roadside green areas indicate how vital these green areas are in the urban ecosystem. Preserving the holistic structure of groves is extremely important, in terms of habitat quality and biodiversity protection. Therefore, the high aggregation index value of this class is promising. Roadside green areas can potentially curb the adverse effects of roads which suppress the urban ecosystem in terms of pollutants such as noise, emissions, and PM. It is crucial for the units belonging to these areas to be clustered and gain a continuous structure as much as possible, in order to be linearly effective.

The results of this research informed us that the ecological indicator–landscape structure relationship, which provides inferences, contains important clues regarding the urban ecosystem. Furthermore, we determined that PM is significantly correlated with metrics that are indicators of landscape structure. For this reason, looking at the relationship between PM and metric values obtained from high-resolution satellite imagery and/or detailed DEM data in future research is expected to enable more detailed interpretations [110].

4. Conclusions

In metropolitan areas such as Istanbul, where the urbanization pressure is intense, research is of vital importance to ensure the continuity of the urban ecosystem. Landscape plans should focus on ecosystem relations and the inclusion of implementation strategies, thus guiding development plans within the sustainability framework. In areas where rapid transformation processes are experienced, it is necessary to produce comprehensive, practical, and up-to-date data on the deterioration/transformation rates. In this sense, the landscape structure, which can be evaluated at wide scale through the use of RS and GIS technologies, provides important clues for the urban ecosystem. Furthermore, the spatial arrangement and structure can be used as indicators, in terms of landscape functions.

In the first stage, we revealed and interpreted the change trends of green areas and artificial surfaces over a 51-year period using freely accessible Landsat (Corona for 1963) satellite imagery with medium-level resolution. After this general evaluation, we examined the spatial relationships of cultural landscapes that shape the urban ecosystem in more detail, using high-resolution Pleiades satellite images. In addition, we measured the PM_{10} concentration (in 2014) at 50 test points representing different cultural landscapes in the research area. Finally, we interpreted the results by comparing them with the general and detailed data obtained for the landscape structure. We also analyzed the correlations between PM_{10} and landscape-level landscape metrics.

The most striking results achieved in this comprehensive and multi-component study are summarized below:

- The two-stage landscape pattern evaluation method, based on the temporal–spatial findings related to the landscape structure of the research area, enabled the interpretation of the spatial arrangement of landscape classes on a more detailed scale and the determination of administrative priorities regarding landscape functions.
- An interpretation of the relationships between landscape structure, particulate matter concentration, and habitat quality provided essential findings for the urban ecosystem.
- Results from the low-resolution data revealed significant correlations between particulate matter concentration and landscape structure indices. Examining these relationships at more detailed scales can significantly contribute to the evaluation of important components, such as habitat quality, biodiversity, and microclimatic relationships in the urban ecosystem.
- Assessing the landscape structure through a detailed holistic approach ensures that the habitat relationships can be evaluated more accurately and comprehensively. Different resolution RS data (satellite images and orthophotos) available on a wide scale facilitate such an evaluation.

- The research was productive in creating an ecological basis within a short time, which is extremely important for the evaluation and management of urban landscapes experiencing a rapid transformation process.
- Associating cultural landscape types with the living environments of indicator species enabled us to establish a bridge between landscape structure and important factors for landscape function, such as water cycle, pollutants, and climate. In this way, the landscape structure could be evaluated as an indicator of landscape functions.
- An alternative model was created, in order to associate species–habitat relations, by looking at landscape structure–ecological indicator interactions.
- We revealed a holistic view of the spatial transformation processes in urban landscapes, which have dynamic drivers at the local, regional, national, and international levels that serve to accelerate urbanization. Such an assessment is crucial for ensuring the sustainability of the urban ecosystem and presenting a model for similar landscapes. Moreover, the proposed framework allows lead landscape planners and managers to better assess cause–effect relationships.

Author Contributions: Conceptualization, G.A.A.; methodology, G.A.A., Ş.T. and N.M.; software, G.A.A., Ş.T., N.M. and E.S.C.; validation, G.A.A., Ş.T., N.M. and E.S.C.; formal analysis, G.A.A., Ş.T., N.M. and E.S.C.; investigation, G.A.A.; resources, G.A.A., Ş.T., N.M. and A.U.; data curation, G.A.A.; writing—original draft preparation, G.A.A. and Ş.T.; writing—review and editing, G.A.A. and N.M.; visualization, G.A.A. and E.S.C.; supervision, G.A.A.; project administration, G.A.A.; funding acquisition, G.A.A. All authors have read and agreed to the published version of the manuscript.

Funding: This research was carried out in the scope of the project titled "Istanbul-Besiktas District an Ecological Planning Approach for an Urban Landscape Plan and Implementation Strategy" funded by The Scientific and Technological Research Council of Turkey (TUBITAK) with the project number 114O341-TOVAG. We are grateful to TUBITAK for their support.

Conflicts of Interest: The authors declare no conflict of interest.

References

1. European Union Environment. Available online: https://europa.eu/european-union/topics/environment_en (accessed on 26 August 2022).
2. UN-Environment Programme. Available online: https://www.unenvironment.org/ (accessed on 26 August 2022).
3. UN–Sustainable Development Goals (SDGs). Available online: https://sdgs.un.org/goals#goals (accessed on 8 September 2022).
4. The European Green Deal, 2021. Available online: https://ec.europa.eu/info/strategy/priorities-2019-2024/european-green-deal_en (accessed on 26 August 2022).
5. Sundseth, K.; Creed, P. *NATURA 2000, Protecting Europe's Biodiversity*; European Commission, Directorate General for the Environment: Brusselles, Belgium, 2008; ISBN 978 92 79 08308 2.
6. COM–Commission of the European Communities. Our Life Insurance, Our Natural Capital: An EU Biodiversity Strategy to 2020. Communication from the Commission to the European Parliament, the Council, the Economic and Social Committee and the Committee of the Regions. Brussels, 3.5.2011; COM (2011) 244 Final. Available online: http://eur-lex.europa.eu/legal-content/EN/TXT/PDF/?uri=CELEX:52011DC0244&from=de (accessed on 20 September 2022).
7. Zonneveld, I.S. The land unit–A fundamental concept in landscape ecology, and its applications. *Landsc. Ecol.* **1989**, *3*, 67–86. [CrossRef]
8. Antrop, M. Landscape change: Plan or chaos? *Landsc. Urban Plan.* **1998**, *41*, 155–161. [CrossRef]
9. Bailey, R.G. *Ecoregion-Based Design for Sustainability*; Springer: New York, NY, USA, 2002; ISBN 0-387-95430-9.
10. Steinhardt, U.; Blumenstein, O.; Barsch, H. *Lehrbuch der Landschaftsökologie (Textbook of Landscape Ecology)*; Springer: Berlin/Heidelberg, Germany, 2005; ISBN 3-8274-1448-2. (In German)
11. Forman, R.T.T. *Urban Regions. Ecology and Planning Beyond the City*; Cambridge University Press: Cambridge, UK, 2008; ISBN 978-0-521-67076-0.
12. Forman, R.T.T. *Urban Ecology: Science of Cities*; Cambridge University Press: Cambridge, UK, 2014; ISBN 978-0-521-18824-1.
13. Wu, J. Urban sustainability: An inevitable goal of landscape research. Editorial. *Landsc. Ecol.* **2010**, *25*, 1–4. [CrossRef]
14. Wu, J. Urban ecology and sustainability: The state-of-the-science and future Directions. *Landsc. Urban Plan.* **2014**, *125*, 209–221. [CrossRef]
15. Potschin, M.; Haines-Young, R. "Rio+10", sustainability science and Landscape Ecology. *Landsc. Urban Plan.* **2006**, *75*, 162–174. [CrossRef]

16. Rees, W.; Wackernagel, M. Urban Ecological Footprints: Why Cities Cannot be Sustainable and Why They are a Key to Sustainability. *Environ. Impact Assess. Rev.* **1996**, *16*, 223–248. [CrossRef]
17. Solarski, M.; Krzysztofik, R. Is the Naturalization of the Townscape a Condition of De-Industrialization? An Example of Bytom in Southern Poland. *Land* **2021**, *10*, 838. [CrossRef]
18. Aksu, G.A.; Musaoğlu, N.; Uzun, A. An auxiliary tool for landscape evaluation Ecological risk analysis based on analytic hierarchy process. *Fresenius Environ. Bull.* **2017**, *26*, 84–92.
19. Aksu, G.A.; Küçük, N. Evaluation of urban topography–biotope–population density relations for Istanbul–Besiktas urban landscape using AHP. *Environ. Dev. Sustain.* **2020**, *22*, 733–758. [CrossRef]
20. Sertel, E.; Topaloğlu, R.; Şallı, B.; Yay Algan, I.; Aksu, G. Comparison of Landscape Metrics for Three Different Level Land Cover/Land Use Maps. *ISPRS Int. J. Geo-Inf.* **2018**, *7*, 408. [CrossRef]
21. Topaloğlu, R.H.; Aksu, G.A.; Ghale, Y.A.G.; Sertel, E. High-resolution land use and land cover change analysis using GEOBIA and landscape metrics: A case of Istanbul, Turkey. *Geocarto Int.* **2021**. [CrossRef]
22. Bentrup, G. *Conservation Buffers. Design Guidelines for Buffers, Corridors, and Greenways*; General Technical Report SRS 109; Department of Agriculture, Forest Service, Southern Research Station: Asheville, NC, USA, 2008.
23. Gartland, L. *Heat Islands: Understanding and Mitigating Heat in Urban Areas*; Earthscan: Oxford, UK, 2008; ISBN 978-1-84407-250-7.
24. Lehmann, I.; Mathey, J.; Rößler, S.; Bräuer, A.; Goldberg, V. Urban vegetation structure types as a methodological approach for identifying ecosystem services–Application to the analysis of micro-climatic effects. *Ecol. Indic.* **2014**, *42*, 58–72. [CrossRef]
25. Boehnke, D.; Krehl, A.; Mörmann, K.; Volk, R.; Lützkendorf, T.; Naber, E.; Becker, R.; Norra, S. Mapping Urban Green and Its Ecosystem Services at Microscale—A Methodological Approach for Climate Adaptation and Biodiversity. *Sustainability* **2022**, *14*, 9029. [CrossRef]
26. Lennon, J. Cultural Landscape Management Practice: Some Australian Case Studies. In *Conserving Cultural Landscapes Challenges and New Directions, Part IV: Confronting the Everyday Challenge of Cultural Landscape Management*; Taylor, K., St Clair, A., Mitchell, N.J., Eds.; Routledge: London, UK, 2015; ISBN 9781315813226.
27. Kaim, D.; Kozak, J.; Ostafin, K.; Dobosz, M.; Ostapowicz, K.; Kolecka, N.; Gimmi, U. Uncertainty in historical land-use reconstructions with topographic maps. *Quaest. Geogr.* **2014**, *33*, 55–63. [CrossRef]
28. Skaloš, J.; Weber, M.; Lipský, Z.; Trpáková, I.; Šantrůčková, M.; Uhlířová, L.; Kukla, P. Using old military survey maps and orthophotograph maps to analyse long-term land cover changes—Case study (Czech Republic). *Appl. Geogr.* **2011**, *31*, 426–438. [CrossRef]
29. Petit, C.; Lambin, E. Impact of data integration technique on historical land-use/land-cover change: Comparing historical maps with remote sensing data in the Belgian Ardennes. *Landsc. Ecol.* **2002**, *17*, 117–132. [CrossRef]
30. Güngöroğlu, C.; Musaoğlu, N.; Türkkan, M.; Yöntem, O.; Yılmaztürk, A.; Çayır, G. *CBS Destekli Uzaktan Algılama Teknikleri Kullanılarak Biyotop Tiplerinin Sınıflandırılması ve Haritalanması (Köprülü Kanyon Milli Parkı Örneği). (Classification and Mapping of Biotope Types Through GIS-based Remote Sensing Techniques–The Example of the Köprülü Kanyon National Park–)*; Ministry of Environment and Forests; South-West Anatolia Forest Research Institute (SAFRI): Antalya, Turkey, 2008. (In Turkish)
31. Odum, E.P.; Barrett, G.W. *Fundamentals of Ecology*; Thomson Brooks/Cole: Boston, MA, USA, 2005; ISBN 978-0534420666.
32. Kutzenberger, H. *Animal Ecology and Landscape Planning, Pilot Project in the Context of Cultural Landscape Research*; Research Area Cultural Landscape, Federal Ministry of Science and Transport: Vienna, Austria, 1998; Volume 3.
33. Kutzenberger, H. Animal Ecological Characterization of Cultural Landscapes as a Basis for Landscape Planning. Ph.D. Thesis, University of Natural Resources and Applied Life Sciences, Vienna, Austria, 2001.
34. Hampicke, U. *Kulturlandschaft–Äcker, Wiesen, Wälder und ihre Produkte. (Cultural Landscape–Fields, Meadows, Forests and their Products)*; Springer: Berlin/Heidelberg, Germany, 2018; (In German). ISBN 978-3-662-57752-3.
35. Türkeş, M.; Musaoğlu, N.; Özcan, O. Assessing the vulnerability of a forest ecosystem to climate change and variability in the western Mediterranean sub-region of Turkey: Future evaluation. *J. For. Res.* **2018**, *29*, 1177–1186. [CrossRef]
36. Yagmur, N.; Tanik, A.; Tuzcu, A.; Musaoglu, N.; Erten, E.; Bilgilioglu, B.B. Opportunities Provided by Remote Sensing Data for Watershed Management: Example of Konya Closed Basin. *Int. J. Eng. Geosci.* **2020**, *5*, 120–129.
37. Tuzcu, A.; Dervişoğşu, A.; Musaoğlu, N.; Tanık, A. Determination of Land-Use/Cover and Water Turbidity of Wetlands by Using High Resolution Satellite Image. In Proceedings of the International Symposium on Applied Geoinformatics (ISAG 2019), Istanbul, Turkey, 7 November 2019; Volume 1, pp. 154–158.
38. Riitters, K.H.; O'Neill, R.V.; Hunsaker, C.T.; Wickham, J.D.; Yankee, D.H.; Timmins, S.P.; Jones, K.B.; Jackson, B.L. A factor analysis of landscape pattern and structure metrics. *Landsc. Ecol.* **1995**, *10*, 23–39. [CrossRef]
39. McGarigal, K. *Encyclopedia of environmetric. Landscape Pattern Metrics*; Wiley: Chichester, UK, 2002; Volume 2, pp. 1135–1142.
40. Southworth, J.; Nagendra, H.; Tucker, C. Fragmentation of a landscape: Incorporating landscape metrics into satellite analyses of land-cover change. *Landsc. Res.* **2002**, *27*, 253–269. [CrossRef]
41. Leitão, A.B.; Miller, J.; Ahern, J.; McGarigal, K. *Measuring Landscapes. A Planner's Handbook*; Island Press: Washington, DC, USA, 2006; ISBN 1-55963-899-0.
42. Cushman, S.A.; McGarigal, K.; Neel, M.C. Parsimony in landscape metrics: Strength, universality, and consistency. *Ecol. Indic.* **2008**, *8*, 691–703. [CrossRef]
43. Uuemaa, E.; Antrop, M.; Roosaare, J.; Marja, R.; Mander, Ü. Landscape metrics and indices: An overview of their use in landscape research. *Living Rev. Landsc. Res.* **2009**, *3*, 1–28. [CrossRef]

44. McGarigal, K.; Cushman, S.A.; Ene, E. FRAGSTATS v4: Spatial Pattern Analysis Program for Categorical and Continuous Maps. Computer software program produced by the authors at the University of Massachusetts, Amherst. 2012. Available online: https://www.scirp.org/(S(lz5mqp453edsnp55rrgjct55))/reference/ReferencesPapers.aspx?ReferenceID=1892846 (accessed on 20 October 2022).
45. Liu, T.; Yang, X. Monitoring land changes in an urban area using satellite imagery, GIS, and landscape metrics. *Appl Geogr.* **2015**, *56*, 42–54. [CrossRef]
46. Szabó, S.; Bertalan, L.; Kerekes, Á.; Novák, T.J. Possibilities of land-use change analysis in a mountainous rural area: A methodological approach. *Int. J. Geogr. Inf. Sci.* **2016**, *30*, 708–726. [CrossRef]
47. Xu, C.; Haase, D.; Su, M.; Wang, Y.; Pauleit, S. Assessment of landscape changes under different urban dynamics based on a multiple-scenario modeling approach. *Env. Plan B Urban Anal City Sci.* **2020**, *47*, 1361–1379. [CrossRef]
48. Zhang, Q.; Chen, C.; Wang, J.; Yang, D.; Zhang, Y.; Wang, Z.; Gao, M. The spatial granularity effect, changing landscape patterns, and suitable landscape metrics in the Three Gorges Reservoir Area, 1995–2015. *Ecol. Indic.* **2020**, *114*, 106259. [CrossRef]
49. Karimi, J.D.; Corstanje, R.; Harris, J.A. Understanding the importance of landscape configuration on ecosystem service bundles at a high resolution in urban landscapes in the UK. *Landsc. Ecol.* **2021**, *36*, 2007–2024. [CrossRef]
50. Batur, A. Association of Particulate Matter (PM_{10}) Concentration with the Urban Green System: Istanbul–Besiktas Sample. Matser's Thesis, Istanbul Commerce University Graduate School of Applied and Natural Sciences, Istanbul, Turkey, 2021.
51. Deng, S.; Ma, J.; Zhang, L.; Jia, Z.; Ma, L. Microclimate Simulation and Model Optimization of the Effect of Roadway Green Space on Atmospheric Particulate Matter. *Environ. Pollut.* **2019**, *246*, 932–944. [CrossRef]
52. Wu, H.; Yang, C.; Chen, J.; Yang, S.; Lu, T.; Lin, X. Effects of Green Space Landscape Patterns on Particulate Matter in Zhejiang Province, China. *Atmos. Pollut. Res.* **2018**, *9*, 923–933. [CrossRef]
53. Kais, K.; Goła´s, M.; Suchocka, M. Awareness of Air Pollution and Ecosystem Services Provided by Trees: The Case Study of Warsaw City. *Sustainability* **2021**, *13*, 10611. [CrossRef]
54. Jaung, W.; Carrasco, L.R.; Shaikh, F.F.E.A.; Tan, P.Y.; Richards, D.R. Temperature and air pollution reductions by urban green spaces are highly valued in a tropical city-state. *Urban For. Urban Green.* **2020**, *55*, 126827. [CrossRef]
55. TUIK. Turkish Statistical Institute, Address Based Population Registration System Results official web site. 2017. Available online: http://tuik.gov.tr (accessed on 26 August 2022).
56. Lillesand, T.; Kiefer, R.W.; Chipman, J. *Remote Sensing and Image Interpretation*, 7th ed.; John Wiley & Sons Inc: Hoboken, NJ, USA, 2015.
57. Foody, G.M. Status of Land Cover Classification Accuracy Assessment. *Remote Sens. Environ.* **2002**, *80*, 185–201. [CrossRef]
58. UHUZAM İTÜ. Available online: http://Www.Cscrs.Itu.Edu.Tr/Assets/Downloads/Pleiadesuserguide.Pdf (accessed on 26 August 2022).
59. Myneni, R.B.; Asrar, G. Atmospheric Effects and Spectral Vegetation Indices. *Remote Sens. Environ.* **1994**, *47*, 390–402. [CrossRef]
60. Richards, J.A.; Jia, X. *Remote Sensing Digital Image Analysis: An Introduction*, 3rd ed.; Springer: Sydney, Australia, 1999.
61. Forman, R.T.T. *Land Mosaics*; Cambridge University Press: Cambridge, UK, 1995; ISBN 978-0-521-479980-6.
62. Antrop, M. The language of landscape ecologists and planners: A comparative content analysis of concepts used in landscape ecology. *Landsc. Urban Plan.* **2001**, *55*, 163–173. [CrossRef]
63. Kupfer, J.A. Landscape ecology and biogeography: Rethinking landscape metrics in a post-FRAGSTATS landscape. *Prog. Phys. Geogr. Earth Environ.* **2012**, *36*, 400–420. [CrossRef]
64. Li, H.; Reynolds, J.F. A simulation experiment to quantify spatial heterogeneity in categorical maps. *Ecology* **1994**, *75*, 2446–2455. [CrossRef]
65. Peng, J.; Wu, J.; Li, W. Evaluating the effectiveness of landscape metrics in quantifying spatial patterns. *Ecol. Indic.* **2010**, *10*, 217–223. [CrossRef]
66. McGarigal, K.; Cushman, S.A. The gradient concept of landscape structure. In *Issues and Perspectives in Landscape Ecology*; Wiens, J., Moss, M., Eds.; Cambridge University Press: Cambridge, UK, 2005; pp. 112–119.
67. Liu, M.; Hu, Y.; Li, C. Landscape metrics for three-dimensional urban building pattern recognition. *Appl. Geogr.* **2017**, *87*, 66–72. [CrossRef]
68. Amiri, B.J.; Asgarian, A.; Sakieh, Y. Introducing landscape accuracy metric for spatial performance evaluation of land use/land cover change models. *Geocarto Int.* **2017**, *32*, 1171–1187. [CrossRef]
69. Lamine, S.; Petropoulos, G.P.; Singh, S.K.; Szabó, S.; Bachari, N.E.I.; Srivastava, P.K.; Suman, S. Quantifying land use/land cover spatio-temporal landscape pattern dynamics from Hyperion using SVMs classifier and FRAGSTATS®. *Geocarto Int.* **2018**, *33*, 862–878. [CrossRef]
70. Leitão, A.B.; Muge, F. The Role of Landscape Metrics in Environmental Planning and Management of Mining Activities. In Proceedings of the APCOM'2001-29th International Symposium on Computer Applications in the Minerals Industries, Beijing, China, 25–27 April 2001; Heping, X., Wang, Y., Jiang, Y., Eds.; Balkema Publishers: Rotterdam, The Netherlands, 2001; pp. 713–718.
71. Colson, F.; Bogaert, J.; Ceulemans, R. Fragmentation in the Legal Amazon, Brazil: Can landscape metrics indicate agricultural policy differences? *Ecol. Indic.* **2011**, *11*, 1467–1471. [CrossRef]
72. Aguilera, F.; Valenzuela, L.M.; Botequilha-Leitão, A. Landscape metrics in the analysis of urban land use patterns: A case study in a Spanish metropolitan area. *Landsc. Urban Plan.* **2011**, *99*, 226–238. [CrossRef]

73. Vaz, E.; De Noronha, T.; Nijkamp, P. Exploratory Landscape Metrics for Agricultural Sustainability. *Agroecol. Sustain. Food Syst.* **2014**, *38*, 92–108. [CrossRef]
74. Ozcan, O.; Aksu, G.A.; Erten, E.; Musaoglu, N.; Çetin, M. Degradation monitoring in Silvopastoral systems: A case study of the Mediterranean region of Turkey. *Adv. Space Res.* **2019**, *63*, 160–171. [CrossRef]
75. Li, E.; Endter-Wada, J.; Li, S. Dynamics of Utah's agricultural landscapes in response to urbanization: A comparison between irrigated and non-irrigated agricultural lands. *Appl. Geogr.* **2019**, *105*, 58–72. [CrossRef]
76. Abedini, A.; Khalili, A.; Asadi, N. Urban Sprawl Evaluation Using Landscape Metrics and Black-and White Hypothesis (Case Study: Urmia City). *J. Indian Soc. Remote Sens.* **2020**, *48*, 1021–1034. [CrossRef]
77. Wolff, S.; Lakes, T. Characterising Agricultural Landscapes using Landscape Metrics and Cluster Analysis in Brandenburg, Germany. *GI_Forum* **2020**, *1*, 89–98. [CrossRef]
78. Wolff, S.; Hüttel, S.; Nendel, C.; Lakes, T. Agricultural Landscapes in Brandenburg, Germany: An Analysis of Characteristics and Spatial Patterns. *Int. J. Environ. Res.* **2021**, *15*, 487–507. [CrossRef]
79. Zhang, L.; Zhai, Z.; Zhou, Y.; Liu, S.; Wang, L. The Landscape Pattern Evolution of Typical Open-Pit Coal Mines Based on Land Use in Inner Mongolia of China during 20 Years. *Sustainability* **2022**, *14*, 9590. [CrossRef]
80. Bao, S.; Yang, F. Spatio-Temporal Dynamic of the Land Use/Cover Change and Scenario Simulation in the Southeast Coastal Shelterbelt System Construction Project Region of China. *Sustainability* **2022**, *14*, 8952. [CrossRef]
81. Miller, W.; Collins, M.G.; Steiner, F.R.; Cook, E. An approach for greenway suitability analysis. *Landsc. Urban Plan.* **1998**, *42*, 91–105. [CrossRef]
82. Tischendorf, L.; Fahrig, L. How should we measure landscape connectivity? *Landsc. Ecol.* **2000**, *15*, 633–641. [CrossRef]
83. Tischendorf, L. Can landscape indices predict ecological processes consistently? *Landsc. Ecol.* **2001**, *16*, 235–254. [CrossRef]
84. Ramachandra, T.V.; Bharath, H.A.; Sreekantha, S. Spatial Metrics based Landscape Structure and Dynamics Assessment for an emerging Indian Megalopolis. *Int. J. Adv. Res. Artif. Intell.* **2012**, *1*, 1. [CrossRef]
85. Scolozzi, R.; Geneletti, D. A multi-scale qualitative approach to assess the impact of urbanization on natural habitats and their connectivity. *Environ. Impact Assess. Rev.* **2012**, *36*, 9–22. [CrossRef]
86. Fan, C.; Myint, S. A comparison of spatial autocorrelation indices and landscape metrics in measuring urban landscape fragmentation. *Landsc. Urban Plan.* **2014**, *121*, 117–128. [CrossRef]
87. Weber, N.; Haase, D.; Franck, U. Assessing modelled outdoor traffic induced noise and air pollution around urban structures using the concept of landscape metrics. *Landsc. Urban Plan.* **2014**, *125*, 105–116. [CrossRef]
88. Prastacos, P.; Chrysoulakis, N.; Kochilakis, G. Urban Atlas, land use modelling and spatial metric techniques. *Cybergeo Eur. J. Geogr. Reg. Urban Plan.* **2017**, *815*. [CrossRef]
89. Wang, H.; Li, C. Analysis of scale effect and change characteristics of ecological landscape pattern in urban waters. *Arab. J. Geosci.* **2021**, *14*, 569. [CrossRef]
90. Szabó, S.; Csorba, P.; Szilassi, P. Tools for Landscape Ecological Planning–Scale, and Aggregation Sensitivity of the Contagion Type Landscape Metric Indices. *Carpathian J. Earth Environ. Sci.* **2012**, *7*, 127–136.
91. Singh, S.K.; Srivastava, P.K.; Szabó, S.; Petropoulos, G.P.; Gupta, M.; Islam, T. Landscape transform and spatial metrics for mapping spatiotemporal land cover dynamics using Earth Observation data-sets. *Geocarto. Int.* **2017**, *32*, 113–127. [CrossRef]
92. Wiens, J.A. Spatial Scaling in Ecology. *Funct. Ecol.* **1989**, *3*, 385–397. [CrossRef]
93. Forman, R.T.T.; Godron, M. *Landscape Ecology*; John Wiley & Sons: New York, NY, USA, 1986; ISBN 0-471-87037-4.
94. Ranney, J.W.; Bruner, M.C.; Levenson, J.B. The importance of edge in the structure and dynamics of forest islands. In *Forest Island Dynamics in Man-Dominated Landscapes*; Burgess, R.L., Sharpe, D.M., Eds.; Springer: New York, NY, USA, 1981; pp. 67–95.
95. Moore, N.W.; Hooper, M.D. On the number of bird species in British woods. *Biol. Conserv.* **1975**, *8*, 239–250. [CrossRef]
96. Whitcomb, R.F.; Robbins, C.S.; Lynch, J.F.; Whitcomb, B.L.; Klimkiewicz, K.; Bystrak, D. Effects of forest. In *Forest Island Dynamics in Man-Dominated Landscapes*; Burgess, R.L., Sharpe, D.M., Eds.; Springer: New York, NY, USA, 1981; pp. 125–205.
97. Gyenizse, P.; Bognár, Z.; Czigány, S.; Elekes, T. Landscape shape index, as a potencial indicator of urban development in Hungary. *Landsc. Amp. Env.* **2014**, *8*, 78–88.
98. Buechner, M. Are small-scale landscape features important factors for field studies of small mammal dispersal sinks? *Landsc. Ecol.* **1989**, *2*, 191–199. [CrossRef]
99. Andrén, H. Effects of Habitat Fragmentation on Birds and Mammals in Landscapes with Different Proportions of Suitable Habitat: A Review. *Oikos* **1994**, *71*, 355–366. [CrossRef]
100. He, H.S.; DeZonia, B.E.; Mladenoff, D.J. An aggregation index (AI) to quantify spatial patterns of landscapes. *Landsc. Ecol.* **2000**, *15*, 591–601. [CrossRef]
101. Abhijith, K.V.; Kumar, P.; Gallagher, J.; McNabola, A.; Baldauf, R.; Pilla, F.; Pulvirenti, B. Air pollution abatement performances of green infrastructure in open road and built-up street canyon environments—A review. *Atmos. Environ.* **2017**, *162*, 71–86. [CrossRef]
102. Viecco, M.; Vera, S.; Jorquera, H.; Bustamante, W.; Gironás, J.; Dobbs, C.; Leiva, E. Potential of particle matter dry deposition on green roofs and living walls vegetation for mitigating urban atmospheric pollution in semiarid climates. *Sustainability* **2018**, *10*, 2431. [CrossRef]
103. Garcia, J.N.P.M.; Cerdeira, R.S.; Tavares, N.A.; Coelho, L.M.R. Studying street geometry influence in PM10 concentration. *Environ. Pollut.* **2012**, *50*, 283–292. [CrossRef]

104. Roselund, M. Environmental Factors in Cardiovascular Disease. Ph.D. Thesis, The Institute of Environmental Medicine, Karolinska Institutet, Stockholm, Sweden, 2005.
105. Vlachokostas, C.; Achillas, C.; Michailidou, A.V.; Moussiopoulos, N. Measuring combined exposure to environmental pressures in urban areas: An air quality and noise pollution assessment approach. *Environ. Int.* **2012**, *39*, 8–18. [CrossRef] [PubMed]
106. Li, S.; Lu, X.X.; He, M.; Zhou, Y.; Li, L.; Ziegler, A.D. Daily CO2 partial pressure and CO2 outgassing in the upper Yangtze River basin: A case study of the Longchuan River, China. *J. Hydrol.* **2012**, *466–467*, 141–150. [CrossRef]
107. Aksu, G.A. Evaluation of Roadside Green Areas in Urban Landscapes in the Framework of Urban Ecosystem. *Eur. J. Sci. Technol.* **2021**, *25*, 736–748.
108. Gandy, M. *Natura Urbana: Ecological Constellations in Urban Space*; MIT Press: Cambridge, MA, USA, 2022; ISBN 9780262046282.
109. Kırca, S.; Çolak, A.H.; Rotherdam, I.D. The Ancient Woodland Concept as a Practical Conservation Tool: The Turkish Experience. In *Ancient Woodlands and Trees: A Guide for Landscape Planners and Forest Managers*; Çolak, A.H., Kırca, S., Rotherham, I.D., Eds.; IUFRO World Series: Vienna, Austria, 2018; Volume 37, pp. 214–241.110.
110. Śliwiński, D.; Konieczna, A.; Roman, K. Geostatistical Resampling of LiDAR-Derived DEM in Wide Resolution Range for Modelling in SWAT: A Case Study of Zgłowiączka River (Poland). *Remote Sens.* **2022**, *14*, 1281. [CrossRef]

sustainability

MDPI

Article

Assessment of Urban Green Development Efficiency Based on Three-Stage DEA: A Case Study from China's Yangtze River Delta

Qi Yang *, Zhonggen Sun and Hubiao Zhang

School of Public Administration, Hohai University, Nanjing 211100, China
* Correspondence: qyang1990@hhu.edu.cn

Abstract: With the march of global urbanization, there are looming problems including environmental degradation and remediation all over the world. In this case, urban green development is the key to overcoming climate crisis, biodiversity loss and pollution. In this paper, a three-stage DEA model was employed to study the urban green development efficiency (GDE), with cities in the Yangtze River Delta (YRD) as the object. In the study, the regional economic foundation, urbanization level, industrial structure and government planning were used as external environmental variables, and the impact of objective external environmental factors was tested empirically, thereby eliminating the adverse environmental impact and statistical noise to obtain more truthful GDE. According to the results, first, the influence of external environmental factors and stochastic disturbance on GDE was effectively removed by virtue of the three-stage DEA model, and the GDE of the YRD was measured in a true and objective manner. The GDE of the YRD in Stage III was notably higher than that in Stage I since the GDE in Stage I was underestimated under the influence of objective environmental variables. Second, the GDE level showed heterogeneity in different cities, which behaved better in coastal and southeastern regions than in central, western and northern regions. Third, regarding the impact of external environmental variables, the GDE was enhanced by increasing the proportion of the tertiary industry and the green area of built districts but weakened when the area of built districts (ABD) reflecting urban construction was expanded. The index gross regional product (GRP) reflects local economic development level, the impact of which on GDE was not determined in this paper. As a consequence, in the process of urban development, it is suggested to focus on the innovation and application of green technology, upgrade the industrial structure, cultivate green talents, and formulate reasonable green transformation policies.

Keywords: urban green development efficiency; Yangtze River Delta; three-stage DEA analysis; China

Citation: Yang, Q.; Sun, Z.; Zhang, H. Assessment of Urban Green Development Efficiency Based on Three-Stage DEA: A Case Study from China's Yangtze River Delta. *Sustainability* **2022**, *14*, 12076. https://doi.org/10.3390/su141912076

Academic Editors: Lu Zhang, Bing Kuang and Bohan Yang

Received: 3 September 2022
Accepted: 21 September 2022
Published: 24 September 2022

1. Introduction

As the global economy develops, the scale of cities has continued to expand and the urban population has risen sharply, impacting the environment in many aspects. The area of cities only accounts for 3% of the world's land, and cities contribute 80% of the gross world product (GWP) at the expense of 70% of the world's resources and 75% of the global greenhouse gas emissions [1,2]. With global urbanization, human beings consumed natural resources and energy over the past 100 years, reaching an unprecedented level in human history. Accordingly, resource consumption and greenhouse gas emissions have sharply deteriorated the global ecological environment. Moreover, urban environmental problems are no longer limited to cities, but environmental problems involving all regions and all countries. As Anwarul K. Chowdhury, the Chairman of the Global Forum on Human Settlements (GFHS) and former Deputy Secretary-General and High Representative of the United Nations, said, "The world is undergoing a process of urbanization, and a new urban age has come. It is conceivable that the global urbanization level will be as high as 70% in the next 40 years. Sustainable urban development is one of the most serious challenges for human society in the 21st century. As a growing number of people settle in cities, cities will face the greatest challenges in the world at all levels, so concerted efforts and sincere cooperation are required from all over the world". On 18 November 2021, the United Nations Environment Program (UNEP) and UN-Habitat jointly released the Global Environment Outlook for Cities report, calling for urgent action to achieve net-zero recycling cities that are resilient, sustainable, inclusive and equitable, thus providing feasible solutions for the construction of environmentally friendly and green cities. Urban green development is the key to overcoming climate crisis, biodiversity loss and pollution, and it is also an important way to build urban ecological civilization and promote economic transformation by improving the green development level [3]. Not only is green development an idea describing a green environment from all aspects, but it also puts a premium on coordinated economic, social and environmental development, which is a comprehensive model aiming at efficiency, development and sustainability [4]. The economic vitality, innovation and competitiveness serving high-quality urban development are closely linked to green development. In the absence of green development, economic development will lose driving force and vitality, and similarly, there will be a lack of foundation and support concerning innovation and competitiveness [5].

The green development level can be measured by two main methods including the comprehensive index system and green development efficiency (GDE). For the first method, the regional green development status is evaluated by constructing an index system. Zhang et al. (2021) established an index system to measure the GDE in the Yangtze River Delta (YRD) on the strength of four dimensions: social development, economic development, energy consumption and ecological environment. Yang et al. (2019) evaluated the green development of resource-based cities in China and found that the green development level in the east of China was higher than that in the west [6]. The second method is usually realized by parametric stochastic frontier analysis (SFA) and nonparametric data envelopment analysis (DEA)) [7]. SFA is commonly applied to a single output scenario and requires the estimation of specific functional forms, but incorrect results may be caused by an incorrect functional form [8,9]. As a linear programming technique, DEA is widely used in the evaluation of the relative efficiency of homogeneous decision-making units, especially for multiple input-output scenarios [10–12]. Hence, there are an increasing number of scholars using DEA and its extended models to evaluate regional GDE. For instance, Wu et al. (2020) analyzed the GDE of 30 provinces in China in 2015 using a multi-objective DEA model from the perspective of resource allocation [13]. According to the annual cross-sectional data of different regions, Yang et al. (2015) employed the super-efficiency DEA model and the Malmquist index model to calculate the GDE of 31 regions in China during 2008–2012 [14].

Notwithstanding, environmental variables and statistical noise bring about considerable impacts in the traditional DEA, so the estimation of results may be biased and inaccurate [15]. In order to solve this problem, Fried et al. proposed a three-stage DEA model, that is, after calculating the efficiency value with the traditional DEA, the changes in the environment, statistical noise and management efficiency were analyzed with the help of the SFA model, the original input variables were adjusted, and then a second DEA calculation was performed to obtain the real efficiency value [16,17]. The three-stage DEA model has been applied by many scholars to calculate the efficiency of different subjects in different fields, and the results obtained are superior to those obtained through the traditional DEA model [18–20]. At present, there is still little information on GDE at the city level in a region since GDE is calculated using the traditional DEA in most of the existing studies.

To fill this gap, in this paper, based on the three-stage DEA model, cities in the YRD were selected as the object of the study. In the study, the regional economic foundation, urbanization level, industrial structure and government planning were served as external environmental variables, and the impact of objective external environmental factors was tested empirically, thereby eliminating environmental factors and statistical noise to obtain a more truthful GDE, as well as policy suggestions for improving the urban environment. The other parts of the paper are organized as follows: The second part contains the scope of the study, variable selection and description, and computation model description. The GDE calculation is conducted in the third part, and further discussion on the results is revealed in the fourth part. Finally, the conclusions are summarized, and some suggestions and implications for the sustainable ecological development of cities in the future are put forward.

2. Materials and Methods

2.1. Scope of Study

China is a developing country with the largest energy consumption and carbon dioxide emissions in the world, where sustainable urban development faces severe challenges. Benefited from the policy dividends of reform and opening up and the high attention of the State, the YRD is one of the regions with the most active economic development, the highest degree of openness, and the strongest innovation capability in China, which holds a pivotal strategic position in the national modernization and all-round opening-up pattern. According to the Outline of the Integrated Regional Development of the YRD approved by the State Council in 2019, the YRD covers an area of 358,000 km^2, including Shanghai municipality and three provinces, i.e., Jiangsu, Zhejiang, and Anhui provinces, as shown in Figure 1.

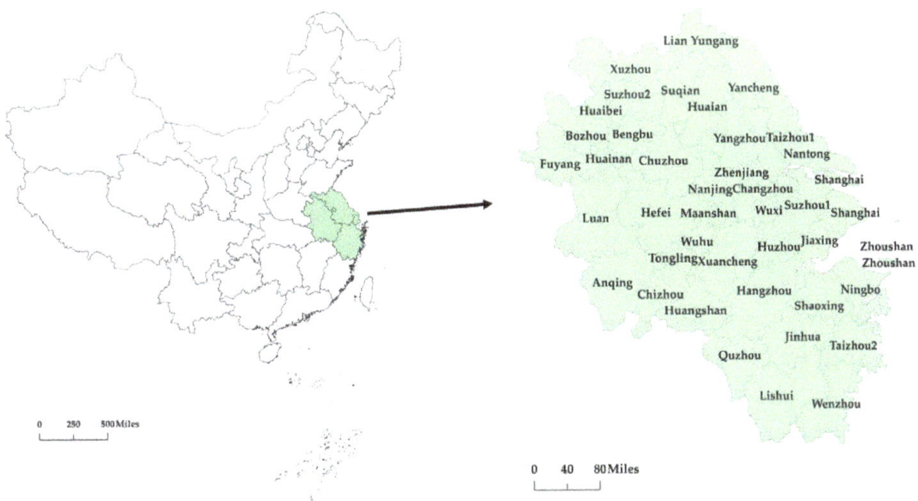

Figure 1. Distribution of 41 Cities in the YRD.

2.2. Variable Selection and Description

The GDE indexes shall be selected in accordance with the connotation of GDE. Based on previous studies, in this paper, GDE was defined as a fact that the maximum economic and social benefits are obtained with the minimum factor input and the minimum environmental output, so as to achieve a win–win situation of "economy–society–ecology". Comparatively, this definition better reflects the connotation of the social level than the previous definitions, which is completely consistent with the concept of urban green development.

For input indexes, the general input factors mainly include capital, labor, resources and technology [15,18]. Referring to the multilayer evaluation indexes on urban development systems of Feng and Xu (1999), Su et al. (2019), Zhang et al. (2021) [21–23], the investment in fixed assets (IFA) represents the capital input factor, the employment in the management of water conservancy and environment (EMWCE) indicates the elements of labor input, the annual electricity consumption (AEC) of the entire society stands for the input of energy factors, and the expenditure for education, science and technology (EEST) denotes the technical input factors. According to relevant study results and the availability of data, the total retail sales of consumer goods (TRSCG) were used as the desired output to represent the economic and social levels of a city. The volume of industrial wastewater discharged (VIWD) and volume of industry sulfur dioxide produced (VISDP) were selected to comprehensively investigate the environmental pollution factors.

Environmental variables in this study refer to factors that can affect GDE but cannot be controlled or changed by samples subjectively [24,25]. In this paper, the indexes gross regional product (GRP) [26], area of built districts (ABD) [22], the tertiary industry as a percentage of GRP (TIP) [27] and green covered area of complete area (GCA) [28] were selected as the environmental variables to indicate the economic development, urban construction, industrial structure and government planning, respectively.

With 41 cities in the YRD as the object of the study, the GDE there during 2009–2018 was evaluated, and corresponding data were obtained from the China Statistical Yearbook, China City Statistical Yearbook and official websites of the Bureau of Statistics of various cities. Table 1 presents the evaluation index system, where four inputs, three desirable outputs and four environmental variables are listed, and descriptive statistics of the selected data are exhibited in Table 2.

Table 1. Evaluation index system of GDE.

Variable	No.	Index	Unit
Input Variables	I1	AEC	10,000 kwh
	I2	IFA	10,000 yuan
	I3	EEST	10,000 yuan
	I4	EMWCE	person
Output Variables	O1	VIWD	10,000 tons
	O2	VISDP	ton
	O3	TRSCG	10,000 yuan
Environmental Variables	E1	GRP	10,000 yuan
	E2	ABD	sq. km
	E3	TIP	%
	E4	GCA	hectare

Source: Authors' work.

Table 2. Descriptive statistics.

Variable	Number	Mean Value	Standard Deviation	Min.	Max.
I1	410	1,828,000	3,031,000	67,166	31,820,000
I2	410	21,730,000	18,060,000	2,352,000	112,400,000
I3	410	1,022,000	1,572,000	64,104	13,440,000
I4	410	9873	12,761	455	93,600
O1	410	11,624	13,123	486	80,468
O2	410	43,067	45,387	1407	496,377
O3	410	13,540,000	16,900,000	791,784	126,700,000
E1	410	35,780,000	44,200,000	1,331,000	326,800,000
E2	410	176.5	186.4	31	1238
E3	410	0.42	0.0825	0.234	0.793
E4	410	7925	10,934	1256	139,427

Source: Authors' work.

For indexes of desirable output, environmental factors are always considered undesirable outputs [7]. Given that the outputs of the DEA model are generally desirable, it is unreasonable to select the three-stage DEA method when environmental pollutants are undesirable outputs. Some scholars treat undesirable outputs as inputs [29,30], which only requires information on whether the data should be minimized or maximized but cannot reflect the real production process. Therefore, the above-required indexes should be converted accordingly. The data conversion function processing method is an ideal efficiency evaluation method proposed by Seiford and Zhu (2002), containing negative output, linear and nonlinear data conversion and other types. In this study, the method was specially selected for data conversion of the environmental pollutant indexes. The specific formula is $Y_i = -Y_i + D$, where D represents a very large vector to ensure that all converted output data are positive. Referring to the existing study results, the C value was set to 1.1 times the maximum value in the sample area.

Under the application conditions of the DEA model, the pollution emission index was transformed and processed. The industrial wastewater and sulfur dioxide emissions were reduced to a comprehensive index, and the pollution index was converted by the data conversion function processing method. The linear data conversion method for reinforcing the environmental pollutants after conversion can reasonably solve the problem of the undesirable outputs in the three-stage DEA model for efficiency evaluation, effectively maintaining the convex and linear relationship.

2.3. Computation Model Description

Leveraging the three-stage DEA model, the true GDE was calculated as per the steps below:

Stage I: The traditional DEA model was applied. Charnes, Cooper and Rhodes introduced a DEA method, also called the CCR model, to calculate the relative effectiveness of decision-making units (DMUs) under constant returns to scale [12]. Later, Banker, Charnes and Cooper decomposed the comprehensive technical efficiency in the CCR model into PTE (pure technical efficiency) and SE (scale efficiency) which have been used to measure the effectiveness of DMUs under variable returns to scale, also known as the BCC model [31]. This paper employed the BCC model to estimate the initial effectiveness of 41 cities in the study area, and the calculation process is expressed as follows:

$$\min_{\theta,\lambda} = \left[\theta - \left(e^t s^- + e^t s^+\right)\right] \tag{1}$$

$$\sum_{k=1}^{n} \lambda_i y_{rk} - s^+ = y_{0k} \tag{2}$$

$$\sum_{k=1}^{n} \lambda_i y_{rk} + s^- = \theta x_{0k} \tag{3}$$

where $i = 1, 2, \ldots, m$ and $r = 1, 2, \ldots, s$. n indicates the number of measuring units, m represents the number of input indexes and s denotes the number of output indexes. x_{ik} $(i = 1, 2, \ldots, m)$ refers to the i_{th} input element of the k_{th} measuring unit, y_{rk} $(r = 1, 2, \ldots, s)$ stands for the r_{th} output element of the k_{th} measuring unit and θ indicates the valid value of DMUs. If $\theta = 1$ and $s^+ = s^- = 0$, the measuring unit is of DEA efficiency; if $\theta = 1$ and $s^+ \neq s^- \neq 0$, the measuring unit is of weak DEA efficiency; if $\theta < 1$, the measuring unit is of non-DEA efficiency.

Stage II: In the second stage, the input slacks in Stage I were decomposed with the SFA model for eliminating the influence of uncontrollable effects on efficiency. It was a regression equation with input slacks as the explained variable and environmental variables as the explanatory variable [32]. Input slack refers to the difference between the input of the i_{th} measuring unit and the optimal efficiency of a certain actual input in Stage I. According to the study by Fried et al. (2002), the input slacks in Stage I was decomposed into three components including the influence of environmental effects, managerial inefficiencies, and stochastic disturbance. In the case of n DMUs, every DMU contains p observable environmental variables $Z_i = \left[Z_{1i}, \ldots Z_{pi} \right]$. Input slacks can be decomposed into the following form:

$$s_{ik} = f^i\left(z_k; \beta^i\right) + v_{ik} + u_{ik} \tag{4}$$

where is the slack value for the i_{th} input of the k_{th} DMU and $f^i(z_k ; \beta^i)$ marks the environmental effects, which is denoted as $f^i(z_k ; \beta^i) = z_k \times \beta^i$. $v_{ik} + u_{ik}$ stands for the mixed error term, v_{ik} is the stochastic error term, and μ_{ik} refers to the managerial inefficiency. If $v_{ik} \sim N(0, \theta^2_{vi})$, $v_{ik} \sim N^+(u^i, \sigma^2_{ui})$, v_{ik} and u_{ik} are independent of each other. $\gamma = \sigma^2_{ui}/(\sigma^2_{ui} + \sigma^2_{vi})$ is defined. When γ tends to 1, the influence of managerial factors is dominant, and when γ tends to 0, the difference in efficiency is mainly attributed to stochastic disturbance.

To adjust the measuring unit to the same external environment and stochastic factor state based on the most effective measuring unit, the unknown parameters were estimated by the maximum likelihood method, and then the original input was adjusted according to the formula below.

$$\overset{\wedge}{x_{ik}} = x_{ik} + \left[\max_k\left\{z_k\overset{\wedge}{\beta^i}\right\} - z_k\overset{\wedge}{\beta^i}\right] + \left[\max_k\left\{\overset{\wedge}{v_{ik}}\right\} - \overset{\wedge}{v_{ik}}\right] \tag{5}$$

$$i = 1, 2, \ldots, m;\ k = 1, 2, \ldots, n;$$

where $\hat{x_{ik}}$ is the adjusted input variable, and x_{ik} is the original input variable. The first square bracket indicates that the environment of DMU is adjusted to the same level, and the second indicates that the statistical noise of DMU is adjusted to the same situation. According to the above formula, statistical noise and managerial inefficiency shall be separated first. The statistical noise condition was estimated as:

$$\hat{E}[v_{ik}|v_{ik} + u_{ik}] = s_{ik} - z_k\,\overset{\wedge}{\beta^i} - \hat{E}[u_{ik}|v_{ik} + u_{ik}] \tag{6}$$

Fried et al. (2002) failed to provide an estimation formula for management inefficiency, but recommended the formula proposed by Jondrow et al. (1982), i.e., $\hat{E}\,[u_{ik} \mid v_{ik} + u_{ik}]$, to estimate the managerial inefficiency. However, the estimation formula by Jondrow et al. (1982) was based on the stochastic frontier production function, and the DEA model by Fried et al. (2002) was on the basis of the stochastic frontier cost function [33]. Some scholars failed to notice this point and misused the formula, resulting in low credibility of results [32,34,35]. Instead, the estimation formula of managerial inefficiency in the three-stage DEA model should be derived according to their methods. Luo (2012) proposed an estimation formula for the managerial inefficiency of the three-stage DEA model based on the assumption of uniform distribution, earning a more reasonable construction of the DEA model.

$$E(u|\varepsilon) = \sigma_*\left[\frac{\phi(\lambda\frac{\varepsilon}{\sigma})}{\Phi(\frac{\lambda\varepsilon}{\sigma})} + \frac{\lambda\varepsilon}{\sigma}\right] \tag{7}$$

where $\sigma_* = \sigma_u\sigma_v/\sigma$, $\sigma = \sqrt{\sigma^2_u + \sigma^2_v}$, and $\lambda = \sigma_u/\sigma_v$, $\phi(.)$, $\Phi(.)$ refer to the density and distribution functions of the standard normal distribution, respectively.

Stage III: The adjusted input variable and the original output variable were put into the BCC model again, obtaining the efficiency value without the influence of environmental effects, managerial inefficiencies, and stochastic disturbance. Comparatively, this efficiency value was more objective and scientific than that obtained in Stage I.

See Figure 2 for the process framework of the whole model.

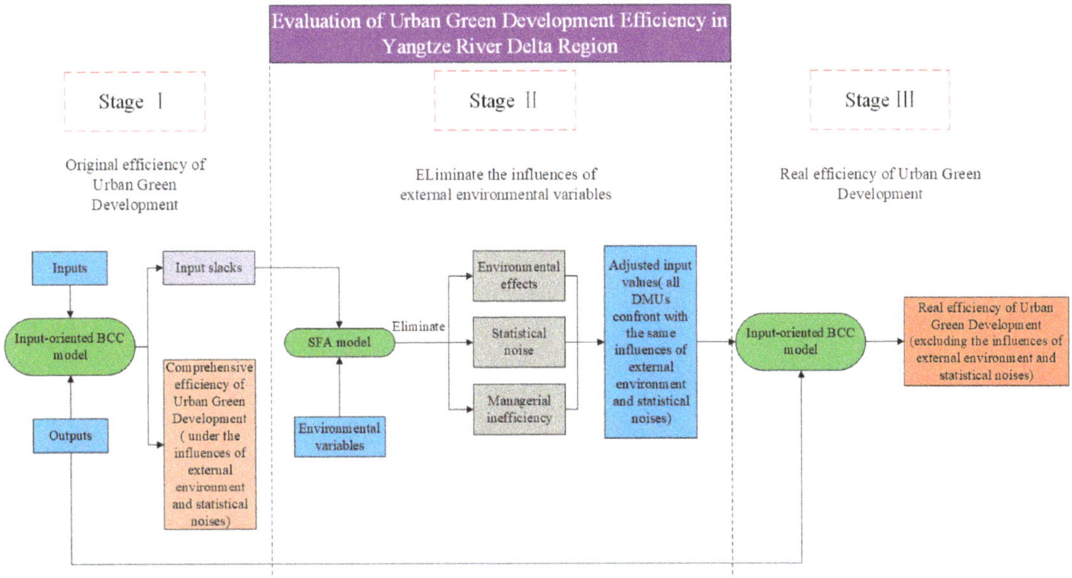

Figure 2. Calculation process of the three-stage DEA model for GDE in the YRD.

3. Results

3.1. Stage I: Comprehensive Technical Efficiency from the BCC Model

The GDE of 41 cities in the YRD during 2009–2018 was obtained using the input-oriented BCC model (Table 3). The mean comprehensive efficiency was 1 in both Wenzhou and Jinhua cities, being the highest among the 41 cities. The mean comprehensive efficiency was lower than 0.7 in Suqian, Wuhu and Huai'an cities, showing poor performance, which was 0.560, 0.593 and 0.655, respectively. The mean comprehensive efficiency was higher than 0.98 and lower than 1 in Nanjing, Quzhou, Huaibei, Lishui, Taizhou[2], Huangshan and Shanghai, showing excellent performance, which was 0.996, 0.996, 0.994, 0.992, 0.992, 0.985 and 0.980, respectively. The mean comprehensive efficiency of the other cities, Chuzhou, Lianyungang, Anqing, Suzhou[2], Lu'an, Ma'anshan, Taizhou[1], Hefei, Xuzhou, Bengbu, Zhenjiang, Ningbo, Yangzhou and Huainan, was lower than the mean of the overall mean comprehensive efficiency of the 41 cities in the YRD (0.859), which was 0.707, 0.709, 0.722, 0.734, 0.751, 0.770, 0.788, 0.792, 0.794, 0.797, 0.833, 0.837, 0.840 and 0.845, respectively. The mean comprehensive efficiency of Suzhou[1], Yancheng, Jiaxing, Zhoushan, Shaoxing, Nantong, Xuancheng, Hangzhou, Tongling, Changzhou, Bozhou, Wuxi, Fuyang, Huzhou and Chizhou was higher than the mean of the overall mean comprehensive efficiency of the 41 cities in the YRD (0.859), which was 0.861, 0.863, 0.868, 0.878, 0.881, 0.887, 0.891, 0.893, 0.909, 0.933, 0.939, 0.946, 0.95, 0.952 and 0.956, respectively. From 2009 to 2018, the GDE of some of the 41 cities in the YRD fluctuated greatly. For example, the GDE of Chizhou was 0.647 in 2016 and above 0.9 in the other years; the GDE of Wuxi was 0.740 in 2015 and above 0.9 in the other years; the GDE of Xuancheng was all above 0.9 from 2009 to 2015 and dropped to 0.646, 0.668 and 0.691 respectively in 2016–2018; the GDE of Yancheng was all above 0.9 from 2009 to 2014 and dropped to 0.631, 0.792, 0.682 and 0.634 respectively in 2015–2018. The above results showed that the changing trend of GDE is unstable, presenting big fluctuations in this area. In consequence, more measures should be taken to improve GDE steadily.

Table 3. GDE of 41 cities (2009–2018) in Stage I.

City	2009	2010	2011	2012	2013	2014	2015	2016	2017	2018	Mean	Ranking
Shanghai	0.898	0.934	1.000	1.000	1.000	1.000	1.000	0.968	1.000	1.000	0.980	I
Nanjing	1.000	1.000	1.000	1.000	1.000	1.000	0.956	1.000	1.000	1.000	0.996	I
Wuxi	0.973	0.926	1.000	0.968	0.953	0.903	0.740	1.000	1.000	1.000	0.946	II
Xuzhou	0.783	0.721	0.615	1.000	0.622	0.741	0.666	1.000	0.824	0.965	0.794	III
Changzhou	1.000	0.956	0.889	0.938	0.925	0.923	0.695	1.000	1.000	1.000	0.933	II
Suzhou [1]	0.838	0.819	0.766	0.759	0.762	0.808	0.860	1.000	1.000	1.000	0.861	III
Nantong	1.000	0.951	0.841	1.000	0.785	0.778	0.739	0.935	0.902	0.943	0.887	II
Lian Yungang	0.839	0.766	0.681	0.636	0.654	0.580	0.549	1.000	0.660	0.721	0.709	IV
Huaian	0.694	0.632	0.587	0.581	0.614	0.636	0.547	0.736	0.689	0.832	0.655	IV
Yancheng	0.970	0.952	0.984	0.987	1.000	1.000	0.631	0.792	0.682	0.634	0.863	III
Yangzhou	0.966	0.958	0.888	0.884	0.867	0.853	0.585	0.801	0.765	0.832	0.840	III
Zhenjiang	0.859	0.857	0.854	0.806	0.804	0.900	0.664	0.920	0.833	0.830	0.833	III
Taizhou [1]	0.846	0.819	0.889	0.811	0.783	0.868	0.601	0.764	0.730	0.764	0.788	IV
Suqian	0.547	0.498	0.546	0.522	0.509	0.782	0.505	0.570	0.559	0.561	0.560	IV
Hangzhou	0.816	0.788	0.842	0.902	0.951	0.953	0.865	0.934	1.000	0.877	0.893	II
Ningbo	0.768	0.768	0.803	0.813	0.852	0.858	0.852	0.869	0.904	0.886	0.837	III
Wenzhou	1.000	1.000	1.000	1.000	1.000	1.000	1.000	1.000	1.000	1.000	1.000	I
Jiaxing	0.929	0.888	0.831	0.844	0.839	0.820	0.981	1.000	0.781	0.767	0.868	III
Huzhou	0.886	0.986	0.972	1.000	0.995	0.871	0.918	1.000	0.925	0.963	0.952	II
Shaoxing	0.954	0.976	0.988	0.947	0.840	0.789	0.714	1.000	0.827	0.773	0.881	II
Jinhua	1.000	1.000	1.000	1.000	1.000	1.000	1.000	1.000	1.000	1.000	1.000	I
Quzhou	1.000	0.956	1.000	1.000	1.000	1.000	1.000	1.000	1.000	1.000	0.996	I
Zhoushan	0.929	1.000	0.953	0.918	0.878	0.781	0.664	0.728	0.925	1.000	0.878	III
Taizhou [2]	0.954	0.961	1.000	1.000	1.000	1.000	1.000	1.000	1.000	1.000	0.992	I
Lishui	0.929	1.000	1.000	1.000	1.000	1.000	1.000	1.000	0.993	1.000	0.992	I
Hefei	0.948	0.837	0.750	0.650	0.690	0.688	0.708	0.666	1.000	0.979	0.792	IV
Wuhu	0.766	0.665	0.563	0.526	0.509	0.470	0.519	0.529	0.622	0.760	0.593	IV
Bengbu	0.985	0.881	0.745	0.756	0.710	0.711	0.786	0.597	0.919	0.880	0.797	III
Huainan	1.000	0.954	0.761	0.852	0.763	0.929	0.714	1.000	0.707	0.766	0.845	III
Maanshan	0.890	0.778	0.644	0.629	0.587	0.736	0.717	0.953	1.000	0.767	0.770	IV
Huaibei	0.939	1.000	1.000	1.000	1.000	1.000	1.000	1.000	1.000	1.000	0.994	I
Tongling	1.000	1.000	1.000	0.916	0.896	0.912	0.791	0.749	0.821	1.000	0.909	II
Anqing	0.625	0.652	0.650	0.699	0.706	0.706	0.865	0.674	0.840	0.805	0.722	IV
Huangshan	1.000	1.000	1.000	1.000	1.000	1.000	1.000	0.846	1.000	1.000	0.985	I
Chuzhou	0.744	0.722	0.831	0.696	0.741	0.740	0.800	0.710	0.573	0.508	0.707	IV
Fuyang	1.000	1.000	1.000	1.000	1.000	1.000	1.000	0.856	0.926	0.720	0.950	II
Suzhou [2]	0.817	0.834	0.686	0.678	0.737	0.713	0.621	0.736	0.634	0.883	0.734	IV
Luan	0.767	0.746	0.691	0.789	0.726	1.000	0.725	0.558	0.750	0.759	0.751	IV
Bozhou	1.000	1.000	1.000	1.000	1.000	1.000	0.992	0.718	0.847	0.834	0.939	II
Chizhou	1.000	1.000	1.000	1.000	1.000	0.995	0.918	0.647	1.000	1.000	0.956	I
Xuancheng	1.000	1.000	1.000	1.000	1.000	0.909	1.000	0.646	0.668	0.691	0.891	II
YRD	0.893	0.877	0.860	0.866	0.846	0.862	0.802	0.851	0.861	0.871	0.859	

Source: Authors' work. I II III IV represents the ranking 1~10, 11~20, 21~30, 31~41, respectively. Suzhou [1] and Taizhou [1] belong to Jiangsu Province. Suzhou [2] belongs to Anhui province, and Taizhou [2] belongs to Zhejiang provinces.

In this study, 41 cities were divided into Shanghai Municipality, Jiangsu, Zhejiang and Anhui provinces by provincial level, and the GDE time series in the YRD was further analyzed, as shown in Figure 3. The GDE for the whole region was between 0.802 and 0.893 during the study period, with a zero cut-off point in 2015. Distinctively, the GDE of Shanghai Municipality rose from 2009 to 2011 and remained at the forefront almost throughout the subsequent period, the GDE of Zhejiang Province ranked second overall, and the GDE showed a trend of declining first and then rising in both Jiangsu and Anhui provinces. The foregoing results demonstrated that the GDE needs to be promoted constantly in the YRD and stabilized in Jiangsu and Anhui provinces.

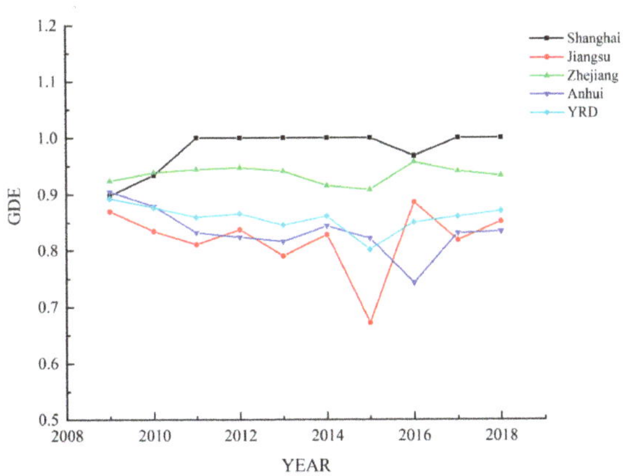

Figure 3. Time series of GDE in the YRD in Stage I.

3.2. Stage II: SFA Model

In this part, an SFA regression model was built, of which the slack value of each input variable served as the explained variable and GRP, ABD, TIP and GCA were taken as explanatory variables and an SFA regression model was built with the software Frontier 4.1 to estimate the impact of environmental variables [36]. The SFA regression results are listed in Table 4.

Table 4. The results of SFA regression.

Explanatory Variable	Slacks of Input Variables			
	AEC	IFA	EEST	EMWCE
Constant term	−177,242.07 −140,141.30	−2,391,731.60 −1,891,088.14	−49,187.69 −751.78	−816.14 −71.51
GRP	−6226.56 (−511.36) ***	−2,017,080.80 (−165,654.65) ***	−30,534.40 (−319.20) ***	385.84 (1.44) *
ABD	363,134.99 (23,017.63) ***	18,522,799.00 (1,174,083.64) ***	499,887.82 (40,106.30) ***	7736.39 (17.37) ***
TIP	−55,091.87 (−5258.59) ***	−1,120,127.10 (−106,917.67) ***	−38,150.40 (−234.72) ***	−531.48 (−11.99) ***
GCA	−328,483.35 (−21,419.04) ***	−15,710,684.00 (−1,024,428.59) ***	−455,881.82 (−19,887.36) ***	−7669.25 (−111.83) ***
γ	1.00	1.00	0.98	1.00
Log likelihood function	−561.07559	−675.04933	−529.03493	−357.3225
LR test	35.5	34.6	37.3	28.7

Notes: *, and *** indicate the significance level at 10%, 5% and 1%, respectively.

According to Table 4, the four models were subject to the LR test and the value of γ was 1 or close to 1, indicating that in the mixed error term, the management inefficiency has a much greater impact on the input slack than the stochastic error term. In the case of a negative regression coefficient, the increase of the explanatory variable reduced the slack of the input variable, narrowing the gap between the actual and ideal value of the input variable. Hence, the increase of the explanatory variable was conducive to the enhancement of GDE. On the contrary, when the regression coefficient was greater than 0, the increase of the explanatory variables was adverse to the improvement of GDE. As shown in Table 4, GRP had a significant negative relationship with AEC, IFA and EEST and a significant positive relationship with EMWCE, ABD significantly had a positive impact on the slack variables of the four inputs, and both TIP and GCA had significant negative impacts on the slack variables of the four inputs (below 1%), which are discussed in the next part.

In accordance with Formulas (4)–(6), u_i the management inefficiency term was separated and calculated next, so that the measuring unit was adjusted to the same external environment and stochastic factor state, thereby adjusting the original data to obtain new input variables. The calculation process was complicated, and it was omitted herein due to the limited space.

3.3. Stage III: Actual GDE in the YRD

Table 5 lists the actual GDE in the YRD based on the adjusted input value (2009~2018). As shown in the table, the actual mean GDE was 1 in Nanjing, Wenzhou and Jinhua cities, notably superior to that in other cities. The actual mean GDE of Huai'an, Wuhu and Suqian was dramatically lower than that of other regions, which was 0.754, 0.734 and 0.705, respectively, showing poor performance. Besides, the actual mean GDE was higher than 0.98 in Huaibei, Shanghai, Tongling and Taizhou[2], being 0.999, 0.990, 0.990 and 0.982, respectively, which was better than that in other cities. The overall mean of the actual GDE of the 41 cities in the YRD was 0.908. In addition to the aforementioned cities, the actual mean GDE was lower than the overall mean (0.908) in Taizhou[1], Lianyungang, Suzhou[2], Xuzhou, Anqing, Zhenjiang, Maanshan, Zhoushan, Hefei, Jiaxing, Ningbo, Yangzhou, Yancheng, Lishui and Shaoxing, which was 0.828, 0.832, 0.850, 0.853, 0.858, 0.866, 0.871, 0.872, 0.875, 0.885, 0.887, 0.888, 0.901, 0.903 and 0.907, respectively. Meanwhile, the actual mean GDE was higher than the overall mean (0.908) in Lu'an, Chuzhou, Hangzhou, Suzhou[1], Xuancheng, Nantong, Chizhou, Quzhou, Huangshan, Huzhou, Bengbu, Huainan, Bozhou, Changzhou, Wuxi and Fuyang, which was 0.908, 0.909, 0.915, 0.918, 0.921, 0.926, 0.937, 0.938, 0.941, 0.941, 0.946, 0.953, 0.955, 0.959, 0.965 and 0.977, respectively. From the overall trend, the actual GDE of many cities dropped dramatically in 2016. For example, the actual GDE of Chizhou was 0.378 in 2016 and above 0.995 in the other years. The actual GDE of Xuancheng was 0.559 in 2016 and above 0.8 in the other years. The actual GDE of Bengbu was 0.581 in 2016 and above 0.9 in the other years. The actual GDE trend of most other cities was relatively stable. For example, the actual GDE of Huzhou exceeded 0.9 in the ten years from 2009 to 2018.

Table 5. GDE of 41 cities (2009–2018) in Stage III.

City	2009	2010	2011	2012	2013	2014	2015	2016	2017	2018	Mean	Ranking
Shanghai	0.926	0.969	1.000	1.000	1.000	1.000	1.000	1.000	1.000	1.000	0.990	I
Nanjing	1.000	1.000	1.000	1.000	1.000	1.000	1.000	1.000	1.000	1.000	1.000	I
Wuxi	0.959	0.914	0.992	0.938	0.991	0.968	0.887	1.000	1.000	1.000	0.965	I
Xuzhou	0.817	0.755	0.664	1.000	0.724	0.867	0.832	1.000	0.891	0.983	0.853	IV
Changzhou	1.000	0.923	0.914	0.935	0.994	1.000	0.820	1.000	1.000	1.000	0.959	I
Suzhou [1]	0.853	0.833	0.849	0.772	0.877	1.000	1.000	1.000	1.000	1.000	0.918	II
Nantong	1.000	0.961	0.809	1.000	0.844	0.843	0.915	0.997	0.942	0.951	0.926	II
Lian Yungang	0.915	0.824	0.848	0.736	0.793	0.839	0.705	1.000	0.864	0.853	0.838	IV
Huaian	0.767	0.681	0.704	0.681	0.749	0.828	0.715	0.677	0.842	0.896	0.754	IV
Yancheng	1.000	1.000	0.950	1.000	1.000	1.000	0.774	0.864	0.741	0.682	0.901	III
Yangzhou	0.997	0.984	0.877	0.878	0.950	0.954	0.743	0.770	0.867	0.862	0.888	III
Zhenjiang	0.928	0.903	0.864	0.808	0.868	0.908	0.756	0.833	0.912	0.875	0.866	IV
Taizhou [1]	0.881	0.840	0.851	0.816	0.873	0.906	0.766	0.696	0.818	0.837	0.828	IV
Suqian	0.616	0.552	0.757	0.638	0.683	0.978	0.765	0.562	0.775	0.728	0.705	IV
Hangzhou	0.837	0.813	0.855	0.931	1.000	1.000	0.938	0.905	1.000	0.869	0.915	III
Ningbo	0.783	0.801	0.856	0.880	0.951	0.787	0.928	0.913	1.000	0.975	0.887	III
Wenzhou	1.000	1.000	1.000	1.000	1.000	1.000	1.000	1.000	1.000	1.000	1.000	I
Jiaxing	0.932	0.913	0.787	0.805	0.861	0.845	1.000	1.000	0.883	0.825	0.885	III
Huzhou	0.913	0.990	0.927	0.946	0.989	0.931	0.990	0.909	0.902	0.913	0.941	II
Shaoxing	0.949	0.999	0.883	0.943	0.909	0.844	0.810	1.000	0.910	0.826	0.907	III
Jinhua	1.000	1.000	1.000	1.000	1.000	1.000	1.000	1.000	1.000	1.000	1.000	I

Table 5. *Cont.*

City	2009	2010	2011	2012	2013	2014	2015	2016	2017	2018	Mean	Ranking
Quzhou	0.986	0.947	0.905	0.960	0.961	0.975	0.912	1.000	0.846	0.883	0.938	II
Zhoushan	0.843	0.942	0.961	0.898	0.910	0.929	0.799	0.490	0.952	1.000	0.872	IV
Taizhou [2]	0.940	0.966	0.983	0.940	1.000	1.000	1.000	0.986	1.000	1.000	0.982	I
Lishui	0.846	1.000	0.922	0.909	1.000	0.968	1.000	0.668	0.865	0.851	0.903	III
Hefei	0.987	0.879	0.837	0.741	0.777	0.831	0.898	0.804	1.000	1.000	0.875	III
Wuhu	0.849	0.755	0.798	0.626	0.717	0.890	0.668	0.588	0.664	0.787	0.734	IV
Bengbu	1.000	0.938	0.967	1.000	0.980	1.000	0.991	0.581	1.000	1.000	0.946	II
Huainan	1.000	0.925	0.937	1.000	0.930	0.938	0.897	1.000	0.926	0.974	0.953	II
Maanshan	1.000	0.787	0.818	0.723	0.811	0.878	0.793	0.961	1.000	0.937	0.871	IV
Huaibei	1.000	1.000	0.992	1.000	1.000	1.000	1.000	1.000	1.000	1.000	0.999	I
Tongling	1.000	1.000	1.000	0.915	1.000	1.000	1.000	1.000	0.981	1.000	0.990	I
Anqing	0.674	0.723	0.840	0.863	0.889	0.964	1.000	0.655	0.978	0.995	0.858	IV
Huangshan	1.000	1.000	1.000	1.000	1.000	1.000	1.000	0.422	0.995	0.993	0.941	II
Chuzhou	0.984	0.917	0.867	0.998	0.990	0.974	0.988	0.784	0.822	0.769	0.909	III
Fuyang	1.000	1.000	1.000	1.000	1.000	1.000	1.000	0.882	0.978	0.905	0.977	I
Suzhou [2]	0.906	0.918	0.843	0.870	0.892	0.936	0.875	0.654	0.791	0.812	0.850	IV
Luan	0.901	0.824	0.855	0.977	0.921	1.000	0.999	0.600	1.000	1.000	0.908	III
Bozhou	1.000	1.000	1.000	1.000	1.000	1.000	0.965	0.580	1.000	1.000	0.955	II
Chizhou	1.000	1.000	1.000	1.000	1.000	0.995	1.000	0.378	1.000	1.000	0.937	II
Xuancheng	1.000	1.000	1.000	1.000	1.000	1.000	1.000	0.559	0.849	0.801	0.921	II
YRD	0.924	0.906	0.900	0.906	0.923	0.946	0.906	0.822	0.927	0.922	0.908	

Source: Authors' work. I, II, III and IV represent the ranking of 1~10, 11~20, 21~30 and 31~41, respectively. Suzhou [1] and Taizhou [1] belong to Jiangsu Province. Suzhou [2] belongs to Anhui province, and Taizhou [2] belongs to Zhejiang provinces.

Figure 4 illustrates the time series of the actual GDE in the YRD. According to Figure 4, the GDE in the YRD generally tends stable, with the highest level of 0.946 in 2014 and the lowest level of 0.822 in 2016, presenting a downward trend from 2014 to 2016. The actual GDE had been kept at 1 in Shanghai Municipality since it rose from 0.926 in 2009 to 1 in 2011, which fluctuated continuously in Jiangsu, Zhejiang and Anhui provinces from 2009 to 2018. Among them, the actual GDE of Jiangsu and Anhui provinces dropped sharply in 2015 and 2016, respectively. On the whole, the fluctuation range of GDE in Zhejiang Province was relatively small, and the GDE of Jiangsu Province was at the lowest level during 2009–2018. The above results indicated that although the influence of external environmental factors is removed, there are still regional differences in the actual GDE of cities in the YRD.

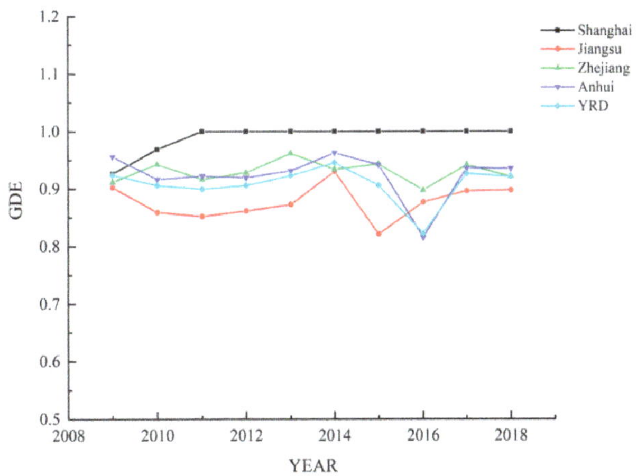

Figure 4. Time series of GDE in the YRD in Stage III.

4. Discussion

4.1. GDE Analysis in the YRD

As shown in Figure 4, the actual mean GDE of the YRD was 0.908 (2009~2018). After adjustment, the GDE of the YRD in Sage III was obviously higher than that in Stage I, which testified that objective environmental factors cause people to underestimate GDE. Comparing the initial GDE (Table 3) and the actual GDE (Table 5) in the YRD, except for Chizhou, Huangshan, Huzhou, Quzhou, Taizhou[2] and Zhoushan, the GDE was enhanced upon the removal of external environmental factors and the mixed error term. From 2009 to 2018, the mean GDE of 41 cities in the YRD was 0.859 when external factors were taken into account, which was increased to 0.908 when external environmental factors were excluded. In stage III, the GDE of Nanjing, Jinhua and Wenzhou cities reached the optimal level, while only Jinhua and Wenzhou cities maintained this efficiency level in Stage I, which showed that external environmental factors negatively affected the GDE of Nanjing City. Consequently, there is substantial potential to improve the external environment.

Since it was a national strategy of China to integrate the YRD, the development of the YRD has been constantly concerned by all walks of life. In this case, the ecological environment is also one of the inevitable problems in the development process, and the main causes of excessive resource consumption and environmental pollution can be explored by virtue of effective environmental efficiency measurement, so as to improve environmental governance policies. There are many GDE calculation methods, of which the use of the three-stage DEA model enables obtains more objective and accurate efficiency by separating environmental variables such as management inefficiency and statistical noise. Before the environmental interference factors were excluded, that is, in Stage I of this study, the obtained GDE in the YRD showed a trend of falling first and then rising, and the zero cut-off point appeared in 2015. Wang et al. (2019) measured the GDE in the YRD from 2005 to 2015 using the Super-SBM model and concluded that there was a downward trend, thus predicting that the efficiency would increase after 2015 [37], which was verified in this study. Nevertheless, when the environmental factors and stochastic disturbance were removed, that is, when the actual GDE was obtained, no matter from the perspective of the entire region or the four provinces, the GDE in the entire time series was improved to a certain extent compared with that in Stage I, indicating that objective factors may cause people to underestimate the GDE. Consistent with the results herein, Guo et al. (2018) also came to the conclusion that the mean environmental efficiency of the central, eastern and western regions, as well as the whole country, was underestimated during the three-stage measurement of environmental efficiency in China [38].

For further analysis, the GDE calculated in the previous parts was divided into five levels, and the GDE spatial distribution maps in the YRD in 2009, 2012, 2015 and 2018 were drawn using ArcGIS, as shown in Figure 5. Thereout, the differences in spatial distribution, as well as the spatial characteristics, were determined intuitively, and it was visibly that spatial heterogeneity existed in the GDE of the YRD and changed with time. For example, the GDE in the northeast coastal region changed from a high level at the beginning to a low level in the later period, while the GDE in the western region experienced the opposite process. In addition, the GDE was maintained at a high level in some cities in the northwest, southeast and middle of China, as well as a small area formed around Shanghai Municipality, presenting certain clustering characteristics, and the GDE there was better than that in other regions.

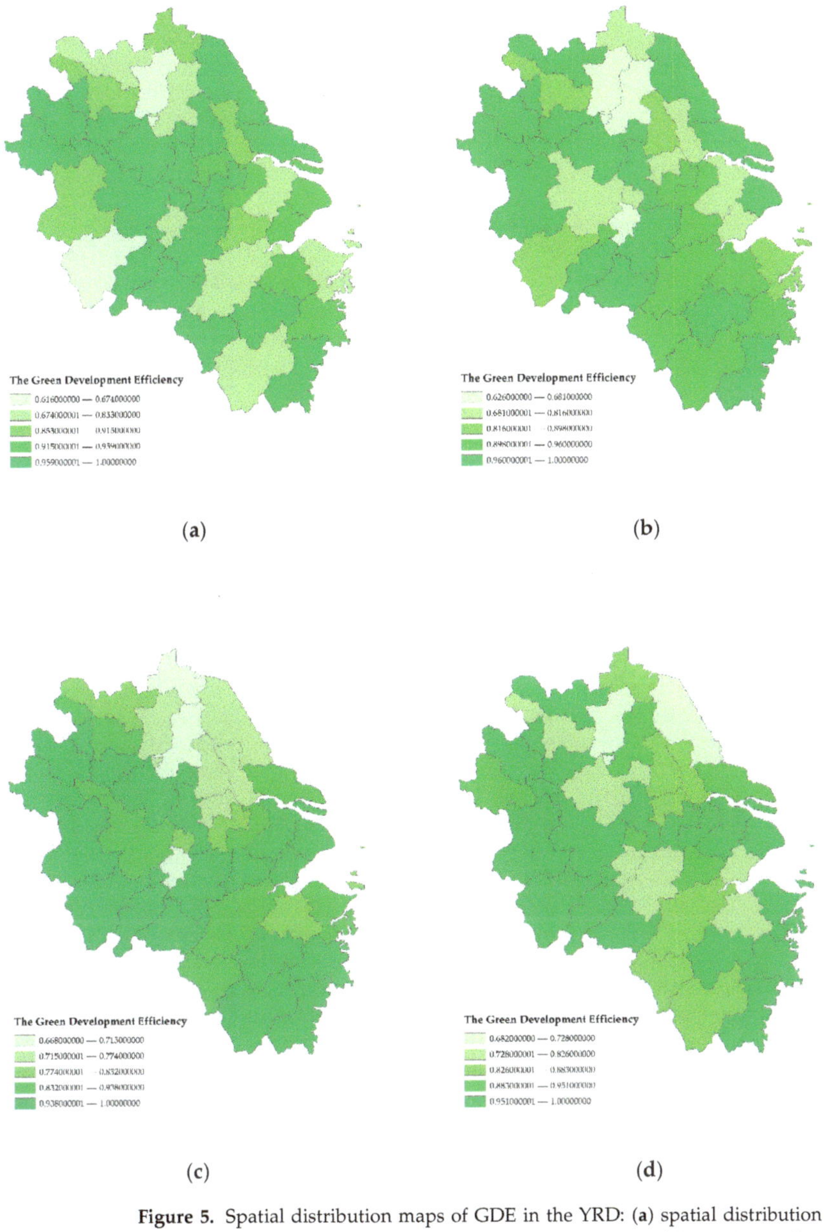

Figure 5. Spatial distribution maps of GDE in the YRD: (**a**) spatial distribution of GDE in 2009; (**b**) spatial distribution of GDE in 2012; (**c**) spatial distribution of GDE in 2015; (**d**) spatial distribution of GDE in 2018.

Based on the spatial distribution maps drawn from the actual GDE results, the GDE in the YRD showed certain spatial heterogeneity, not only different from the conclusion of Wang et al. (2018) who calculated that the GDE is low in the east and high in the west of the YRD [37], but also different from the conclusion of Deng et al. (2021) who found that the GDE in the eastern region is significantly higher than that in the western region [39]. In this study, it was discovered that the GDE showed high clustering characteristics to some extent in several cities in the northwest, southeast, and middle of China, as well as a small area around Shanghai Municipality. Combined with the efficiency decomposition diagram in Figure 5, the GDE tends to have high-high clustering in cities with good economic development levels and low-low clustering in cities with relatively low economic development levels. In this sense, environmental protection was better implemented in the eastern region with better economic development than in the western region. Feng et al. (2020) believed that economic development is correlated with green development, but there is not a complete positive correlation [33]. As shown in this paper, after excluding the objective factor of economic development, the GDE was high in some economically underdeveloped regions. For example, surrounded by mountains and rivers, some cities in central China such as Xuancheng, Huangshan, and Quzhou are famous for tourism and mainly develop tourism and service industries, and there are few industries characterized by high pollution, presenting high green development levels. Topography affects industries, thereby affecting local green development. Accordingly, it is necessary to comprehensively consider the factors affecting GDE from various aspects [40].

4.2. SFA Regression Analysis

In Stage II, SFA regression was performed on the input slack variables obtained in Stage I and four environmental variables, and some meaningful information was obtained.

(1) GRP is negatively correlated with the slacks of electricity consumption, fixed asset investment and scientific education investment in the whole society, while it has a positive correlation with the slacks of water conservancy and environment practitioners. It indicates that the increase in GRP makes the electricity consumption, fixed asset investment and scientific education investment rationalized on the one hand, and on the other hand, it shows inefficiency in the input of water conservancy and environmental practitioners. GRP represents the local economic development level, and the classic environmental Kuznets curve shows that the quality of the environment will first decline and then rise with the development of the economy [41]. From the perspective of input, it is the rationality and waste of these different input factors that explain the complexity of the mechanism of the relationship between economic development and green development.

(2) ABD has a significant positive correlation with the slacks of the four input variables, proving that the increase in the urban construction area will increase the input slacks, which goes against the GDE. With the expansion of urban space, among the land cover types in the urban fringe area, land types with less interference from human activities such as cultivated land, forest land and orchards have been greatly reduced and replaced by high-density urban land. The impervious area in the urban center area has been increased, and the natural green area has been reduced, replaced by squares and roads covered with cement and asphalt. The increase of impervious area and the reduction of the green area have seriously caused problems for the water environment and atmospheric environment in cities [22,42]. On the contrary, in terms of geographic space, cities surrounded by mountainous terrain and famous for tourism enjoy high GDE as large-scale construction may not be applicable, such as Xuancheng, Huangshan, Quzhou and Jinhua, which is consistent to the conclusion of Li et al. (2022) [40].

(3) TIP has a negative impact on the slacks of the four input variables, that is, the higher the proportion of the tertiary industry, the more beneficial to input slack reduction and GDE improvement, which fully reveals that the optimization of industrial structure is conducive to local green development. According to the *13th Five-Year Plan for Economic and Social Development of the People's Republic of China (2016–2020)*, green development can be achieved through industrial restructuring. Many studies also support that the high proportion of the tertiary industry is conducive to the protection of the ecological environment [27,43]. Guo et al. (2020) concluded that the secondary industry is adverse to green development, which in turn supports the aforesaid statement [44]. The rise of the tertiary industry, on the one hand, compresses the secondary industry supported by a large amount of fixed capital investment, which is conducive to reducing the waste of capital input; on the other

hand, it has a strong ability to absorb labor employment and reduces the slack degree of labor equivalent. In this sense, industrial structure optimization is the appropriate path for green development.

(4) GCA is negatively correlated with the slacks of the four input variables. The increase in the urban green area supports the social fixed asset input, social electricity consumption, scientific education investment and water conservancy and environmental management personnel to reach the ideal input value, suggesting that the increase in urban green area is helpful to the rational use of energy, capital, labor and innovation input. Similar to the formulation of environmental policies and the investment in pollution control, the planning of urban green areas demonstrates the active intervention of the government in green development planning and plays an essential role in urban sustainable development [45]. Today, governments are gradually conscious that the construction of green spaces has become a vital issue for high-quality economic development [22]. Many policies have been issued by governments at all levels, such as the *Regulations of China on Urban Greening, Shanghai Greening Regulations*, specifying that urban greening should keep pace with urban development. In line with the study results of this paper, in the process of urban expansion, reasonable planning and investment should be carried out on the coverage of the urban green area, in a bid to prevent the expansion of urban construction area from reducing the GDE and causing a series of environmental problems.

4.3. GDE Decomposition Analysis in the YRD

To better understand the GDE in the YRD, the software DEAP 2.1 was applied to divide the actual GDE into two types, that is, the comprehensive technical efficiency (TE) was decomposed into pure technical efficiency (PTE) and scale efficiency (SE) [31]. PTE reflected the production efficiency of DMUs at certain input factors at an exact scale (usually optimal) and explained how to effectively apply green technologies to achieve maximum efficiency, and SE presented the realization degree of scale effects on green development [46].

The efficiency decomposition scatters diagram of 41 cities was drawn according to PTE and SE, which was divided into four quadrants according to the mean value to represent four categories of high-high, low-high, low-low and high-low, respectively (Figure 6). It can be seen that after the environmental factors and stochastic disturbance were eliminated, the mean PTE rose from 0.883 to 0.966, while the mean SE slightly dropped from 0.974 to 0.939. On the whole, environmental factors affected the real performance of PTE, causing GDE to be underestimated.

The high-high category contained areas with high PTE and high SE. According to the efficiency decomposition in Stage I, 13 cities were included in this category, and Shanghai, Changzhou, Tongling, Huainan, and Bozhou cities became new members of this category in Stage III, implying that the SE level of these cities was improved dramatically with consideration of the environmental disturbance factors. Among them, both PTE and SE of Nanjing, Jinhua and Wenzhou cities were 1, indicating that these cities achieved a high-efficiency level.

The high-low category contained cities with high SE and low PTE. According to the calculation results of Stage I, 10 cities were classified into this category, and the number was reduced to 8 in Stage III. Among them, Huangshan, Nantong, Huzhou and Shaoxing cities fell from the original high-high category to the high-low category, reflecting that the utilization efficiency there in the exogenous environment should be strengthened. It is necessary to invest more in green technology innovation and application, enhance the quality of green talents and raise energy utilization to improve the overall GDE locally.

The low-high category contained areas with low SE and high PTE. Nantong, Huzhou and Shaoxing cities, which were originally included in the high-high category in Stage I, were assigned to this category in Stage III, indicating that the SE of the three cities was overestimated. For these cities, the SE should be enhanced by increasing the green investment, conducting green transformation and strengthening green talent aggregation.

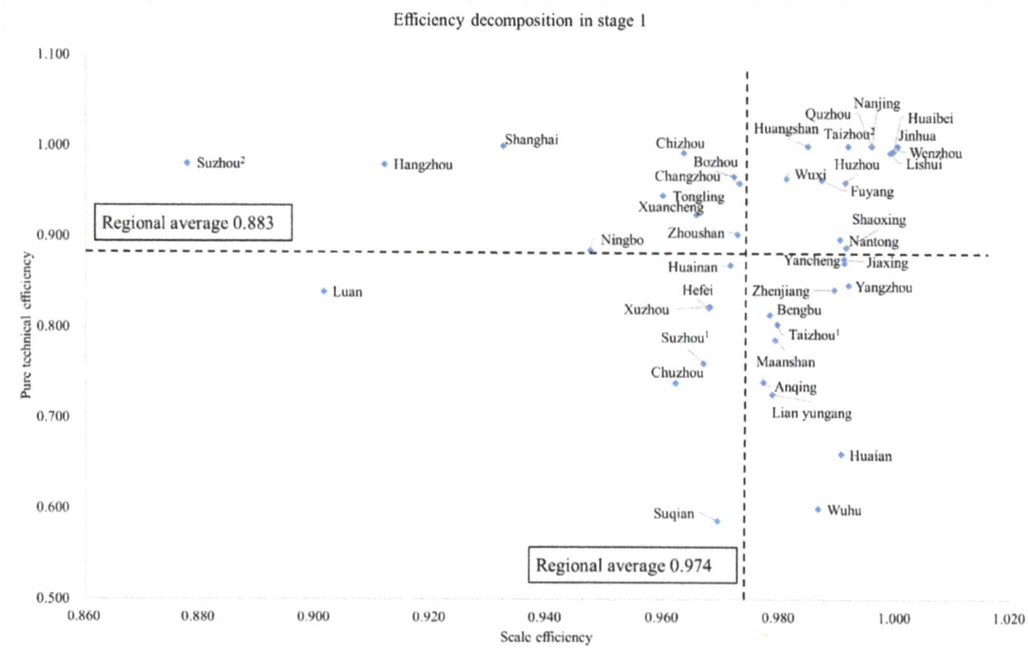

(**a**)

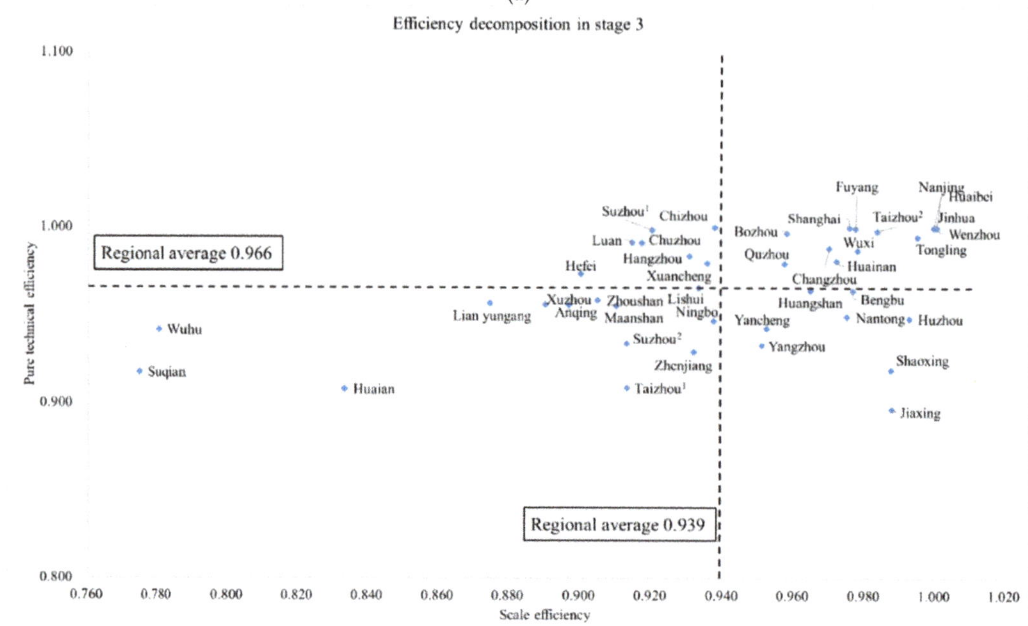

(**b**)

Figure 6. GDE decomposition of 41 cities: (**a**) Stage I (**b**) Stage III.

The low-low category contained areas with low PTE and low SE. Comparing Stage I with Stage III, the number of cities in this category was increased from 7 to 12, and all cities included in this category presented poor GDE. Xuzhou, Huai'an and Suqian cities were contained in the low-low category no matter whether the influence of external environmental factors and statistical noise was excluded. Most cities in this category were weak in economic foundations and dominated by traditional energy-intensive industries. In consequence, for cities in this category, it is not only necessary to advance technological innovation during green transformation, but also to consider the stimulation of scale effect. In the meantime, the priority should be given to the construction of a green economy and a sound green development foundation.

5. Conclusions

At present, green development is a considerable environmental management issue in China, aiming to improve the status of regional environmental development in light of energy saving, emission reduction and pollutant control [13]. Since the integration of the YRD was brought into the national strategy, the development of the region has received continuous attention from all sectors of society. As a result, the ecological environment is one of the inevitable issues in the development process. In this study, panel data from 2009 to 2018 of 41 cities in the YRD were selected and the three-stage DEA model was applied to calculate the objective GDE in this region. Besides, a comprehensive analysis was performed on the grounds of the empirical results. After the adjustment, the GDE in each city changed considerably, which proved that it is objective and accurate to measure GDE after eliminating environmental factors and stochastic disturbance. On this basis, the following suggestions and implications were drawn:

(1) The GDE in the YRD adjusted in Stage III was clearly higher than that in Stage I, mainly because the GDE was underestimated under the influence of objective environmental variables. The GDE levels of different cities showed heterogeneity upon the removal of external environmental factors and stochastic disturbances. The GDE developed out of balance in the four provincial administrative regions and generally behaved better in the coastal and southeastern areas than that in the central, western and northern regions in terms of spatial distribution. As a national central city, Shanghai Municipality serves as the center of the international economy, finance, trade, shipping and technological innovation in China, which is required to not only maintain high-quality development as a leader in the YRD but also to focus on the balanced development of cities in the YRD as a whole. For other regions, it is necessary to control the industrial scale, actively use foreign capital to improve production technologies, achieve clean production and reduce energy consumption.

(2) In terms of external environmental variables, the ABD reflecting urban construction has a negative impact on GDE since urban construction requires the improvement of urban governance infrastructure, which will inevitably lead to an increase in investment in pollution control. Consequently, the faster the urban construction process is, the more capital, labor, energy and resources will be required, which partly generates redundant inputs, thus reducing GDE. Moreover, industrial structure adjustment and green covered area are conducive to GDE, so it is necessary to sequentially strengthen the development of the tertiary industry, reduce the idle employees caused by labor aggregation, and improve the regional economic level while improving the capital utilization efficiency. Besides, the government should increase green investment and carry out rational layouts of urban green spaces to prevent the reduction of the green development level in the process of urban expansion and construction.

(3) The GRP reflects the local economic development level, the impact of which on GDE was not determined in this paper. In spite of this, it is believed that in the new era emphasizing high-quality development, more emphasis should be put on innovation and ecology, which are beneficial to the healthy and sustainable development of cities. In the future, the government still needs to play an active role in pollution control and urban green planning. While accelerating the process of urbanization, it is necessary to promote clean production, control pollution emissions, eliminate passive terminal control, pay attention

to the excessive consumption of resources and energy in urban construction, keep abreast of the speed of urbanization, adhere to quality-oriented policies, and create a new spatial pattern of intensive and efficient urbanization. Additionally, it is necessary to cultivate new growth points for cities, give play to regional advantages pursuant to different orientations, and realize the coordinated development of urban agglomerations, cities and industries.

Suggestions for future study: First and foremost, when measuring the GDE of 41 cities in YRD from 2009 to 2018, the time lag effect and delayed utility between inputs and outputs have been neglected to some extent. The digestion and absorption of inputs often take time to produce effective outputs, which means that green development inputs will not be converted into relevant outputs in an instant, and further verification is required. Secondly, although the influence mechanism of four objective environmental variables has been involved in this paper, there are still some unconsidered factors, such as urban resources, culture, society, etc. [22,23], and more attention can be paid to the correlations between the factors and the green ecology of cities in future studies. Last but not least, though complex, in-depth research is required to reveal the impact of economic development on urban green development.

Author Contributions: Conceptualization, Z.S. and H.Z.; Data curation, Q.Y.; Formal analysis, Q.Y.; Methodology, Q.Y.; Project administration, Z.S.; Resources, Q.Y.; Software, Q.Y.; Supervision, Z.S. and H.Z.; Writing — original draft, Q.Y.; Writing — review & editing, Z.S. and H.Z. All authors have read and agreed to the published version of the manuscript.

Funding: This research received no external funding.

Data Availability Statement: All data are from Chinese official institutions, please contact the corresponding author if necessary.

Conflicts of Interest: The authors declare no conflict of interest.

References

1. Ruggieri, R.; Ruggeri, M.; Vinci, G. Efficient energy and electric transport in a Smart City: Evaluation of sustainability and competitivness. In Proceedings of the 2020 IEEE International Conference on Environment and Electrical Engineering and 2020 IEEE Industrial and Commercial Power Systems Europe (EEEIC/I&CPS Europe), Madrid, Spain, 9–12 June 2020; pp. 1–4.
2. Sait, M.A.; Chigbu, U.E.; Hamiduddin, I.; De Vries, W.T. Renewable energy as an underutilised resource in cities: Germany's 'Energiewende'and lessons for post-brexit cities in the United Kingdom. *Resources* **2019**, *8*, 7. [CrossRef]
3. Kumar, R.R.; Stauvermann, P.J.; Patel, A. Exploring the link between research and economic growth: An empirical study of China and USA. *Qual. Quant.* **2016**, *50*, 1073–1091. [CrossRef]
4. Sun, C.; Tong, Y.; Zou, W. The evolution and a temporal-spatial difference analysis of green development in China. *Sustain. Cities Soc.* **2018**, *41*, 52–61. [CrossRef]
5. Zhang, H.; Geng, Z.; Yin, R.; Zhang, W. Regional differences and convergence tendency of green development competitiveness in China. *J. Clean. Prod.* **2020**, *254*, 119922. [CrossRef]
6. Yang, Y.Y.; Guo, H.X.; Chen, L.F.; Liu, X.; Gu, M.Y.; Ke, X.L. Regional analysis of the green development level differences in Chinese mineral resource-based cities. *Resour. Policy* **2019**, *61*, 261–272. [CrossRef]
7. Chen, L.; Jia, G. Environmental efficiency analysis of China's regional industry: A data envelopment analysis (DEA) based approach. *J. Clean. Prod.* **2017**, *142*, 846–853. [CrossRef]
8. Kuosmanen, T.; Kortelainen, M. Stochastic non-smooth envelopment of data: Semi-parametric frontier estimation subject to shape constraints. *J. Product. Anal.* **2012**, *38*, 11–28. [CrossRef]
9. Dyson, R.G.; Shale, E.A. Data envelopment analysis, operational research and uncertainty. *J. Oper. Res. Soc.* **2010**, *61*, 25–34. [CrossRef]
10. Cook, W.D.; Tone, K.; Zhu, J. Data envelopment analysis: Prior to choosing a model. *Omega* **2014**, *44*, 1–4. [CrossRef]
11. Cook, W.D.; Zhu, J. Classifying inputs and outputs in data envelopment analysis. *Eur. J. Oper. Res.* **2007**, *180*, 692–699. [CrossRef]
12. Charnes, A.; Cooper, W.W.; Rhodes, E. Measuring the efficiency of decision making units. *Eur. J. Oper. Res.* **1978**, *2*, 429–444. [CrossRef]
13. Wu, J.; Lu, W.; Li, M. A DEA-based improvement of China's green development from the perspective of resource reallocation. *Sci. Total Environ.* **2020**, *717*, 137106. [CrossRef] [PubMed]
14. Yang, Q.; Wan, X.Z.; Ma, H.M. Assessing Green Development Efficiency of Municipalities and Provinces in China Integrating Models of Super-Efficiency DEA and Malmquist Index. *Sustainability* **2015**, *7*, 4492–4510. [CrossRef]
15. Zhou, X.; Xu, Z.; Chai, J.; Yao, L.; Wang, S.; Lev, B. Efficiency evaluation for banking systems under uncertainty: A multi-period three-stage DEA model. *Omega* **2019**, *85*, 68–82. [CrossRef]

16. Fried, H.O.; Lovell, C.; Schmidt, S.S.; Yaisawarng, S. Accounting for Environmental Effects and Statistical Noise in Data Envelopment Analysis. *J. Product. Anal.* **2002**, *17*, 157–174. [CrossRef]
17. Lee, J.Y. Application of the three-stage DEA in measuring efficiency–an empirical evidence. *Appl. Econ. Lett.* **2008**, *15*, 49–52. [CrossRef]
18. Jia, S.; Wang, C.; Li, Y.; Zhang, F.; Liu, W. The urbanization efficiency in Chengdu City: An estimation based on a three-stage DEA model. *Phys. Chem. Earth Parts A/B/C* **2017**, *101*, 59–69. [CrossRef]
19. Hu, Y.; Wu, Y.; Zhou, W.; Li, T.; Li, L. A three-stage DEA-based efficiency evaluation of social security expenditure in China. *PLoS ONE* **2020**, *15*, e0226046. [CrossRef]
20. Li, H.; He, H.; Shan, J.; Cai, J. Innovation efficiency of semiconductor industry in China: A new framework based on generalized three-stage DEA analysis. *Socio-Econ. Plan. Sci.* **2019**, *66*, 136–148. [CrossRef]
21. Feng, S.; Xu, L.D. Decision support for fuzzy comprehensive evaluation of urban development. *Fuzzy Sets Syst.* **1999**, *105*, 1–12. [CrossRef]
22. Su, Y.; Xue, H.; Liang, H. An evaluation model for urban comprehensive carrying capacity: An empirical case from Harbin city. *Int. J. Environ. Res. Public Health* **2019**, *16*, 367. [CrossRef] [PubMed]
23. Zhang, Y.; Liu, X.; Ding, M. Energy-Consumption-Based Comprehensive Evaluation of Urban Green Development-Cases of 27 Cities in the Integrated Central Area of the Yangtze River Delta, China. *IEEE Access* **2021**, *9*, 79105–79117. [CrossRef]
24. Yang, X.; Shang, G. Smallholders' Agricultural Production Efficiency of Conservation Tillage in Jianghan Plain, China—Based on a Three-Stage DEA Model. *Int. J. Environ. Res. Public Health* **2020**, *17*, 7470. [CrossRef]
25. Zhang, C.; Liu, H.; Bressers, H.; Buchanan, K.S. Productivity growth and environmental regulations-accounting for undesirable outputs: Analysis of China's thirty provincial regions using the Malmquist–Luenberger index. *Ecol. Econ.* **2011**, *70*, 2369–2379. [CrossRef]
26. Xu, J.; Li, A.; Li, D.; Liu, Y.; Du, Y.; Pei, T.; Ma, T.; Zhou, C. Difference of urban development in China from the perspective of passenger transport around Spring Festival. *Appl. Geogr.* **2017**, *87*, 85–96. [CrossRef]
27. Zhu, B.Z.; Zhang, M.F.; Zhou, Y.H.; Wang, P.; Sheng, J.C.; He, K.J.; Wei, Y.M.; Xie, R. Exploring the effect of industrial structure adjustment on interprovincial green development efficiency in China: A novel integrated approach. *Energy Policy* **2019**, *134*, 12. [CrossRef]
28. Van Herzele, A.; Wiedemann, T. A monitoring tool for the provision of accessible and attractive urban green spaces. *Landsc. Urban Plan.* **2003**, *63*, 109–126. [CrossRef]
29. Hailu, A. Non-parametric Productivity Analysis with Undesirable Outputs: An Application to the Canadian Pulp and Paper Industry. *Am. J. Agric. Econ.* **2011**, *85*, 1075–1077. [CrossRef]
30. Dyckhoff, H.; Allen, K. Measuring ecological efficiency with data envelopment analysis (DEA). *Eur. J. Oper. Res.* **2001**, *132*, 312–325. [CrossRef]
31. Banker, R.D.; Charnes, A.; Cooper, W.W. Some models for estimating technical and scale inefficiencies in data envelopment analysis. *Manag. Sci.* **1984**, *30*, 1078–1092. [CrossRef]
32. Jondrow, J.; Lovell, C.K.; Materov, I.S.; Schmidt, P. On the estimation of technical inefficiency in the stochastic frontier production function model. *J. Econom.* **1982**, *19*, 233–238. [CrossRef]
33. Feng, M.; Li, X. Evaluating the efficiency of industrial environmental regulation in China: A three-stage data envelopment analysis approach. *J. Clean. Prod.* **2020**, *242*, 118535. [CrossRef]
34. Xie, B.-C.; Duan, N.; Wang, Y.-S. Environmental efficiency and abatement cost of China's industrial sectors based on a three-stage data envelopment analysis. *J. Clean. Prod.* **2017**, *153*, 626–636. [CrossRef]
35. Liu, X.; Liu, J. Measurement of low carbon economy efficiency with a three-stage data envelopment analysis: A comparison of the largest twenty CO2 emitting countries. *Int. J. Environ. Res. Public Health* **2016**, *13*, 1116. [CrossRef] [PubMed]
36. Coelli, T.J. *A Guide to Frontier Version 4.1: A Computer Program for Stochastic Frontier Production and Cost Function Estimation*; CEPA Working Papers; CEPA: Armidale, Australia, 1996; Volume 7, pp. 1–33.
37. Wang, Z.; Wang, X.; Liang, L. Green economic efficiency in the Yangtze River Delta: Spatiotemporal evolution and influencing factors. *Ecosyst. Health Sustain.* **2019**, *5*, 20–35. [CrossRef]
38. Guo, S.; Tong, M.; Guo, J.; Han, Y. Measurement and influencing factors of inter-provincial real environmental efficiency based on three-stage DEA model. *China Popul. Resour Environ.* **2018**, *28*, 106–116.
39. Deng, Q.; Qin, Y.; He, X. Recognition of the Spatio-temporal Heterogeneity Characteristics and Influencing Factors about Inclusive Green Development in the Yangtze River Economic Belt. *J. Cent. South Univ. For. Technol.* **2021**, *15*, 1–9.
40. Li, D.; Shangguan, Z.; Huang, M.; Zhang, X.; Tang, L. Impacts of urban development on regional green development efficiency—A case of the Yangtze River Delta in China. *Energies* **2022**, *15*, 4785. [CrossRef]
41. Dasgupta, S.; Laplante, B.; Wang, H.; Wheeler, D. Confronting the environmental Kuznets curve. *J. Econ. Perspect.* **2002**, *16*, 147–168. [CrossRef]
42. Musakwa, W.; Van Niekerk, A. Monitoring sustainable urban development using built-up area indicators: A case study of Stellenbosch, South Africa. *Environ. Dev. Sustain.* **2015**, *17*, 547–566. [CrossRef]
43. Chen, L.; Zhang, X.; He, F.; Yuan, R. Regional Green Development Level and Its Spatial Relationship under the Constraints of Haze in China. *J. Clean. Prod.* **2018**, *210*, 376–387. [CrossRef]

44. Guo, Y.H.; Tong, L.J.; Mei, L. The effect of industrial agglomeration on green development efficiency in Northeast China since the revitalization. *J. Clean. Prod.* **2020**, *258*, 13. [CrossRef]
45. Droste, N.; Hansjürgens, B.; Kuikman, P.; Otter, N.; Antikainen, R.; Leskinen, P.; Pitkänen, K.; Saikku, L.; Loiseau, E.; Thomsen, M. Steering innovations towards a green economy: Understanding government intervention. *J. Clean. Prod.* **2016**, *135*, 426–434. [CrossRef]
46. Mirmozaffari, M.; Alinezhad, A. Window analysis using two-stage DEA in heart hospitals. In Proceedings of the 10th International Conference on Innovations in Science, Engineering, Computers and Technology (ISECT-2017), Dubai, United Arab Emirates, 17–19 October 2017; pp. 44–51.

sustainability

Article

The Efficiency of Urban–Rural Integration in the Yangtze River Economic Belt and Its Optimization

Gubu Muga [1,2], Shougeng Hu [1,2], Zhilan Wang [1,2], Luyi Tong [1,2], Zongnan Hu [1,2], Hui Huang [1,2] and Shijin Qu [1,2,*]

1 School of Public Administration, China University of Geosciences, Wuhan 430074, China
2 Key Laboratory for Rule of Law Research, Ministry of Natural Resources, Wuhan 430074, China
* Correspondence: qusj@cug.edu.cn

Abstract: China has entered a new stage of integrated development of urban and rural areas under the constraints of scarce land resources and the need for high-quality economic and social development. While there is concern about the state and speed of urban–rural integrated development (URID), increasing attention is being paid to efficiency improvement. This paper comprehensively measures the efficiency of URID from the input–output perspective, taking into account the impact of carbon emissions; it also studies the efficiency of URID and its developmental spatiotemporal characteristics in 73 cities within three major city clusters in the Yangtze River Economic Belt (YREB) from 2010 to 2019, and analyzes the input–output optimization strategies for URID within each of these major urban systems. The results show that (1) the comprehensive efficiency evaluation system constructed by the study can more objectively reflect the state and trends of URID. From 2010 to 2019, the efficiency of URID in the three major city clusters in the YREB showed a downward trend; in cities with better economic development, the efficiency of URID was lower than in cities with average economic development, where carbon emission indicators showed a significant impact. (2) The spatial distribution of URID efficiency in the three major city clusters in the YREB follows an inverted "U" shape; URID efficiency in the urban agglomeration in the middle reaches of the Yangtze River (MRYRUA) is higher than in the Chengyu urban agglomeration (CYUA), where it is higher than in the Yangtze River Delta urban agglomeration (YRDUA). (3) The input redundancy rates are high in the indicators for culture, sports and media, energy conservation and environmental protection, urban and rural communities, and housing security expenditures. Carbon emission redundancy has a negative impact on efficiency in URID. Based on the high redundancy rates of each input–output indicator, this paper proposes methods to optimize the efficiency of URID in each of the three major city clusters and provides directional guidance for promoting the high-quality development of regional urban–rural integration.

Keywords: integrated urban–rural development; efficiency; spatiotemporal evolution; carbon emissions; urban agglomerations

Citation: Muga, G.; Hu, S.; Wang, Z.; Tong, L.; Hu, Z.; Huang, H.; Qu, S. The Efficiency of Urban–Rural Integration in the Yangtze River Economic Belt and Its Optimization. *Sustainability* **2023**, *15*, 2419. https://doi.org/10.3390/su15032419

Academic Editors: Lu Zhang, Bing Kuang and Bohan Yang

Received: 2 December 2022
Revised: 16 January 2023
Accepted: 19 January 2023
Published: 29 January 2023

1. Introduction

Urban–rural integrated development (URID) is seen as the model for coordinated urban–rural development in China's new era and is no longer equivalent to the one-directional emphasis on industrial development feeding the agricultural sector in "promoting agriculture with industry". URID is committed to the preservation of the characteristics of urban and rural areas, respectively, and the establishment of a new type of urban–rural relationship characterized by comprehensive integrated development to replace the previous rural–urban dichotomy [1,2]. The integrated development of urban and rural areas has always been an important goal for China to achieve urban and rural coprosperity. In the past 70 years, urban–rural relations have changed from the initial division to the current stage of integration and development, and although significant achievements have been made, integration still faces problems, such as unbalanced urban–rural development, inadequate rural development, an inadequate two-way flow of factors, and unreasonable allocation of public resources, which hinder further integration and development [3–5]. Improving the distribution of basic public services and the reasonable allocation of public resources in urban and rural areas is of great practical significance to the success of the strategy for urban–rural integration and development [6]. At the same time, it also puts the governance capacity of administrative departments at all levels to a severe test. For local governments, as the main enforcer of urban–rural integration policy, obtaining the best output efficiency while controlling input costs has become important to effectively promoting urban–rural integration strategies. Therefore, it is necessary to measure the efficiency of the current implementation of integrated urban–rural development in government departments across China and propose measures to improve efficiency.

Data envelopment analysis (DEA) is used to assess the relative validity of decision units in a "multiple input, multiple output" model [7,8]. This method has been applied to multiple fields of research. For example, in government management, scholars have conducted an overall assessment of service efficiency by examining the public services provided by local governments in Portugal and Norway [9,10]; in corporate management, scholars have analyzed the technical efficiency of American Airlines from 1970 to 1990 and the relationship between the stock market and the technical efficiency of the company [11]. In bank management, scholars have constructed a bank efficiency evaluation system to measure the efficiency of Swedish banking services as well as the average efficiency level of the industry, based on the concept of service efficiency [12]; in agricultural production, researchers measured the efficiency of agricultural production in 18 developing countries from 1961 to 1985 and confirmed that the results were consistent with the findings of previous studies that agricultural production efficiency was declining in developing countries [13].

Due to different national conditions, international research on urban–rural integration is still lacking. Most foreign countries explore the definition of urban and rural patterns, influencing factors, and policy recommendations. For example, for the definition of urban–rural patterns, some European countries use urban–rural typology for the definition of urban–rural spatial patterns. In Denmark, a study has compared urban–rural typologies from OECD, Eurostat, and ESPON, and reduced them to the level of Danish municipalities; the reduced typologies are largely consistent in terms of overall spatial patterns, and their urban–rural patterns are more diverse than the original typologies, providing a clearer picture of the urban–rural structure in Denmark [14]. In terms of influencing factors, a researcher used migration patterns to analyze urban–rural relationships. The study elaborated the spatial distribution of types of in-migration and the relation to selected location determinants in the metropolitan area of Copenhagen for the years 1986–2011 [15]. In addition to this, there are studies in Spain that have used an integrated approach based on statistical and cartographic techniques, incorporating socioeconomic and land use variables using a multivariate statistical framework to explore the processes of change in urban–rural relations in Spain [16]. In terms of policy recommendations, in Europe, policy documents at national and regional levels are increasingly emphasizing urban–rural interdependence, moving toward regionalization and shifting the focus of development more toward functional regions rather than towns and villages [17,18]. Research on urban–rural integration in China has focused on theoretical analysis [19–21], level measurement [22–26], assessment of implementation [27,28], and research on policy tools to manage it [29,30], while relatively little research has been conducted on the efficiency of urban–rural integrated development (URID). The existing studies mainly use the data envelopment analysis (DEA) method to study the efficiency of URID from the input–output perspective without considering undesired outputs, and mainly involve static studies at the provincial and municipal levels in a single year at the spatial and temporal scales. For example, at the provincial level, the DEA model, combining analytical hierarchical processes (APH) and DEA methods, was used to measure the efficiency of URID in 30 Chinese provinces; it was found that there was a gradient of higher efficiency in the eastern region than in the central region and higher efficiency in the central region than in the western region [31,32]. Expanding and refining the URID index system and analyzing the efficiency measurement of urban and rural planning overall with spatial differentiation laws for 30 provinces in China, the results show that the eastern region still has the highest efficiency and that regional socioeconomic development is not related to the overall efficiency of urban–rural development [33,34]. There are obvious differences in natural conditions and the human geographic environment in different regions of China. Therefore, it is necessary to conduct research on urban–rural relations at the regional level and formulate regional urban–rural integration policies according to local conditions. After measuring the urban–rural integration efficiency of different prefecture-level cities in Gansu and Jiangsu Provinces and analyzing the spatial divergence pattern and influencing factors at the regional level, researchers found that 14 prefecture-level cities in Gansu Province showed spatial distribution characteristics of high efficiency in the west and low efficiency in the east, with a north-south divergence in 2009; 13 prefecture-level cities in Jiangsu Province showed low overall efficiency in 2015, with a spatial distribution pattern of south Jiangsu > north Jiangsu > middle Jiangsu [35,36]. Considering the undesirable output of regional carbon emissions, the efficiency of URID and its dynamic evolutionary characteristics in 27 cities in the Yangtze River Delta region from 2008–2017 were analyzed using a superefficient epsilon-based measure (super EBM) model, including total factor productivity changes and driving factors. Researchers found that efficiency is low across the delta and the efficiency of URID in economically developed cities is lower than in less economically developed cities. The redundancy of undesirable indicators of carbon emissions has a greater impact on the loss of URID efficiency, but the overall trend in total factor productivity is improving [37]. The above shows that China and Europe differ in their research directions and approaches to urban–rural integration. European countries focus on developing toward functional areas rather than towns and villages, while China focuses on urban–rural parity and tends to develop villages. In future research, the methodology and indicator construction of European countries can be used to make studies more comprehensive.

In summary, systematic and mature cases of research into the efficiency of URID are still lacking, especially in city clusters with rapid economic development and obvious urban–rural differences. At the same time, most existing studies are static studies on a single year, lacking dynamic monitoring and an analysis of the variance in efficiency at different time scales. Furthermore, most of the current research on urban–rural integration efficiency mostly measure using traditional DEA methods, and less consideration is given to the influence of unexpected values, especially carbon emissions, leading to an overestimation of efficiency. In this regard, the three major city clusters of the Yangtze River Economic Belt (YREB) (hereafter referred to as the three major city clusters), which span three major regions of east, central, and west China, are targeted for research in this paper. The low-carbon concept is introduced using the EBM superefficiency model, taking carbon emissions into consideration as an undesired output. Based on the panel data of 73 cities in the three major city clusters that carried out urban–rural integration from 2010 to 2019, the efficiency, characteristics of spatiotemporal evolution, and the correlation between URID efficiency, carbon emissions and efficiency improvement are all analyzed. The study purpose is to reveal the spatial and temporal patterns of URID efficiency of the three major urban agglomerations in the YREB, to provide a basis for policy formulation on URID efficiency of the urban agglomerations in the YREB, and provide direction for promoting the high-quality development of regional urban–rural integration.

2. Materials and Methods

2.1. Study Area

The YREB spans the three major regions of China's east, center, and west, covering 9 provinces and 2 municipalities directly under the Central Government, with a total area of approximately 2.05 million km^2; its population and GDP exceed 40% of that of the country. In these three regions of the YREB, the Yangtze River Delta urban agglomeration

(YRDUA), the urban agglomeration in the middle reaches of the Yangtze River (MRYRUA), and the Chengyu urban agglomeration (CYUA) are the strategic core areas of economic growth [38,39] and are located in the lower, middle, and upper reaches of the Yangtze River (Figure 1). The Yangtze River Delta region is one of the regions with the most active economic development, the highest degree of openness, and the strongest innovation capacity in China, and it has a pivotal strategic position in the general plan for national modernization and the overall pattern for economic opening. Promoting the integrated development of the Yangtze River Delta, enhancing the innovation and competitiveness of the Yangtze River Delta region, and improving the efficiency of economic agglomeration, regional connectivity, and policy synergy are all highly significant to leading the country's development of a high-quality modern economic system. The city cluster in the middle reaches of the Yangtze River is an important part of the YREB and is also a key area in the strategy to promote the rise of the central region, deepen reform and opening, and promote new urbanization in all aspects. The central region also occupies an important position in the pattern of regional development in China. The CYUA is an important ecological barrier in the upper reaches of the Yangtze River, as a comprehensive transportation hub that integrates the east and the west and connects the north and the south in southwest China. As the connection point between the "Belt and Road" and the YREB, the region has the substantial responsibility to integrate and promote the development of the YREB. The development of the YREB must prioritize ecological and green development, and the Chengdu-Chongqing city cluster plays a leading role in this green development. The three major city clusters are important engines to support and lead the high-quality and integrated development of the YREB, and they are also important functional areas in the strategic pattern of China's regional development. To this end, the efficiency and spatial and temporal evolutionary characteristics of URID in the three major city clusters were scientifically analyzed. This study provides a basis for the YREB to achieve high-quality development and to collaboratively promote the policy guidelines for URID.

2.2. Materials

The data in the paper were mainly obtained from the China City Statistical Yearbook, the statistical yearbooks of provinces and municipalities in the Yangtze River Economic Zone, and the Final Statement of General Public Budget Expenditure in each city. Some data were calculated based on the yearbook data, and the missing data for individual years were made up by linear interpolation. Carbon emission data were obtained from Oda et al. [40] and counted by ArcGIS software.

2.3. Methods

For this article, we constructed an index system that can calculate a comprehensive coefficient to measure the development efficiency of urban–rural integration using the EBM superefficiency model. The degree of development of urban–rural integration in this index system is calculated using another index system and the vertical and horizontal scatter degree method. In addition, the temporal and spatial variations of URID efficiency in the YREB are analyzed by the trend surface method.

2.3.1. Evaluation of URID Efficiency

1. Efficiency measurement index system

As an indicator to measure the maximum efficiency of inputs and outputs between urban and rural areas, the efficiency of urban–rural integration refers to the efficiency of the allocation of capital, technology, talent, land, and other factors between urban and rural areas. Maximum efficiency is the optimal combination of factor inputs to produce the "best" product mix, so that the allocation of input and output resources between urban and rural areas is optimized.

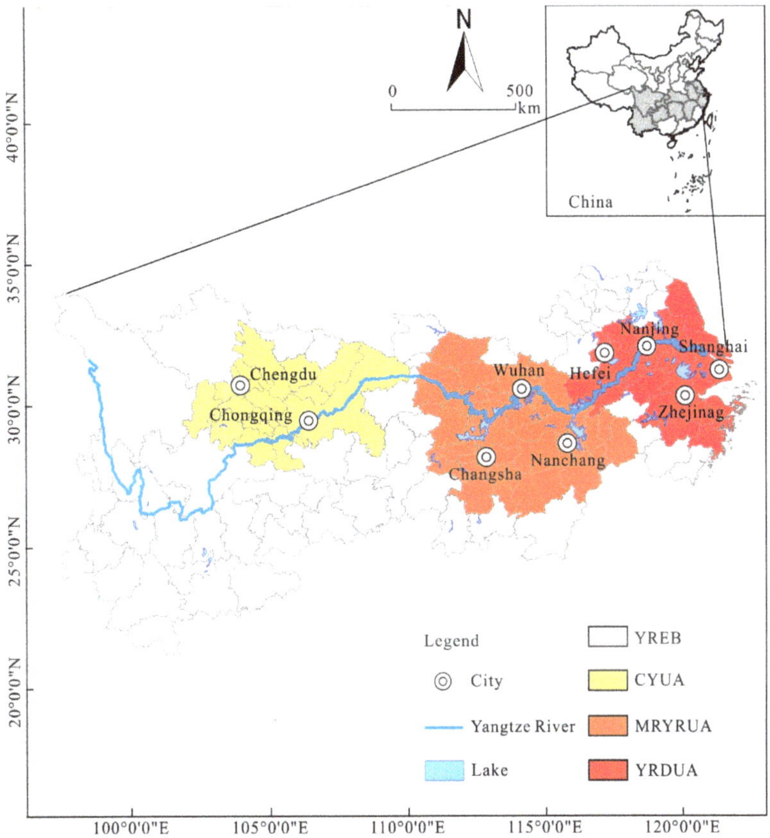

Figure 1. Location diagram of the three major city clusters in the YREB. Note: YREB = Yangtze River Economic Belt, CYUA = Chengyu Urban Agglomeration, MRYRUA = Urban Agglomeration in the Middle Reaches of Yangtze River, YRDUA = Yangtze River Delta Urban Agglomeration.

The study combined the connotations of urban–rural integration referred to in the literature [34,37], and followed the principles of objectivity, systematicity, comparability, and operability to evaluate URID efficiency in the three major city clusters in the YREB for both inputs and outputs (Table 1). URID, as an important public project led by the government, has intricate and complex forms and structures of inputs, which are difficult to refine. At the same time, various elements measuring URID efficiency are derived from financial inputs and transformations. Combined with general public service expenditure from government finance, 11 indicators, such as education, science and technology, culture, sports, and media, were selected. In terms of outputs, the level of urban–rural integration and carbon emission efficiency were selected as the expected values, and total carbon emissions were selected as the unexpected value. Carbon emissions, as an important indicator reflecting the quality of URID, are closely related to urban and rural social and economic activities. Taking carbon emissions into consideration can more objectively examine whether URID is performing as expected, reflecting low-carbon and sustainable urban–rural development. For example, traditional productivity measures that ignore carbon emissions and other undesirable outputs will lead to overestimation of the true efficiency of urban–rural integration.

Table 1. Input–output index system of efficiency for URID.

Index Attribute	Index Selection	ID
Input indicators	Education (100 million yuan)	Ip[1]
	Science and technology (100 million yuan)	Ip[2]
	Culture, sports, and media (100 million yuan)	Ip[3]
	Social security and employment (100 million yuan)	Ip[4]
	Hygiene and health (100 million yuan)	Ip[5]
	Energy conservation and environmental protection (100 million yuan)	Ip[6]
	Urban and rural communities (100 million yuan)	Ip[7]
	Agriculture, forest, and water (100 million yuan)	Ip[8]
	Public transportation (100 million yuan)	Ip[9]
	Business services (100 million yuan)	Ip[10]
	Expenditure on housing security (100 million yuan)	Ip[11]
Output indicators	The level of integrated urban and rural development	Op[1]
	Carbon emission efficiency (ton/10,000)	Op[2]
	Carbon emissions (10,000 tons)	Op[3]

2. The EBM superefficiency model

The traditional DEA model cannot measure slack variables, while the slack-based measure (SBM) model loses the proportional information between the actual value of inputs and outputs and the target value. Aiming at these shortcomings, Tone et al. [41,42] proposed a hybrid model: an epsilon-based measure (EBM) model that includes both radial and SBM distance functions. This model can measure not only the improvement ratio between the target value and the actual value, but also the gap between the target value and the actual value by solving the nonradial values of each input–output so that the efficiency of the decision-making unit (DMU) can be measured more accurately. The conventional EBM model cannot compare multiple input DMUs at the frontier, but the superefficiency EBM model can make up for this deficiency. In view of this, this paper uses MaxDEA9 software, selects the EBM model to be nonoriented, sets the superefficiency option, and calculates the efficiency of URID. The expressions are as follows:

$$r^* = \min \frac{\theta - \varepsilon^- \sum_{i=1}^{m} \frac{\omega_i^- s_i^-}{x_{i0}}}{\varphi + \varepsilon^+ \left(\sum_{r=1}^{s} \frac{\omega_r^+ s_r^+}{y_{r0}} + \sum_{p=1}^{q} \frac{\omega_p^{u-} s_p^{u-}}{u_{p0}} \right)} \tag{1}$$

$$s.t. \sum_{j=1}^{n} x_{ij}\lambda_j + s_i^- = \theta x_{i0} \tag{2}$$

$$\sum_{j=1}^{n} y_{rj}\lambda_j - s_r^+ = \varphi y_{r0} \tag{3}$$

$$\sum_{j=1}^{n} u_{pj}\lambda_j + s_p^- = \varphi u_{p0} \tag{4}$$

In the formula, $r^*(0 \leq r^* \leq 1)$ is the optimal efficiency value and x_{i0}, y_{1r0}, u_{p02}, and s_i^- are the DMU_0 input, expected output, and undesired output, respectively, followed by DMU_0 as the input slack. s_r^+ and s_p^- are the expected output and undesired output slack, respectively; ω_i^-, ω_r^+, ω_p^{u-} are the input and expected output of each indicator, respectively, followed by the importance of undesired outputs. θ is the efficiency value under radial conditions; ε is the core parameter of the importance degree of the nonradial part when $0 \leq \varepsilon \leq 1$ is satisfied.

2.3.2. Evaluation of the URID Level

1. Level measurement index system

Above, the expected value output is used as an indicator of the level of URID efficiency to measure the level of urban–rural integration in a scientific way. Seventeen indicators were selected from five dimensions, including integration, cultural integration, spatial integration, and ecological integration (Table 2) [22,43–47]. In addition, the indicators of the URID level are divided into comprehensive and comparative categories, where the comprehensive category mainly reflects the overall development of regional urban and rural areas while the comparative category mainly reflects the differences between urban and rural areas. The two are indispensable and complement each other. If there is a lack of comparative indicators, the level measurement results will deviate from the stated research goals and become an evaluation of the comprehensive development level of the region; similarly, if there is a lack of comprehensive indicators, the measurement results will also deviate from the research goals.

Table 2. Indicator system for measuring the level of URID.

Dimensionality	Indicator Name	Indicator Calculation and Description	Attribute	Category
Economic integration	Per capita GDP	GDP/regional resident population (yuan)	+	Comprehensive
	Disposable income ratio of urban and rural residents	Per capita disposable income of urban residents/per capita disposable income of rural residents (%)	−	Comparison
	Per capita consumption ratio of urban and rural households	Per capita consumption expenditure of urban residents/per capita consumption expenditure of rural residents (%)	−	Comparison
	Engel's coefficient ratio between urban and rural areas	Urban Engel's coefficient/rural Engel's coefficient (%)	+	Comparison
	Binary contrast coefficient	(Output value of primary industry/employees of primary industry)/(output value of secondary and tertiary industries/employees of secondary and tertiary industries) (%)	+	Comparison
Social integration	Urban and rural cultural, educational and entertainment comparison coefficient	Per capita expenditure on cultural, educational and recreational services for urban residents/per capita expenditure on cultural, educational and recreational services for rural residents (%)	−	Comparison
	Teacher–student ratio in basic education	Number of elementary education teachers/number of elementary education students (%)	+	Comprehensive
	Contrast coefficient of medical care per capita between urban and rural areas	Per capita health care expenditure of urban residents/per capita health care expenditure of rural residents (%)	−	Comparison
	Urban and rural unemployment insurance coverage	Number of urban and rural residents covered by unemployment insurance/number of permanent residents (%)	+	Comprehensive
Population integration	Urban and rural population contrast coefficient	Urban population/rural population (%)	+	Comparison
	The ratio of nonagricultural employment to agricultural employment	Number of employees in the secondary and tertiary industries/number of employees in the primary industry/(%)	+	Comparison
	Population urbanization level	Total urban population/total population (%)	+	Comprehensive
Ecological integration	Vegetation index	Urban and rural NDVA (normalized difference vegetation index)	+	Comprehensive
	Urban and rural sewage treatment	Centralized treatment rate of sewage treatment plant (%)	+	Comprehensive
	Urban and rural domestic waste treatment	Harmless treatment rate of domestic waste (%)	+	Comprehensive
Space integration	Road network density	Highway operating mileage/total land area (km/km2)	+	Comprehensive
	Urban and rural internet user rate	Number of internet users in urban and rural areas/total number of households at the end of the year (%)	+	Comprehensive

Note: 1. An index with an attribute of "+" means that the larger the index value is, the more conducive it is to improving URID; an index with an attribute of "−" means that the larger the index value is, the less conducive it is to improving URID.

2. Vertical and horizontal scatter degree method

There are many methods for measuring the level of URID, such as the commonly used principal component analysis and entropy value methods, but these methods are difficult to evaluate dynamically. The comprehensive evaluation method of a three-dimensional time series can not only reflect the difference of the evaluation objects at certain time section, but can also show the distribution of the evaluation objects longitudinally over time and has strong objectivity [48,49].

$$H_t = A(t)''^T A(t)'' \quad (5)$$

$$H = \sum_{t=1}^{N} H_t \quad (6)$$

$$e^2 = \sum_{t=1}^{T}\sum_{i=1}^{m}(y_i(t) - \overline{y})^2 = \sum_{t=1}^{T}\sum_{i=1}^{m}(y_i(t))^2 = \sum_{t=1}^{T} W^T H_t W = W^T H W \quad (7)$$

$$gti = \mathrm{w}1X_{til} + \mathrm{w}2X_{til}'' + \cdots + \mathrm{w}nX_{til}'' \quad (8)$$

In the formula, T is the research year, m is the number of cities, N is the number of indicators, and the eigenvector u corresponding to the largest eigenvalue of the matrix H is the weight determination vector. After u is obtained, normalization is performed. At this time, e^2 takes the maximum value. This determines the weight vector u, where u(w1, w2, w3, w4...wn). Among these terms, t is the research year, n is the research city, n is the number of indicators in the study, and gti is the urban–rural integration degree of the ith city in the tth year. The weight of each index is multiplied by the corresponding standardized index value of the city in the current year to obtain the urban–rural integration degree of the ith city in the tth year.

2.3.3. Evaluation of the URID Level

A trend surface is a semiquantitative study of geographic data from a large area based on spatial data and simulated spatial surfaces using mathematical fitting, which can be used to explore the spatial trends and distribution patterns of research objects [9]. In this paper, the characteristics of spatial and temporal variation in urban–rural integration in the three major city clusters since 2010 are simulated by means of trend surface analysis with the value of URID efficiency. Let (x_i, y_i) be the spatial location of the ith municipality; then, Z_i (x_i, y_i) is the trend function of the ith municipality, where the X-axis represents the east–west direction and the Y-axis represents the north–south direction.

3. Results

3.1. General Change Characteristics in the Efficiency of URID

Selecting the vertical and horizontal scatter degree method and EBM superefficiency model and using MATLAB and MaxDEA9 software, the URID level and efficiency of 73 cities in the three major city clusters were obtained, as shown in Figure 2a. The level of urban–rural integration has been increasing linearly over time, with an average annual growth rate of 5%. Total carbon emissions show an overall upward trend, as shown by a rapid rise from 2010 to 2014 and a small fluctuation from 2014 to 2019 of "first falling and then rising"; the efficiency of URID shows an overall decreasing trend over time, as shown by a gradual decrease in efficiency from 2010 to 2015 and a small fluctuation from 2015 to 2019 of "gradually rising and then falling", which is the opposite of the trend in carbon emissions.

Further analysis of the change in characteristics of different city clusters shows that, as shown in Figure 2b–d, the URID level of each city cluster is on the rise as a whole, and YRDUA > MRYRUA > CYUA. In the past 10 years, URID has maintained a trend of growth; the overall carbon emissions of the three major city clusters have also shown an upward trend, with YRDUA > CYUA > MRYRUA, and the carbon emissions of YRDUA are on average 3–4 times higher than the other two city clusters. There are obvious differences in the efficiency of URID of the three major city clusters, with MRYRUA > CYUA > YRDUA. The overall efficiency declines from 2010–2015, followed by small fluctuations in 2015–2019, among which the most significant are found in CYUA, "rising first and then falling".

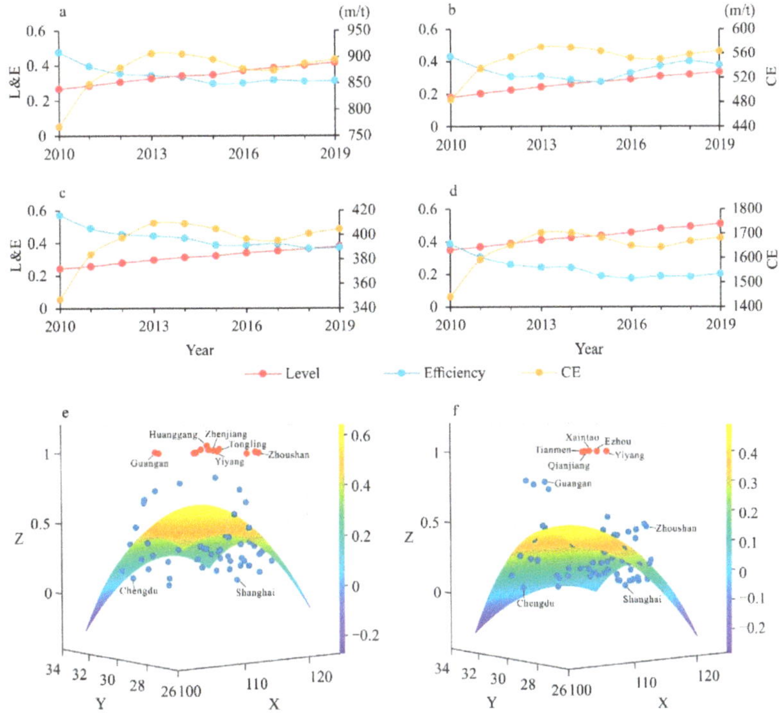

Figure 2. Changes in the level, efficiency, and surface trends of URID in the three major city clusters in the YREB. Note: (**a**) denotes the three major city clusters, (**b**) denotes CYUA, (**c**) denotes MRYRUA, (**d**) denotes YRDUA, (**e**,**f**) denote trend surface analysis of three major city clusters. Dots are efficiency values and red dots are $\geq$1 (effective efficiency). L&E denotes the level and efficiency of URID, CE denotes carbon emissions, Level (red line) denotes the level of URID, Efficiency (blue line) denotes the efficiency of URID, and CE (orange line) denotes carbon emissions.

The trend surface analysis method (Figure 2e,f) helps to reveal the spatial divergence in URID efficiency in the three major city clusters in the YREB. On the whole, from 2010 to 2019, the efficiency of URID in the three major city clusters roughly shows spatial divergence in an inverted U-shape in the east–west and north–south directions: MRYRUA > CYUA > YRDUA. The trend results are the same as those in Figure 2b–d. In 2010 (Figure 2e), the efficiency of URID in the east–west direction increased significantly from Shanghai, Suzhou, and Hangzhou in the eastern Yangtze River Delta city cluster to Qianjiang, Xiantao, and Ezhou in the midstream city cluster, and fell back again in the Chengdu-Chongqing city cluster, with cities such as Guang'an, Chengdu, and Chongqing. The north–south direction shows increases in efficiency from Yiyang, Tongling, and Zhoushan in the south to Huanggang, Ezhou, and Zhenjiang in the middle; it then decreases to Yancheng, Chuzhou, and Mianyang in the north. There are 13 cities with an effective urban–rural integration efficiency > 1, which are mainly concentrated in the central MRYRUA, represented by cities such as Huanggang, Qianjiang, and Yiyang. The change in the URID efficiency trend east–west in 2019 (Figure 2f) is significantly different from that in 2010, gradually increasing from the eastern YRDUA to the western CYUA and decreasing from south to north. The effective coefficient of urban–rural integration efficiency for the total region has been reduced to 5, with representative cities concentrated in the central MRYRUA, with cities such as Ezhou, Xiantao, and Yiyang.

3.2. Evolution of the Spatiotemporal Pattern of URID Efficiency at the City Level

In this paper, the city-level URID efficiency of the three major city clusters from 2010–2019 was divided into five hierarchical gradients by the natural breakpoint method in ArcGIS software, and the spatial distribution diagram for each year was drawn (Figure 3). On the whole, the efficiency of URID in each of the three major city clusters shows a decreasing trend over time in the order MRYRUA > CYUA > YRDUA, which corresponds to the results in Figure 2b–d. The first and second gradients are mainly concentrated in the midstream city cluster, the third and fourth gradients are mainly concentrated in the Chengyu city cluster, and the fifth gradient is mainly concentrated in the Yangtze River Delta city cluster.

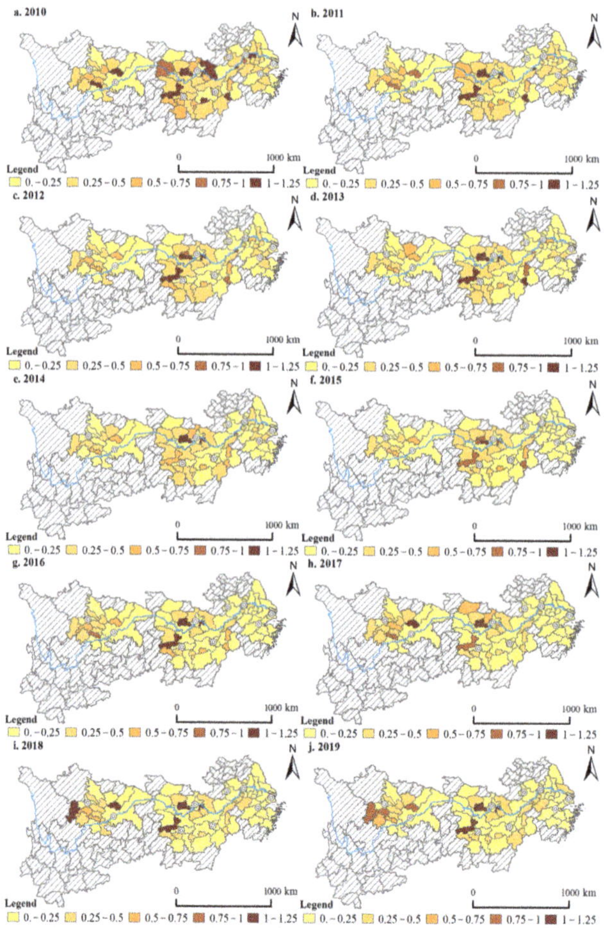

Figure 3. Schematic diagram of the spatial distribution of the efficiency of URID in the three major city clusters. Note: (**a–j**) denotes the spatial distribution of the efficiency of URID in the three major city clusters from 2010–2019, respectively. The efficiency of urban-rural integration development is divided into five categories, which are indicated by different colors.

Combined with the spatiotemporal trends of each city, the URID efficiency of the eastern cluster, the YRDUA, shows a decreasing trend from 2010 to 2019, with an overall mean value of 0.24. In 2010, the average value of URID efficiency was 0.39, and there were three cities with an effective efficiency in the first gradient, including Zhenjiang, Zhoushan, and Tongling; in 2011, the average value of URID efficiency of the first gradient was 0.31, and the cities with effective efficiency were reduced to 1, Tongling, while Zhoushan and Zhenjiang dropped to the third and fourth gradients, respectively. In 2012–2019, the average annual values of URID efficiency were 0.26, 0.244, 0.24, 0.19, 0.18, 0.19, 0.18, and 0.20, dominated by the fourth and fifth gradients, in which the relatively economically developed cities of Shanghai, Suzhou, Nanjing, and Hangzhou show low efficiency. From 2010 to 2019, the central midstream city group had a decreasing trend in URID efficiency, with an overall mean value of 0.43, the highest among the three city groups. In 2010, the average value of URID efficiency was 0.57, the highest of the 10 years measured, and there were eight cities with effective efficiency values in the first gradient, namely, Ezhou, Huanggang, Xiantao, Qianjiang, Tianmen, Yiyang, Yingtan, and Xinyu. In 2011, the average value of URID efficiency was 0.49 and the cities with effective efficiencies in the first gradient decreased to 6, namely, Ezhou, Xiantao, Qianjiang, Tianmen, Yiyang, and Yingtan. Huanggang and Xinyu dropped to the third and fourth gradients. The average values of URID efficiency in 2012–2019 were 0.45, 0.44, 0.43, 0.39, 0.39, 0.40, 0.37, and 0.37, mainly representing cities in the third and fourth gradients. The cities of Xiantao, Ezhou, Xiantao, Qianjiang, and Tianmen fluctuated little and remained stable in the first gradient, while the cities of Yiyang and Yingtan were more volatile and showed instability. The same low efficiency also appeared in Wuhan, Changsha, Nanchang, and other relatively economically developed cities. The overall average value of URID efficiency in western CYUA from 2010 to 2019 is 0.43, with annual averages showing a "U" shape, first decreasing, then increasing. In 2010, the average URID efficiency was 0.43, and there were two cities with effective efficiencies in the first gradient, namely, Guang'an and Zigong. From 2011 to 2015, the URID efficiency showed a decreasing trend and from 2015 to 2019, an increasing trend. The city with effective efficiency in the first gradient is Guang'an; Ya'an was added in 2018, bringing the number of cities with effective efficiency in the first gradient to two. Similar to the first two city clusters, Chengdu, Chongqing, Mianyang, and other relatively developed cities are generally inefficient.

4. Discussion

4.1. Analysis of the Changing Law of URID Efficiency

4.1.1. Analysis of the Overall Laws of Change for URID Efficiency

The results of measuring the level and efficiency of URID in the three major city clusters (Figure 2a) reveal that the level of URID, efficiency, and carbon emissions show different development trends over time. The level of URID rises linearly over time, mainly because the central government attached importance to the "three rural issues" that are part of the "urban–rural integration" and "new socialist countryside construction" that were proposed and implemented in the early stage (2003–2011). In the later stage (2012–present), "urban–rural integration", "precise poverty alleviation", "rural revitalization", and "new socialist countryside construction" policies were proposed. A series of policy strategies such as "URID" have strongly promoted the rapid development of rural areas, gradually narrowing the gap with urban areas and promoting the improvement of urban–rural integration. The overall trend of total carbon emissions is upward, showing a rapid rise from 2010 to 2014 and a temporary dip followed by a rise from 2014 to 2019. The gradual slowdown in emissions after 2014 is closely related to the transformation of the industrial structure and large-scale application of low-carbon technologies following the 2014 declaration of a "new normal" defining the new economic development stage proposed by the central government. By contrast, the efficiency of URID has generally shown a decreasing trend with the passage of time, specifically in 2010–2015. In general, cities increase their efficiency input indicators year by year, which promotes social, economic, and environmental development, and to a certain extent, improves the URID level. However, in the process of rapid urbanization and industrialization, a large number of low-end industries with crude production methods have resulted in the ineffective use of a large number of resource inputs and failed to play a practical role in the integrated development of urban and rural areas. In addition, carbon emissions grew rapidly during this period, indicating that these crude low-end industries relied on resources to a high degree, resulting in regional environmental pollution and increased urban and rural energy consumption. This is also opposite of the general trend of carbon emissions, mainly because the economy entered a new normal in 2015, with economic growth shifting from high-speed to medium-speed, from sloppy upscaling and acceleration to an intensive focus on improving quality and efficiency, and from factor investment-driven growth to quality-, innovation-, and efficiency-driven growth. The effective use of various resources resulted in a steady improvement in the level of urban–rural integration and a slowdown in the growth of carbon emissions.

4.1.2. Analysis of the Change Pattern of URID Efficiency at the City Cluster Level

There are obvious differences in the URID efficiency among city clusters. The URID efficiency of MRYRUA is the highest from 2010 to 2019, which is mainly influenced by the "Rise of Central China", the "YREB Development Strategy", and other plans. However, it is also worth noting that the efficiency values of these regions show a decline, probably due to the fact that the traditional industries, such as steel, automobiles, and transportation equipment manufacturing, are dominated by high dependence on resources and underutilization of resource inputs, resulting in increased carbon emissions and lower efficiency. From 2010 to 2015, the three major urban agglomerations showed a decreasing trend with little difference between them; from 2015 to 2019, the efficiency difference with MRYAUA and YRDUA gradually increased and showed an increasing trend (Figure 2b). The possible reason for this is that although the "Western Development" strategy proposed by the government in the early stage has promoted urban–rural integration in CYUA to a certain extent, on the whole, the infrastructure and industrial development of CYUA is weak and the resources inputs are not fully utilized, resulting in a decrease in efficiency and an increase in carbon emission. After 2015, the strong cooperation between Chongqing and Sichuan has, to a certain extent, contributed to the transformation and upgrading of their industrial structures, effective utilization of resource inputs, and improvement of efficiency. YRDUA had the lowest efficiency of URID but the highest level of URID and carbon emissions from 2010–2019 (Figure 2d). The possible reason for this is that the Yangtze River Delta region has paid more attention to URID in the past decade and its relatively high investment has contributed to the progress of its urban–rural economic, social, and environmental dimensions. However, at the same time, it must be acknowledged that the development pattern of a large number of urban and rural low-end industries in the Yangtze River Delta region is still relatively crude, resulting in a large number of less efficient inputs.

From the analysis of urban–rural function, YRDUA has developed social economy and a high urbanization rate, and the urban and rural areas play their respective functional advantages to form complementary urban–rural functions, which leads to the improvement of the level of urban–rural integrated development. There is no longer a single rural area supporting the urban area and the urban area feeding the rural area, but rather a large number of low-end and rough industrial gatherings in the rural area, which leading to s resource inputs not being ineffectively used and resulting in a large amount of carbon dioxide emission and low efficiency of urban–rural integration development. To avoid this, these areas should reasonably plan their industrial layout and increase scientific and technological innovation. CYUA and MRYRUA are relatively less developed socio-economically, with low urbanization rates, large urban–rural gaps, and better urban functions and lack of rural functions, creating single urban areas feeding rural areas and a low level of urban–rural integrated development. However, the resource inputs from urban functions to rural areas are effectively utilized, improving of the efficiency of urban–rural integration development. The future relies on the rural revitalization strategy to promote rural development, promoting the two-way flow of urban and rural elements and the mutual promotion of urban and rural functions.

4.2. Correlation between Carbon Emissions and URID Efficiency

In this paper, the efficiency of URID in 73 cities in three major city clusters was analyzed in comparison with carbon emissions; the results showed cities in three categories: positive correlation, inverse correlation, and insignificant correlation (Figure 4). There are three cities that were positively correlated, accounting for 4%. As shown in Figure 4a, the efficiency of URID in cities represented by Dazhou gradually increases with carbon emissions, which this paper believes may be due to weak socioeconomic development in cities with small economic volume, low industrialization, limited amounts of resource inputs, and low carbon emissions. Consequently, the efficiency value gradually increases with carbon emissions. However, it is worth noting that, although it becomes positive, the efficiency value fluctuates and does not form a stable trend of growth, There are 54 cities with a reverse correlation with carbon emissions, distributed in each city group, accounting for 74%, the majority of total cities. As shown in Figure 4b, the efficiency of URID represented by Wuhu decreases with the gradual increase of carbon emissions, While carbon emissions increased rapidly in 2010–2014, the efficiency value decreased, probably because of rapid urbanization and industrialization, with a large number of low-end industries with crude production methods making poor use of resource allocation and resource inputs. From 2014 to 2019, as the economy entered the "new normal", all inputs effectively improved, and the efficiency value increased with the reduction of carbon emissions. There are 16 cities, accounting for 22% of the total, that are not significantly correlated with carbon emissions; they are also distributed among all urban clusters. As shown in Figure 3c, the efficiency of URID in cities represented by Ningbo gradually flattens out with increasing carbon emissions, generating a nonsignificant correlation. Among these cities, the more socioeconomically developed the city is, the flatter the urban–rural integration efficiency is, and the larger the carbon emissions are (Figure 3d–f). It may be that resource allocation is unreasonable and the various resources inputs are used only in a limited way, resulting in the waste of some resources, high carbon emissions, and relatively low efficiency values.

4.3. Improvement of URID Efficiency

The EBM model is able to measure the redundancy of inputs, the shortfall of desired outputs, and the redundancy of undesirable outputs in terms of proportional improvement values and slack improvement values. The sum of both the proportional improvement value and the slack improvement value is the overall redundancy value [32]. The analysis of the redundancy (deficiency) of each input–output indicator can reflect the causes of efficiency loss and help to provide guidance for the improvement of URID efficiency in the three major city clusters in the YREB. This paper divides the average value of redundancy (deficiency) of each indicator for the 73 cities in the three major city clusters from 2010 to 2019 by the average value of the corresponding input (output) indicator and obtains the input redundancy rate and output insufficiency (redundancy) rate of each indicator. The calculation results are shown in Table 3.

4.3.1. Input–Output Analysis of URID Efficiency

From the perspective of input indicators, the overall redundancy rate in the three major city clusters is high. The redundancy rates of individual input indicators are all above 50%, indicating that the large amount of resource inputs has not played a practical role in promoting the integrated development of urban and rural areas. From the mean values of each input in Table 3, it is found that redundancy in the inputs for cultural, sports and media (Ip^3), energy conservation and environmental protection (Ip^6), urban and rural communities (Ip^7), and housing security expenditures (Ip^{11}) in most cities are the primary influencing factors of their efficiency loss in URID. This indicates that the three major city clusters, as strategic core areas of economic growth in the YREB, still have unbalanced urban–rural development, with inadequate rural development in the process of integrated development. Urban–rural resource allocation remains unreasonable and unbalanced, and urban–rural factor flow and distribution are still mainly one-way, with two-way interactions largely unformed.

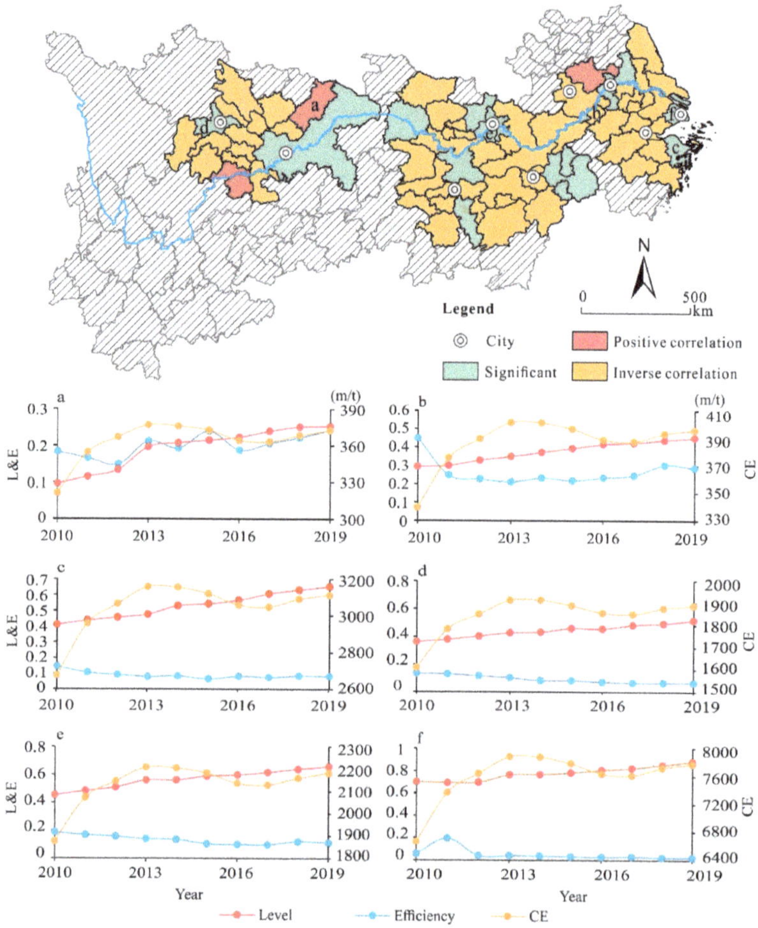

Figure 4. Schematic diagram of the spatial-temporal relationship between URID efficiency and carbon emissions in the three major city clusters. Note: (**a**) (Dazhou) denotes a positive correlation, (**b**) (Wuhu) denotes an inverse correlation, (**c**) (Ningbo) denotes an insignificant correlation, and (**d**–**f**) denote Chengdu, Wuhan, and Shanghai. L&E denotes the level and efficiency of URID, CE denotes carbon emissions, Level (red line) denotes the level of URID, efficiency (blue line) denotes the efficiency of URID, and CE (orange line) denotes carbon emissions.

In recent years, the state has strongly supported the deepening of cultural system reform, promoting the development of cultural industries especially in the countryside, building a large number of cultural centers, rural bookstores, and fitness venues, which have enriched and improved rural cultural life. There is a disconnect between the content of rural cultural construction and the needs of rural residents, a lack of innovation in cultural system construction, a shortage of cultural service personnel, and a lack of investment in construction funds, which have all hampered the role of rural cultural construction in narrowing the gap with urban areas and promoting URID. There are obvious differences between urban and rural areas in terms of energy conservation and environmental protection (Ip^6), with cities being effectively targeted as the center of gravity for pollution prevention and control, while rural areas are devoid of pollution prevention and control, which also hampers integrated urban–rural development. The urban and rural community (Ip^7) input is mainly used for urban and rural community management affairs, community planning and management, public facilities, community housing, environmental sanitation, and construction market management and supervision, which have played a certain role in promoting urban and rural social life and the improvement of urban and rural living environments. However, due to the lack of strict budgetary management in the use of these funds and the relative absence of consideration of the geographical and regional differences in the development of urban and rural communities, the performance of this indicator for promoting URID is not high. Housing security expenditure input (Ip^{11}) is mainly used to support the construction of secure housing projects and secure housing to accelerate the transformation of shantytowns and the renovation of dilapidated houses in the countryside and, in general, to improve the living conditions of urban and rural people in difficulty. This input helps improve urban and rural living conditions and narrow the gap between urban and rural integration development. However, the method for assigning land for guaranteed housing still needs to be improved, and the lack of unified and standardized planning for housing and shantytown renovation leads to unreasonable construction costs. The waste of resource inputs again inhibits the efficient and high-quality development of urban–rural integration. It is worth noting that the YRDUA is different from other city clusters in terms of investment in science and technology (Ip^2), which has a high redundancy rate, similar to the investment in energy conservation and environmental protection. The long-standing development strategy of "emphasizing urban over rural areas" has caused a serious imbalance in the investment in science and technology between urban and rural areas. The high level of redundancy in this indicator thus narrows the urban–rural development gap and promotes urban–rural integration, but the indicator's actual role in development is limited.

Table 3. Results of the input–output optimization of URID in the three major city clusters.

City	Input Redundancy Rate											Underproduction (Redundancy) Rate		
	Ip^1	Ip^2	Ip^3	Ip^4	Ip^5	Ip^6	Ip^7	Ip^8	Ip^9	Ip^{10}	Ip^{11}	Op^1	Op^2	Op^3
CYUA	−0.69	−0.54	−0.79	−0.71	−0.74	−0.80	−0.78	−0.70	−0.66	−0.63	−0.85	0.54	1.42	−0.62
MRYRUA	−0.57	−0.58	−0.59	−0.59	−0.54	−0.62	−0.68	−0.53	−0.51	−0.53	−0.60	0.45	1.38	−0.54
YRDUA	−0.76	−0.87	−0.82	−0.66	−0.70	−0.76	−0.90	−0.67	−0.68	−0.71	−0.77	0.60	3.91	−0.78
Average value	−0.66	−0.68	−0.71	−0.64	−0.64	−0.71	−0.78	−0.62	−0.60	−0.62	−0.71	0.52	2.29	−0.64

Note: 1. Negative numbers in the table indicate that the input is redundant and positive numbers indicate that the output is insufficient. The redundancy (insufficiency) rate refers to the absolute value of the corresponding value of each indicator. 2. The color blocks in the table represent the top three indicators of the redundancy rate. CYUA = Chengyu Urban Agglomeration, MRYRUA = Urban Agglomeration in the Middle Reaches of the Yangtze River, YRDUA = Yangtze River Delta Urban Agglomeration.

The output deficiency (redundancy) rate varies widely among city clusters, which is related to factors such as resource endowment and industrial structure in different cities. The YRDUA has the highest deficiency rate in desired outputs and the highest redundancy rate in unexpected outputs. This indicates that the efficiency of URID in the YRDUA is also largely constrained by output deficiency in URID level and the excess of carbon emissions, which is consistent with the findings of Figures 2d and 4f above. Comparing the desired output deficiency rate and the unexpected output redundancy rate of each city, it can be found that the impact of excessive carbon emissions on efficiency loss in URID is relatively large.

4.3.2. URID Efficiency Input–Output Path Optimization

In terms of improving the efficiency of integrated urban–rural development, the above indicators have great potential for improving resource utilization. Culture, sports, and media (Ip^3) should be oriented toward normalizing rural cultural services and optimizing its financial input structure. The construction of grassroots cultural teams should be increased; mass cultural workers, folk artists, professional cultural workers, comprehensive law enforcement managers, and other cultural teams should all be trained; the healthy development of rural private culture should be actively guided and encouraged, and rural cultural teams should continue to grow. To make full use of the advantages of rural cultural resources, combined with the actual needs of rural residents, the creation of innovative rural cultural industries, such as the Chongqing Fengjie "navel orange cultural festival" and Hubei Zigui commemorative Qu Yuan "dragon boat race" competition, are needed to enrich local cultural life and simultaneously promote the development of local tourism. Rural grassroots cultural institutions should be reformed and improved, as should mechanisms to improve the effectiveness of rural public cultural services. Townships are the link between rural and urban areas; the cultural construction of townships has an influential effect on the surrounding rural areas, which can be promoted through "new urbanization". A stable growth mechanism for financial investment in rural energy conservation and environmental protection is needed (Ip^6), and the financial investment structure for energy conservation and environmental protection should be optimized. Green and clean energy should be promoted, such as the construction of natural gas pipelines, photovoltaic power generation, and other green and clean energy to support rural households. The "toilet revolution" should be accelerated, fully popularizing rural public toilets, connecting domestic sewage treatment, and promoting the effective treatment or resource utilization of toilet sewage and manure. Rural household garbage collection should be connected with village collection, transportation to local waste facilities, and the district treatment system. The concept of urban and rural communities should be strengthened (Ip^7). Budget management should be funded and special personnel for budget management should be hired to refine fund management and optimize the fund input structure for urban and rural communities. A strict budget monitoring and assessment system should be established and developed, with a budget information feedback system; the budget implementation should be widely publicized so that residents are more willing to accept public supervision and reporting. The housing security expenditure (Ip^{11}) focuses on the construction of secure housing projects, promoting the construction of secure housing while increasing investment in the countryside, carrying out scientific and reasonable architectural planning and renovation according to the resource endowments of different villages, reducing unnecessary resource investments, and narrowing the gap between urban and rural areas.

Further analysis of the input–output path optimization of each city cluster was conducted. The input–output results of the CYUA show that the redundancy of housing security expenditure (Ip^{11}) inputs is the primary cause of efficiency loss. Housing security (Ip^{11}) in the Chengdu-Chongqing city cluster can be improved through the following measures. First, liaisons for housing security in the Chengdu-Chongqing city cluster should be established with regular joint meetings and collaboration to promote the improvement of the housing security system. Second, each housing and urban–rural development management authority in the Chengdu-Chongqing city cluster should adhere to integration and coordination plans to promote the construction of guaranteed rental housing and improve the accuracy of guaranteed housing and the efficient utilization of public rentals, further standardizing public rental housing. Third, the common construction and sharing of housing security information in the CYUA should be promoted. The housing security policies of each city should be centralized and unified, including the channels for application for residency in each city in the CYUA. Most of the villages in the CYUA are located in mountainous areas with complicated terrain, so reasonable planning and transformation with engineers should be carried out according to local conditions to reduce unnecessary resource inputs. The redundancy of investment in urban and rural communities (Ip^7) is the primary reason for the loss of input–output efficiency in the city clusters in the middle reaches and the YRDUA. By optimizing financial investments, the use of funds in urban and rural communities can be improved. According to the actual needs of residents, various special funds should be budgeted, and the community fund budget should be reasonably arranged and refined. The next step is to conduct in-depth investigation and research on ways to expand the use of community funds, effectively integrating funds and maximizing benefits. Third, the community should strictly allocate funds for special purposes and separate accounts. Fourth, the construction of smart community infrastructure should be promoted, improving the smart community governance system and building an open community service complex. Examples include "Community Access" in Baoshan District, Shanghai [50] and "Garden Digital Village" in Zhejiang [51]. In addition to urban and rural community (Ip^7) input, the energy saving and environmental protection (Ip^6) input results in a large loss of efficiency in the city cluster in the middle reaches of the Yangtze River; these inputs can be assessed with actual needs, and resources can be moved toward rural areas through reasonable budgeting to accelerate the improvement of rural infrastructure for environmental protection. The science and technology input (Ip^2) is a source of a large loss of efficiency for the YRDUA, which should optimize the structure of expenditures for science and technology to reasonably distribute the ratio of technology inputs between urban and rural locations. Rural modern agricultural science and technology research should be improved and the results converted into practices. The advantages of the density of universities, high-tech enterprises, and scientific research institutions in the Yangtze River Delta city cluster should be optimized to accelerate improvement in industry–university research and optimize the training and management of scientific research talent.

It is noteworthy that the high output deficiency (redundancy) rate of the URID level and carbon emissions in Table 3 has a negative impact on URID efficiency. Considering carbon emissions as an important indicator of the quality of URID, as it is closely related to urban–rural social and economic activities, the inclusion of carbon emissions in assessments of URID levels can more objectively review whether low-carbon and sustainable URID has been achieved. Ignoring carbon emissions as a unexpected output will lead to overestimation of the efficiency of URID. In view of the impact of the high rate of carbon emissions on output deficiency (redundancy) leading to loss of URID efficiency, the development of regional urban–rural integration should be guided toward sustainable development with the concept of "innovation, coordination, green, openness and sharing". Under the framework of a top-level design, the high energy-consuming, high-polluting, and high-emission enterprises in the industrial chain should be gradually phased out by increasing the supervision of emission reduction policies and improving the carbon trading market and its operations. With these measures in place, competition should be reasonably introduced to gradually eliminate enterprises with high energy consumption, high pollution, and high emissions in the industrial chain. The removal of these low performing enterprises will eliminate the restraining effect of environmental factors such as carbon emissions on the efficient and high-quality development of urban–rural integration

There are certain limitations to this study. First, the indicator system of URID needs to be improved, because the unavailability of data makes it difficult to reflect URID comprehensively. In future data updates, the indicator system should be improved, for example, to supplement the space integration indicators, and the output indicators should be considered to include indicators such as poverty headcount and Gini coefficient, because they can reflect the differences in income distribution between urban and rural residents. Secondly, the continuous development of satellite remote sensing technology provides some potential data for the study of urban–rural integration development, and relevant technical methods should be strengthened in further research to extract new data to more comprehensively evaluate URID. Finally, future research address efficiency input–output analysis by using quantitative methods to further in-depth analysis of the mechanism.

5. Conclusions

Based on the vertical and horizontal spread method, the EBM superefficiency model, and the trend surface analysis method, and considering the undesired output of carbon emissions, this paper studies the urban and rural areas of 73 cities in the three major city clusters of the YREB from 2010 to 2019. The efficiency of URID and the characteristics of its temporal and spatial evolution are revealed, and correlations are identified for some cities between URID efficiency and carbon emissions. Efficiency improvement analysis is conducted for all three major city clusters. The main conclusions drawn from the study are as follows:

(1) The level of URID in the three major city clusters in the YREB during 2010–2019 showed a steady improvement; the total carbon emissions showed an overall upward trend, as shown by a rapid upward phase from 2010 to 2014, and a temporary decline and recovery from 2014 to 2019. The overall trend of URID efficiency is decreasing.

(2) The URID efficiency of the three major city clusters in the YREB is spatially distributed in decreasing order from MRYRUA > CYUA > YRDUA, and the gap between cities is gradually widening. In cities with better economic development, the URID level is generally higher than that of cities with average economic development, while the URID efficiency is low.

(3) The URID efficiency of the majority of the 73 cities in the three major city clusters is mainly inversely correlated with carbon emissions and decreases inversely with the gradual increase in carbon emissions. The more developed the socioeconomic structure of the cities is, the flatter the urban–rural integration efficiency is, and the larger the carbon emissions are.

(4) Regarding the input–output efficiency in the URID, the overall redundancy rate of the three major city clusters in the YREB is high in each input indicator, though the redundancy rate of each input indicator varies. Among these indicators, the redundancy of inputs into cultural, sports and media support, energy conservation and environmental protection, urban and rural communities, and housing security expenditures is the primary influence on efficiency loss; in addition, the redundancy of carbon emissions also has an impact on efficiency loss in the URID. Based on these results, optimization paths to improve the efficiency of the URID in the three major city clusters are proposed.

Author Contributions: Conceptualization, G.M., S.H. and L.T.; methodology, G.M., S.Q. and Z.H.; data curation, H.H.; writing—original draft preparation, G.M.; writing—review and editing, S.H., Z.W. and L.T. All authors have read and agreed to the published version of the manuscript.

Funding: This work was supported by the Key Project from the National Social Science Foundation of China (grant no. 18ZDA053).

Institutional Review Board Statement: Not applicable.

Informed Consent Statement: Not applicable.

Data Availability Statement: The data used in this study can be obtained by contacting the corresponding author.

Conflicts of Interest: The authors declare no conflict of interest.

References

1. Liu, Y.; Zang, Y.; Yang, Y. China's rural revitalization and development: Theory, technology and management. *J. Geogr. Sci.* **2020**, *30*, 1923–1942. [CrossRef]
2. Li, Y.H.; Jia, L.R.; Wu, W.H.; Yan, J.Y.; Liu, Y.S. Urbanization for rural sustainability-Rethinking China's urbanization strategy. *J. Clean. Prod.* **2018**, *178*, 580–586. [CrossRef]
3. Liu, Y.S.; Li, Y.H. Revitalize the world's countryside. *Nature* **2017**, *548*, 275–277. [CrossRef] [PubMed]
4. Zhang, Z.; Lu, Y. China's urban-rural relationship: Evolution and prospects. *China Agric. Econ. Rev.* **2018**, *10*, 260–276. [CrossRef]
5. Su, C.; Liu, T.; Chang, H.; Jiang, X. Is urbanization narrowing the urban-rural income gap? A cross-regional study of China. *Habitat Int.* **2015**, *48*, 79–86. [CrossRef]

6. Shi, J.G.; Yan, H.M. Implementation efficiency of urban-rural development integration is measured and suggested. *J. Tongji Univ.* **2014**, *25*, 67–74. (In Chinese)
7. Charnes, A.W.; Cooper, W.W.; Rhodes, E.L. Measuring the Efficiency of Decision Making Units. *Eur. J. Oper. Res.* **1979**, *2*, 429–444. [CrossRef]
8. Banker, R.D. Some models for estimating technical and scale inefficiencies in data envelopmentanalysis. *Manag. Sci.* **1984**, *30*, 1078–1092. [CrossRef]
9. Afonso, A.; Fernandes, S. Measuring local government spending efficiency: Evidence for the Lisbon region. *Reg. Stud.* **2006**, *40*, 39–53. [CrossRef]
10. Kalseth, J.; Rattsø, J. Spending and overspending in local government administration: A minimum requirement approach applied to Norway. *Eur. J. Polit. Econ.* **1995**, *11*, 239–251. [CrossRef]
11. Alam, I.M.S.; Sickles, R.C. The Relationship Between Stock Market Returns and Technical Efficiency Innovations: Evidence from the US Airline Industry. *J. Prod. Anal.* **1998**, *9*, 35–51. [CrossRef]
12. Yörük, B.K.; Zaim, O. Productivity growth in OECD countries: A comparison with Malmquist indices. *J. Comp. Econ.* **2005**, *33*, 401–420. [CrossRef]
13. Fulginiti, L.E.; Perrin, R.K. Agricultural productivity in developing countries. *Agric. Econ.* **1998**, *19*, 45–51. [CrossRef]
14. Fertner, C. Downscaling European urban-rural typologies. *Geogr. Tidsskr.* **2012**, *112*, 77–83. [CrossRef]
15. Fertner, C. The Emergence and Consolidation of the Urban-Rural Region: Migration Patterns around Copenhagen 1986–2011. *Tijdschr. Econ. Soc. Geogr.* **2013**, *104*, 322–337. [CrossRef]
16. Serra, P.; Vera, A.; Tulla, A.F.; Salvati, L. Beyond urban–rural dichotomy: Exploring socioeconomic and land-use processes of change in Spain (1991–2011). *Appl. Geogr.* **2014**, *55*, 71–81. [CrossRef]
17. Caffyn, A.; Dahlström, M. Urban-rural interdependencies: Joining up policy in practice. *Reg. Stud.* **2005**, *39*, 283–296. [CrossRef]
18. Lundh, C.; Prado, S. Markets and politics: The Swedish urban-rural wage gap, 1865–1985. *Eur. Rev. Econ. Hist.* **2015**, *19*, 67–87. [CrossRef]
19. Xiao, L.; He, Z.; Wang, Y.; Guo, Q. Understanding urban-rural linkages from an ecological perspective. *Int. J. Sust. Dev. World* **2017**, *24*, 37–43. [CrossRef]
20. Tang, D.; Bethel, B.J. Yangtze River Economic Belt environmental remediation efficiency based on an input-output optimization analysis. *Environ. Sci. Pollut. Res.* **2021**, *28*, 16734–16743. [CrossRef]
21. Chen, K.Z.; Mao, R.; Zhou, Y. Rurbanomics for common prosperity: New approach to integrated urban-rural development. *China Agric. Econ. Rev.* **2022**. *ahead of print*. [CrossRef]
22. Liu, Y.; Schen, C.; Li, Y. Differentiation regularity of urban-rural equalized development at prefecture-level city in China. *J. Geogr. Sci.* **2015**, *25*, 1075–1088. [CrossRef]
23. Liu, Y.; Lu, S.; Chen, Y. Spatio-temporal change of urban-rural equalized development patterns in China and its driving factors. *J. Rural Stud.* **2013**, *32*, 320–330. [CrossRef]
24. Yang, Y.; Bao, W.; Wang, Y.; Liu, Y. Measurement of urban-rural integration level and its spatial differentiation in China in the new century. *Habitat Int.* **2021**, *117*, 102420. [CrossRef]
25. Du, B.; Wang, Y.; He, J.; Li, W.; Chen, X. Spatio-Temporal Characteristics and Obstacle Factors of the Urban-Rural Integration of China's Shrinking Cities in the Context of Sustainable Development. *Sustainability* **2021**, *13*, 4203. [CrossRef]
26. Wu, X.; Cui, P. A Study of the Time-Space Evolution Characteristics of Urban-Rural Integration Development in a Mountainous Area Based on ESDA-GIS: The Case of the Qinling-Daba Mountains in China. *Sustainability* **2016**, *8*, 1085. [CrossRef]
27. Zheng, Y.; Tan, J.; Huang, Y.; Wang, Z. The Governance Path of Urban-Rural Integration in Changing Urban-Rural Relationships in the Metropolitan Area: A Case Study of Wuhan, China. *Land* **2022**, *11*, 1334. [CrossRef]
28. Chen, C.; LeGates, R.; Zhao, M.; Fang, C. The changing rural-urban divide in China's megacities. *Cities* **2018**, *81*, 81–90. [CrossRef]
29. Chen, K.; Long, H.; Liao, L.; Tu, S.; Li, T. Land use transitions and urban-rural integrated development: Theoretical framework and China's evidence. *Land Use Policy* **2020**, *92*, 104465. [CrossRef]
30. Tian, Y.; Qian, J.; Wang, L. Village classification in metropolitan suburbs from the perspective of urban-rural integration and improvement strategies: A case study of Wuhan, central China. *Land Use Policy* **2021**, *111*, 105748. [CrossRef]
31. Yang, L.; Zhao, F.C. Evaluation of urban-rural integration development efficiency based on DEA technology. *Inq. Into Econ. Issues* **2010**, *6*, 8–13. (In Chinese)
32. Cao, Y.; Yu, F.; Kang, Y.F. Urban and rural overall planning evaluation based on integrated AHP/DEA method. *Stat. Decis.* **2011**, *24*, 58–60. (In Chinese)
33. Yan, H.M.; Shi, J.K. Efficiency evaluation of my country's provincial local governments coordinating urban and rural development-based on DEA analysis at the provincial level. *Issues Agric. Econ.* **2014**, *35*, 94–102. (In Chinese)
34. Hu, Y.G.; Liao, C.Q.; Zhang, X.M.; Wang, C. Evaluation of overall urban and rural development efficiency based on data envelopment analysis. *City Plan. Rev.* **2016**, *40*, 46–50. (In Chinese)
35. Pan, J.H.; Yin, J. Evaluation and spatial difference analysis of urban and rural development efficiency in Gansu Province based on DEA-ESDA. *Econ. Geogr.* **2011**, *31*, 1439–1444. (In Chinese)
36. Yin, J.; Tan, Q.M.; Wu, X.L. Evaluation of urban and rural overall planning efficiency and its spatial spillover effect in Jiangsu Province. *Chin. J. Agric. Res. Reg. Plan.* **2018**, *39*, 76–182. (In Chinese)

37. Shi, J.G.; Duan, K.F.; Wu, G.D.; Li, J. Efficiency of urban-rural integration development in the Yangtze River Delta under the background of carbon emission constraint. *Econ. Geogr.* **2021**, *41*, 57–67. (In Chinese)
38. Fang, C.L.; Zhou, C.H.; Wang, Z.B. Strategic issues of sustainable development of urban agglomerations in the Yangtze River Economic Belt and key points of grading and gradient development. *Prog. Geogr.* **2015**, *34*, 1398–1408. (In Chinese)
39. Cai, B.; Shao, Z.; Fang, S.; Huang, X.; Tang, Y.; Zheng, M.; Zhang, H. The Evolution of urban agglomerations in China and how it deviates from Zipf's law. *Geo-Spat. Inf. Sci.* **2022**, 1–11. [CrossRef]
40. Oda, T.; Maksyutov, S.; Andres, R.J. The Open-source Data Inventory for Anthropogenic CO_2, version 2016 (ODIAC2016): A global monthly fossil fuel CO_2 griddede missions data product for tracer transport simulations and surface flux inversions. *Earth Syst. Sci. Data* **2018**, *10*, 87–107. [CrossRef]
41. Tone, K. A slacks-based measure of super-efficiency in data envelopment analysis. *Eur. J. Oper. Res.* **2002**, *143*, 32–41. [CrossRef]
42. Tone, K.; Tsutsui, M. An epsilon-based measure of efficiency in DEA-A third pole of technical efficiency. *Eur. J. Oper. Res.* **2010**, *207*, 1554–1563.
43. Zhou, J.L.; Qin, F.C.; Liu, J.; Zhu, G.L.; Zou, W. Measurement, spatiotemporal evolution and influence mechanism of urban-rural integration level in China from a multi-dimensional perspective. *Chin. J. Popul. Resour. Environ.* **2019**, *29*, 166–176. (In Chinese)
44. Shi, J.G.; Duan, K.F.; Wu, G.D. Measurement of urban-rural integration development level in the Yangtze River Delta and analysis of its spatiotemporal characteristics. *J. Tongji Univ.* **2022**, *33*, 78–89. (In Chinese)
45. Cai, B.; Shao, Z.; Fang, S.; Huang, X.; Huq, M.E.; Tang, Y.; Li, Y.; Zhuang, Q. Finer-scale spatiotemporal coupling coordination model between socioeconomic activity and eco-environment: A case study of Beijing, China. *Ecol. Indic.* **2021**, *131*, 108165. [CrossRef]
46. Liu, Y.; Long, H.; Chen, Y.; Wang, J.; Li Yurui Li Yuheng Yang, Y.; Zhou, Y. Progress of research on urban-rural transformation and rural development in China in the past decade and future prospects. *J. Geogr. Sci.* **2016**, *26*, 1117–1132.
47. Liu, Y.; Zhou, Y.; Li, Y. Rural regional system and rural revitalization strategy in China. *Acta Geogr. Sin.* **2019**, *74*, 2511–2528. (In Chinese)
48. Guo, Y.J. A New Dynamic Comprehensive Evaluation Method. *J. Manag. Sci. China* **2002**, *2*, 49–54. (In Chinese)
49. Yi, P.T.; Zhang, D.N.; Guo, Y.J.; Gao, L.Q. Dimensionless Method in Dynamic Comprehensive Evaluation. *J. China Univ.* **2009**, *30*, 889–892. (In Chinese)
50. Xia, Y.L. Exploring the intelligent path of community governance under the leadership of party building-Taking the "Community Link" in Baoshan, Shanghai as an example. *SHH. Pty. Hist. Pty. Bldg.* **2018**, *11*, 54–57. (In Chinese)
51. Liu, J.X.; Zeng, S. Intellectual property, top-level design and exploration practice of digital governance in rural China. *J. Lanzhou Univ.* **2020**, *48*, 64–71. (In Chinese)

sustainability

Article

Landslide Displacement Prediction Based on CEEMDAN Method and CNN–BiLSTM Model

Zian Lin [1], Yuanfa Ji [2] and Xiyan Sun [2,*]

[1] School of Computer Science and Information Security, Guilin University of Electronic Technology, Guilin 541004, China; 20031102010@mails.guet.edu.cn

[2] Information and Communication School, Guilin University of Electronic Technology, Guilin 541004, China; jiyuanfa@guet.edu.cn

* Correspondence: sunxiyan@guet.edu.cn

Abstract: Landslides are a typical geological disaster, and are a great challenge to land use management. However, the traditional landslide displacement model has the defect of ignoring random displacement. In order to solve this situation, this paper proposes a CNN–BiLSTM model that combines a convolutional neural network (CNN) model and a bidirectional long short-term memory network (BiLSTM) model. In this model, the CEEMDAN method is innovatively proposed to decompose landslide displacement. The GRA–MIC fusion correlation calculation method is used to select the factors influencing landslide displacement, and finally the CNN–BiLSTM model is used for prediction. The CNN–BiLSTM model was constructed to extract the temporal and spatial characteristics of data for landslide displacement prediction. Two new concepts that evaluate the state of a landslide and the trend of the landslide are proposed to improve the performance of the prediction model. Then, we discuss the prediction performance of the CNN–BiLSTM model under four different input conditions and compare it with seven other prediction models. The experimental prediction results show that the model proposed in this paper can be popularized and applied in areas with frequent landslides, and provide strong support for disaster prevention and reduction and land use management.

Keywords: land use management; landslide displacement prediction; complete ensemble empirical mode decomposition with adaptive noise; bidirectional long short-term memory

Citation: Lin, Z.; Ji, Y.; Sun, X. Landslide Displacement Prediction Based on CEEMDAN Method and CNN–BiLSTM Model. *Sustainability* **2023**, *15*, 10071. https://doi.org/10.3390/su151310071

Academic Editors: Lu Zhang, Bing Kuang and Bohan Yang

Received: 6 June 2023
Revised: 22 June 2023
Accepted: 23 June 2023
Published: 25 June 2023

1. Introduction

As the seasons change, the weather warms, human activity expands, and the frequency of natural disasters increases [1]. All kinds of natural disasters, such as soil erosion, floods, volcanic eruptions, earthquakes, and tsunamis, among which landslides are the most destructive and harmful [2], can lead to the severe loss of lives and properties [3–5]. Landslides are geomorphological processes that involve the mobilization of the ground, rocks, debris, and the mud downslope under the action of gravity, causing local erosion problems [6,7]. Human disturbance is also an important triggering mechanism for landslides [8]. In most countries, landslides have caused severe socioeconomic impacts on people, cities, industrial establishments, and lifelines, including highways, railways, and communication network systems [9]. China is among the countries most vulnerable to landslides worldwide [10]. According to China's 2020 China Statistical Yearbook, 4810 landslides occurred in China in 2020, accounting for 61.3 percent of the country's total geological disasters, causing many property losses and casualties [11]. Therefore, it is important to obtain predictions and alerts for landslides, considering their causes and probability of occurrence, to issue timely landslide hazard warnings and prevent similar tragedies [12]. This also plays an important role in the policymaking of land use management.

Over the years, the field of landslide displacement prediction has benefited from the continuous development of artificial intelligence and landslide monitoring technologies. A variety of analytical methods and machine learning models have been used for landslide displacement prediction [13–24]. Chenhui et al. [13] combined a genetic algorithm with the Elman neural network, optimized the weight, threshold, and number of hidden neurons of the Elman neural network, and solved the problem which Elman easily falls into, of local minima and neuron data being difficult to determine. Yong et al. [14] fused the predicted trend series with the sensitivity state to obtain the nonlinear prediction model. S.H. et al. [15] constructed a weighted multi-kernel grey model based on grey theory, multi-kernel learning and weighted learning. Lizhou et al. [16] proposed a nonlinear grey prediction model with background value optimization (BNGM(1, 1, t2)) and compared it with three kinds of grey Bernoulli models, illustrating the advantages of the proposed model. Yanan et al. [17] proposed a new graph convolution network fused with the GRU model (GC-GRU-N) and applied it to landslide displacement prediction. Cheng et al. [18] improved the bootstrap method, used partial neural networks to construct PI, and used a random vector functional link network (RVFLN) instead of ELM as the predictor of the neural network. As a result, excellent landslide displacement interval prediction was achieved. Jingjing et al. [19] proposed the multi-feature fusion transfer learning (MFTL) method, utilizing the knowledge and skills gained from the Baijiapu landslide scenario, to improve the prediction ability of other landslides. Peihong et al. [20] considered the Laowuji landslide to be a research object, studied its dynamic failure mode, and finally decided to use a variety of factors, including geological conditions, rainfall intensity and human activities, as input and used a long short-term memory (LSTM) model to predict landslide displacement. Heming et al. [21] recombined the mutation displacement data to reduce the displacement of the mutation-affected data in the steady state and accurately predicted the displacement of the landslide mutation segment. Shaohong et al. [22] combined dual support vector regression with the Hausdorff derivative operator and adopted the improved salp group algorithm to determine the model parameters, and the new model was successfully used to predict the actual landslide displacement. Lizheng et al. [23] proposed a low-cost landslide displacement prediction method, which used time series measurements of acoustic emission (AE) and rainfall to predict the displacement, and they verified the effectiveness of the proposed method with a landslide that occurred in Hollin Hill, North Yorkshire, UK. Xinli et al. [24] combined empirical and data methods, and a displacement prediction method was constructed based on the Verhulst inverse function (VIF) and the random forest (RF) algorithm. The performance of the model was evaluated using RMSE and MAPE.

Since landslide displacement changes gradually over time, experts have used the time analysis method to analyze the landslide displacement in many studies [25–35]. Because the moving average (MA) method has the advantage of eliminating the accidental change factors and determining the development trend of things, it is used to analyze the landslide displacement in time and decompose the landslide displacement into trend displacement and periodic displacement for forecasting. Rubin et al. [25] built a landslide displacement prediction model by combining the ELM model with the RS-SVR model of random search support vector regression. They used the ELM model and RS-SVR model to predict the trend displacement and periodic displacement, respectively, and then they summed the two results to obtain the predicted total displacement. Yonggang et al. [26] used a cubic polynomial to predict trend displacement and the GRU model to predict periodic displacement and applied it to the Erdaohe landslide, which achieved good results. Beibei et al. [27] selected input data by calculating the grey correlation degree, and the LSTM model predicted the periodic displacement and used the real data of the Baishuihe landslide and Bazimen landslide to simulate and test the performance of the model. Fasheng et al. [28] established a dynamic model based on displacement observations and used GA-SVR to predict periodic and random terms in displacement. Although the prediction accuracy of random terms is not high, the trend can be reflected to a certain extent, which is helpful for landslide prediction. Zhongqiang et al. [29] used three prediction models, GRU, RF, and LSTM, to verify and compare the prediction effects of three landslides in the Three Gorges area, which illustrated the effectiveness of the three models in landslide displacement prediction. Yankun et al. [30] compared five commonly used machine learning prediction models on three landslide datasets. The Hodrick–Prescott filter was used to decompose landslide displacement into trend displacement and periodic displacement, and double exponential smoothing was used to predict trend displacement. The results show that no model is optimal for the three landslides at the same time, and different models should be selected for different landslides. Zian et al. [31] improved the time series analysis method of landslides, using the WMA method to decompose the landslide displacement and the LSTM model to predict the trend displacement, and obtained good results. Subsequently, Zian et al. [32] further analyzed the composition of landslide displacement and improved the theoretical method, using the EWMA method to decompose and the Double-BiLSTM model to predict landslide displacement, which greatly improved the prediction results. Zizheng et al. [33] and Qi et al. [34] both used the variational mode decomposition (VMD) method, which is a data evaluation and decomposition method that adaptively realizes the frequency domain division of the signal and the effective separation of each component. Zizheng et al. [33] used VMD combined with a periodic neural network model, and Qi et al. [34] used the VMD method combined with the WA-GWO-BP model to achieve the accurate prediction of landslide displacement. Shiluo et al. [35] used the EMD method to decompose Baijiabao landslide displacement data, and one-step-ahead prediction and multistep-ahead prediction methods were used for prediction.

However, the method of time series analysis also has its shortcomings. In the analytical process, the existence of random displacement will be ignored because the model cannot accurately predict random displacement. To solve the shortcomings of the time series analysis method, some experts propose the time frequency analysis method, which does not ignore any part of the data and can effectively improve the accuracy of landslide displacement measurements [36–38]. Zhenglong et al. [36] and Chao et al. [37] divided landslide displacement into subsequences with different frequencies based on wavelet transform theory. Faming et al. [38] improved wavelet transform theory by using the DWT discrete wavelet transform to decompose landslide displacement, using chaos theory to reconstruct each frequency, and finally using the ELM model for prediction.

Most of the previous landslide displacement models adopted the time series method, taking landslide displacement as the data changing with time. Although those models based on the time series method can predict landslide displacement, the method has a disadvantage of ignoring random displacement. In addition, when selecting the input variables of the prediction model, only one correlation calculation method is usually used, which leads to insufficient comprehensiveness. And the choice of the displacement prediction model cannot take into account the temporal and spatial attributes of the data.

In this paper, a landslide displacement prediction model based on time-frequency analysis is proposed. This model uses the complete ensemble empirical mode decomposition with adaptive noise (CEEMDAN) method to analyze landslide displacement, which can overcome the defects of the time series method. The model adopts the joint correlation degree calculation method GRA–MIC to select the influencing factors of displacement, and considers the input variables of the model from multiple perspectives. Finally, combining the advantages of the CNN model and the BiLSTM model in data processing, the CNN–BILSTM model is constructed to effectively extract the spatiotemporal characteristics of the data, and finally achieve accurate displacement prediction. The research results in this paper lay a foundation for the technical progress of landslide monitoring and early warning systems in the future as an important part of disaster prevention and land use management.

The main contributions of this paper are described as follows:

1. According to the principle of time frequency analysis, the CEEMDAN method [39–41] is used to decompose the landslide displacement into multiple subsequences. In this method, the original data are decomposed into different frequency data series

with local characteristics, and the data characteristics of each frequency in landslide displacement are highlighted.
2. This paper analyzes the landslide situation in the study area and proposed two new concepts, using the landslide displacement of the previous month to represent the current state of the landslide and quantifying the difference between two consecutive months of displacement data as the trend of landslide change, adding relevant data of landslide prediction and creating conditions for improving the performance of landslide prediction.
3. To consider the factors affecting landslide displacement more comprehensively, this paper combines two correlation degree calculation methods, GRA [42–44] and MIC [45,46], to obtain the GRA–MIC method. This method comprehensively selects the influencing factors from two perspectives, which is helpful to further improve the accuracy of the landslide displacement prediction model.
4. Combined with the ability of the CNN model [47] to extract local features of data and the BiLSTM model [48,49] to process time series data, the CNN–BiLSTM model was constructed to predict landslide displacement. This paper combines the two models to effectively improve the prediction performance [50].

2. Materials and Methods

2.1. Complete Ensemble Empirical Mode Decomposition with Adaptive Noise

Compared with EEMD, the CEEMDAN method adds adaptive white Gaussian noise at each stage in the decomposition process and obtains each modal component by calculating the unique margin signal. The decomposition process is complete, and the reconstruction error is extremely low [39]. The CEEMDAN method can effectively solve the mode aliasing problem of EMD and overcome the problems of low decomposition efficiency of EEMD and difficulty in completely eliminating noise [40].

$E_j(\cdot)$ is defined as the Jth mode functions obtained by the EMD algorithm, $X(t)$ is the original data series, and $n^i(t)$ is the ith added white Gaussian noise satisfying the standard normal distribution. The implementation steps of the CEEMDAN method are given as follows [41]:

(1) Similar to EEMD, the signal $X(t) + \varepsilon_0 n^i(t)$ is decomposed n times by EMD in the CEEMDAN method, and the first mode functions are obtained by mean calculation:

$$\overline{IMF_1(t)} = \frac{1}{N}\sum_{i=1}^{N} IMF_1^i(t) \tag{1}$$

(2) Calculate the first margin signal $r_1(t)$ as

$$r_1(t) = X(t) - \overline{IMF_1(t)} \tag{2}$$

(3) The EMD algorithm is used to decompose the signal $r_1(t) + \varepsilon_1 E_1(n^i(t))$ n times and then obtain the second mode functions as

$$\overline{IMF_2(t)} = \frac{1}{N}\sum_{i=1}^{N} E_1(r_1(t) + \varepsilon_1 E_1(n^i(t))) \tag{3}$$

(4) For $k = 2, \ldots, K$, calculate the kth residual signal as

$$r_k(t) = r_{k-1}(t) - \overline{IMF_k(t)} \tag{4}$$

(5) The calculation process of step (3) is repeated, and the $k + 1$ mode functions are obtained as

$$\overline{IMF_{k+1}(t)} = \frac{1}{N}\sum_{i=1}^{N} E_1(r_k(t) + \varepsilon_k E_k(n^i(t))) \tag{5}$$

(6) Steps (4) and (5) are repeated until the residual signal meets the termination condition of the decomposition, and K mode functions are finally obtained. The final residual signal of the decomposition is

$$R(t) = X(t) - \sum_{k=1}^{K} \overline{IMF_k(t)} \tag{6}$$

Then, the final original data signal can be decomposed into

$$X(t) = \sum_{k=1}^{K} \overline{IMF_k(t)} + R(t) \tag{7}$$

After the CEEMDAN method has been used to decompose the landslide displacement, each displacement component will be predicted separately in this paper.

2.2. Grey Relation Analysis and Maximal Information Coefficient

The basic idea of grey relation analysis theory is to judge the degree of correlation between factors according to the degree of similarity between curves, which can be used to quantitatively analyze the dynamic development process of the system to determine the degree of contribution of factors to a certain behavior or index [42]. In essence, grey correlation analysis is used to find the main relationship between various factors and determine the relevant factors that cannot be ignored to grasp the main contradiction of the development of things. Grey correlation analysis includes the following three elements: the main sequence, subsequence, and correlation degree. When the method is used to analyze the influence degree, the main sequence is generally the main behavior or index used to evaluate the system performance. The subsequence is made up of the various factors that affect the system performance. The correlation degree is the correlation degree between subsequence and main sequence obtained by grey correlation analysis [43]. In this paper, landslide displacement is selected as the main sequence, and four subsequences constitute sequence X. $X = [X_0, X_1, X_2, X_3, X_4]$ = [landslide displacement, precipitation, reservoir level, trend of landslide, state of landslide]. The analysis and selection of the factors influencing landslide displacement are presented in Section 3.2 of this paper. The process of GRA is described as follows [44].

Because the physical interpretation of each type of data is different, resulting in different ranges of resulting data, it is not suitable for direct comparison. Therefore, data normalization needs to be performed in GRA. The following equation is the normalization method for the data:

$$X_i(k)' = X_i(k) / \frac{1}{n} \sum_{i=1}^{N} X_i(k) \tag{8}$$

where $i = 0, 1, \ldots, m; k = 0, 1, \ldots, n$, M is the number of types of influencing factors, and N is the number of data affecting factors. After data normalization, correlation coefficients between landslide displacement and the other four influencing factors and grey relational grade could be calculated as follows:

$$\xi\left(x_0(k)', x_i(k)'\right) = \frac{min_i min_k \left|x_i(k)' - x_0(k)'\right| + \rho min_i min_k \left|x_i(k)' - x_0(k)'\right|}{\left|x_i(k)' - x_0(k)'\right| + \rho max_i max_k \left|x_i(k)' - x_0(k)'\right|} \tag{9}$$

$$r(x_0, x_i) = \frac{1}{n} \sum_{k=1}^{n} \xi\left(x_0(k)', x_i(k)'\right) \tag{10}$$

where $\xi\left(x_0(k)', x_i(k)'\right)$ is the correlation coefficient between x_i and the sequence x_j, ρ is the resolution coefficient, the usual value is 0.5, and $r(x_0, x_i)$ is the final grey relational grade. Generally, factors with GRD > 0.65 are considered to be important influencing factors [51].

The maximal information coefficient (MIC) was proposed by Reshef et al. [45] in 2011, and it is developed based on mutual information (MI). Mutual information can be regarded as the uncertainty of a random variable reduced by the knowledge of another random variable, which is mainly used to measure the degree of correlation between linear or nonlinear variables, and its value range is [0, 1]. If x and y are random variables, the mutual information is defined as

$$I(x;y) = \sum_{x}\sum_{y} p(x,y) log \frac{p(x,y)}{p(x)p(y)} \tag{11}$$

where $I(x;y)$ is the mutual information of variables x and y, $p(x,y)$ is the joint probability density function, and $p(x)$ and $p(y)$ are the marginal density functions. The greater the mutual information between the two variables is, the stronger the correlation is [46]. Compared with mutual information, MIC overcomes the disadvantage that mutual information cannot be used to conveniently calculate continuous variables based on MI, and it has a higher accuracy. MIC is a normalized maximum mutual information with low computational complexity, good robustness, and higher accuracy than mutual information. When sufficient statistical samples are available, MIC can capture a wide range of relationships and better reflect the degree of association between attributes and features [52]. The scatterplot composed of random variables x and y in two-dimensional space is gridded in m columns and n rows, and then the MIC formula is:

$$MIC(x;y) = max_{m*n<B(n)} \left(\sum_{x}\sum_{y} p(x,y) log \frac{p(x,y)}{p(x)p(y)} / \log \min(m,n)\right) \tag{12}$$

where $m * n < B(n)$ represents the constraints on the total number of meshes, and $B(n)$ is usually set to $n^{0.6}$. The greater the MIC value between the two variables is, the stronger the correlation is. Conversely, for the opposite, the weaker the correlation is. Generally, factors with MIC > 0.3 are considered to be important influencing factors [48].

To better select the factors influencing landslide displacement, the factors selected in this study need to meet both GRG > 0.65 and MIC > 0.3.

2.3. CNN–BiLSTM Model

A convolutional neural network (CNN) is a feedforward neural network [53]. A typical CNN model is shown in Figure 1. It includes an input layer, convolutional layer, pooling layer, fully connected layer, and output layer.

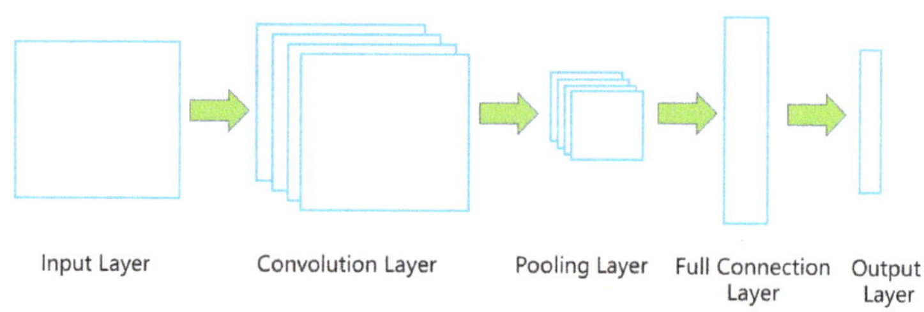

Figure 1. Structure of the CNN model.

The essence of the CNN model lies in the construction of multiple filters that can extract data features, and the hidden topological features among data can be extracted through layer-by-layer convolution and pooling of input data [47]. As the number of layers increases, the extracted features become increasingly abstract. Finally, these abstract features are merged through fully connected layers, and the classification and regression problems are solved by softmax or sigmoid activation functions. One of the characteristics of the CNN model is that it can extract local features of input data [54]. Moreover, the high-level features are abstracted and combined layer by layer, which can effectively realize feature extraction in a complex landslide environment.

The LSTM model is a variant of the RNN model that transmits forward information and processes current information. The LSTM model introduces a new internal state to transmit linear cyclic information, outputs information to the hidden state, and selects to retain or forget information through three control gate units (input gate, forget gate, and output gate) [55]. The input gate controls how much input information needs to be retained at the current time. The forget gate controls how much information needs to be discarded at the last moment. The output gate controls how much information needs to be output to the hidden state at the current time [56]. Although the LSTM model can obtain the feature information over a long distance, the information it obtains is the information obtained before the output time, instead of using the reverse information, while the BiLSTM model can use the past and future information to make more perfect and detailed decisions [49]. The BiLSTM model is an improved version of the LSTM model, which is very suitable for processing time series data [48]. The BiLSTM model is formulated as follows:

$$f_t = \sigma\left(W_f x_t + U_f h_{t-1} + b_f\right) \tag{13}$$

$$\widetilde{c}_t = tanh(W_c x_t + U_c h_{t-1} + b_c) \tag{14}$$

$$i_t = \sigma(W_i x_t + U_i h_{t-1} + b_i) \tag{15}$$

$$c_t = f_t \odot c_{t-1} + i_t \odot \widetilde{c}_t \tag{16}$$

$$o_t = \sigma(W_o x_t + U_o h_{t-1} + b_o) \tag{17}$$

$$\overrightarrow{h}_t = \overrightarrow{LSTM}(h_{t-1}, x_t, c_{t-1}), t \in [1, T] \tag{18}$$

$$\overleftarrow{h}_t = \overleftarrow{LSTM}(h_{t+1}, x_t, c_{t+1}), t \in [T, 1] \tag{19}$$

$$H_t = \left[\overrightarrow{h}_t, \overleftarrow{h}_t\right] \tag{20}$$

where x_t, f_t, i_t, o_t, h_t, C_t, and $\widetilde{C}_t$ denote the input data, forget gate, input gate, output gate, output data, cell state, and temporary state of the cell, respectively; w_f, w_o, w_i, and w_c denote the weight of the forget gate, the weight of the output gate, the weight of the input gate, and the weight of the temporary state, respectively; and b_f, b_i, b_o, and b_c represent the bias of the forget gate, the bias of the input gate, the bias of the output gate, and the bias of the temporary state, respectively. [] is the connection between two vectors, tanh is the tanh function, σ is the sigmoid function, $\odot$ is the matrix product, and $\overrightarrow{h}_t$ and $\overleftarrow{h}_t$ are the outputs of BiLSTM in two directions. H_t is the output of BiLSTM. Figure 2 shows the BiLSTM architecture.

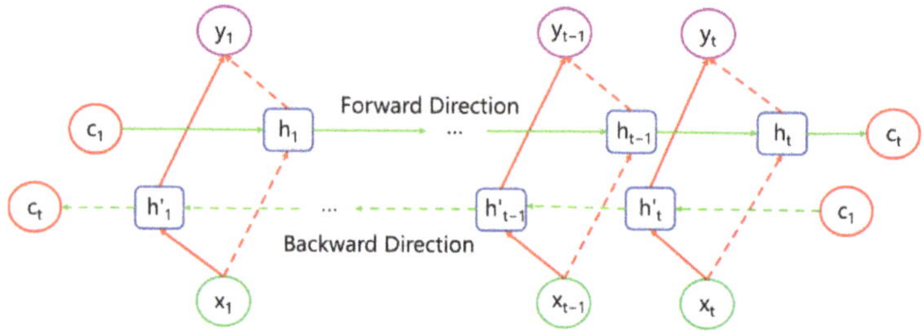

Figure 2. Structure of the BiLSTM model.

CNN and BiLSTM are both mainstream deep learning models. CNN is more suitable for spatial expansion, extracting local data features, and combining and abstracting them into high-level features. BiLSTM is more suitable for time expansion; it has long-term memory function, and it is more suitable for processing time series. In the feature extraction of landslide displacement and environmental factor data, it is necessary to consider not only the spatial relationship between different parameters, but also the change in data in the temporal dimension. Therefore, this paper combines the CNN and BiLSTM models to propose a CNN–BiLSTM model, which enables the model to express features spatiotemporally. The structure of the proposed model is shown in Figure 3.

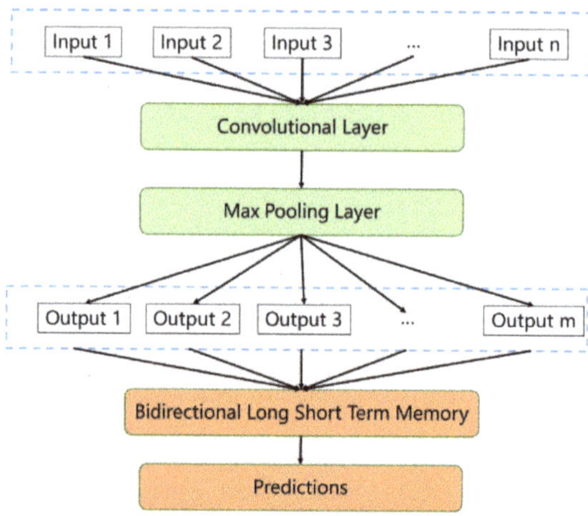

Figure 3. Structure of the CNN–BiLSTM model.

2.4. *Performance Indicators*

To evaluate the prediction effect of different artificial intelligence models, a variety of indicators can be used to verify model performance [57]. In this paper, the root mean square error (*RMSE*), mean absolute error (*MAE*), mean absolute percentage error (*MAPE*), and correlation coefficient R^2 were used to reflect the prediction effect.

$$MSE = \sqrt{\frac{1}{n}\sum_{i=1}^{n}(\hat{y}_i - y_i)^2} \tag{21}$$

$$MAE = \frac{1}{n}\sum_{i=1}^{n}|\hat{y}_i - y_i| \tag{22}$$

$$MAPE = \frac{100\%}{n}\sum_{i=1}^{n}\left|\frac{\hat{y}_i - y_i}{y_i}\right| \tag{23}$$

$$R^2 = 1 - \frac{\sum_{i=1}^{n}(\hat{y}_i - y_i)^2}{\sum_{i=1}^{n}(\overline{y}_i - y_i)^2} \tag{24}$$

where $\hat{y} = \{\hat{y}_1, \hat{y}_2, \ldots, \hat{y}_n\}$ is the predicted value, $y = \{y_1, y_2, \ldots y_n\}$ is the measured value, $\overline{y} = \{\overline{y}_1, \overline{y}_2, \ldots \overline{y}_n\}$ is the average of the measurements, and n is the number of samples. The model is judged according to the results of *RMSE*, *MAE*, *MAPE* and R^2, and the value range of the results is [0, 1]. The closer to 0 the values of the first three evaluation indexes are, the better the prediction performance of the model is. The closer to 1 R^2 is, the better the prediction performance of the model is.

3. Results

3.1. Real Case

The Baishuihe landslide is located on the right bank of the Yangtze River in Zigui County, Three Gorges Reservoir Area, 56 km away from the Three Gorges Dam, 110°32′09″ east longitude, 31°01′34″ north latitude [58]. Surrounded by mountains on three sides and water on one side, it is very conducive to the collection of rainfall. The elevation of the terrain gradually increases from north to south, with a difference of approximately 330 m. The landslide is approximately 700 m wide from east to west and divided by bedrock ridges on both sides. It runs north–south and has a length of approximately 770 m. The overall slope of the landslide ranges from 30° to 35°, the average thickness is 30 m, and the volume is approximately 1260 × 104 m^3. The Baishuihe landslide is an accumulation-type, soil-like landslide with a loose structure; the sliding body is mainly composed of gravel soil and silty clay mixed with gravel, the sliding zone soil is mostly silty clay mixed with gravel or breccias, the underlying bedrock is argillaceous siltstone, mostly in the form of moderate weathering, and the joint and fracture development is relatively obvious. The terrain is stepped, steeper in the upper part and gentle in the middle, creating favorable conditions for the accumulation of colluvial materials. The Baishuihe landslide is a flat transition slide, and the thickness of the sliding body gradually increases from the rear edge to the forward edge, especially in the middle and front of the landslide. Due to the small deformation of the rear part of the landslide, it is in a relatively stable stage, so the main risk of landslide is concentrated in the warning area of the landslide front. There are obvious macroscopic deformations and house cracking on the surface of the landslide in Baishuihe, and the nearby villagers have been relocated. Now, the risk of Baishuihe landslide is mainly a threat to passing boats and roads within the landslide. Eleven Global Positioning System (GPS) monitoring points were installed on the Baishuihe landslide. Since the ZG118 monitoring point was installed in the central area of the whole Baishuihe landslide, it can better reflect the situation of the Baishuihe landslide. Therefore, other studies also use data from the ZG118 monitoring point [59]. The data of two horizontal directions and one vertical direction are monitored for Baishuihe landslide monitoring points, and the final displacement value is a vector calculation value of three directions. Rainfall is based on the data from a local weather station. The reservoir water level is based on the data provided by the Three Gorges hydrology station. The distribution of GPS monitoring points is shown in Figure 4.

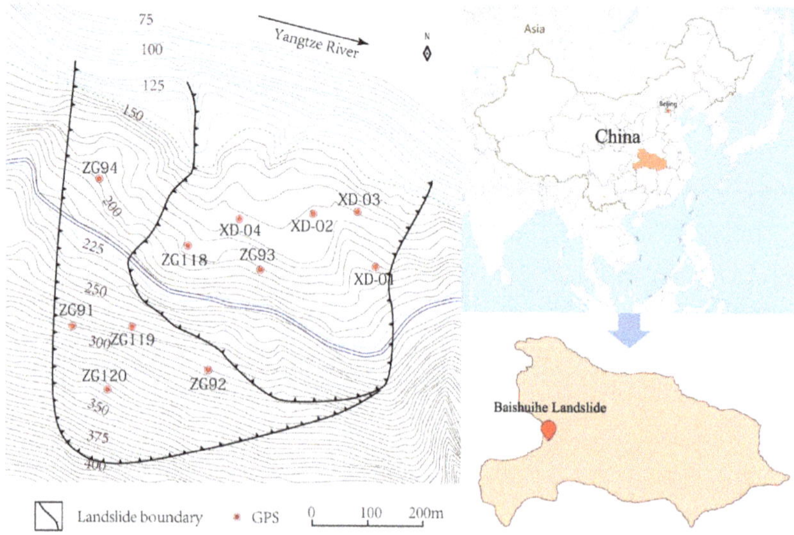

Figure 4. GPS installation positions.

In this paper, the rainfall data and reservoir water level data of the Baishuihe landslide in the same period were monitored and recorded once a month. The time range was from January 2004 to December 2012, with a total of 108 data points, as shown in Figure 5.

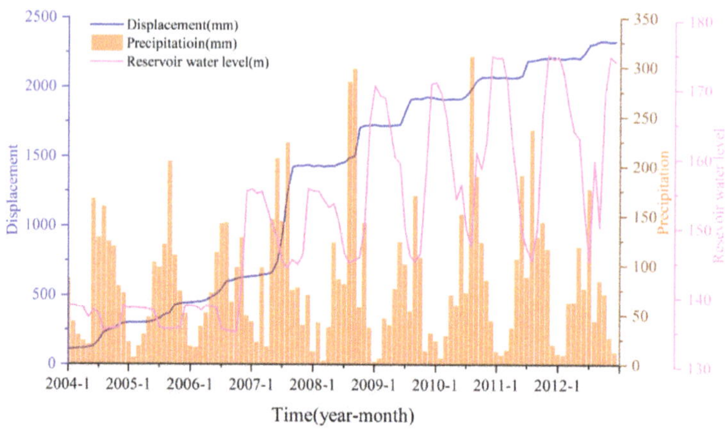

Figure 5. Displacement and environmental data variation in the Baishuihe landslide.

In this paper, the first 96 data points were used as the training data for the model, and the last 12 data points were used as the prediction data for the model test set. The prediction results were compared with the actual measured data to evaluate the prediction performance of the model.

3.2. Analysis of Factors Influencing Landslide Displacement

The Three Gorges area, where the Baishuihe landslide is located, is part of the subtropical monsoon climate zone, with precipitation concentrated from April to August and typically very little rainfall from January to March. Based on the historical data, when rainfall increases, landslide displacement also increases, and when rainfall is scarce, landslide displacement hardly changes. The reason is that a large amount of rain washes the landslide and drives the soil on the slope to slide downward. The rain enters the landslide, increasing the weight of the landslide and increasing the possibility of land sliding. A large amount of rainwater infiltration leads to the saturation of the soil and rock layer on the slope, and even water accumulation on the waterproof layer at the lower part of the slope, thus increasing the weight of the sliding body and reducing the shear strength of the soil and rock layer, resulting in landslides. When studying landslides, many researchers consider rainfall to be one of the influencing factors [16,60], and some scholars also consider rainfall to be the most important influencing factor [61].

Figure 5 shows that the landslide displacement changed the most in 2007, but the rainfall was the largest in 2008 and 2010, which indicates that in addition to rainfall there are other factors that also affect landslide displacement. Because the Baishuihe landslide is on the right bank of the Yangtze River, close to the Three Gorges Dam, it is easily affected by the release of water from the dam. Whenever the Three Gorges Dam opens the sluice to release water, the water level of the reservoir drops and the water level of the Yangtze River rises rapidly, impacting the surface of the Baishuihe landslide, and water flows into the slope, increases the pore water pressure, softens the rock and soil, and increases the bulk density of the landslide. The overall structure of the landslide has an impact that promotes or induces the occurrence of land sliding.

The geological conditions of landslides are complex, and there is no clear and unified standard. According to previous studies, landslides exist in a variety of different states, and the corresponding stability of different states is also different [14,29]. The cumulative displacement–time curve of the Baishuihe landslide presents an obvious ladder-like pattern. In particular, the height of the ladder was highest in 2007. The maximum displacement velocity of the Baishuihe landslide is greater than 26 mm/day, and the average annual deformation rate is also above 250 mm. The displacement of the landslide moved slowly from 2004 to 2006, and the displacement accelerated obviously in 2007. However, the deformation speed dropped again at the beginning of 2008, and the displacement grew slowly. The fastest increase in the deformation rate of the landslide occurred in July 2007, which coincided with the decrease in rainfall and reservoir water level during this period. We believe that the magnitude of landslide displacement is related to the stability of the landslide. When the landslide is in a stable state, it is difficult for external factors to lead to the occurrence of landslide displacement. When the landslide is in an unstable state, relatively minor factors may lead to a more serious landslide collapse phenomenon. Therefore, the landslide displacement can reflect a certain landslide state. This paper intends to use the displacement of the previous month to represent the current state of the landslide and participate in the prediction of landslide displacement.

Due to the influence of the landslide itself and environmental factors, it usually produces a certain displacement every month. If the displacement of the previous month is taken as the current state of the landslide, the difference in the displacement data of two consecutive months is considered to be the change between the two states of the landslide, which reflects the development trend of the landslide to a certain extent. When the change is large, it reflects the development direction of the landslide, indicating that the landslide is in a trend of unstable development and can change violently. To improve the accuracy of landslide displacement prediction, this paper attempts to quantify the difference between two consecutive months of displacement data as the trend of landslide change, which is considered to be one of the inputs of the prediction model.

The selection of influencing factors will directly affect the training and prediction ability of the model [28]. Based on the above analysis, this study believes that the development of landslide displacement is the result of the influence of rainfall, reservoir water level, landslide trend and landslide state. Therefore, this paper considers these four factors to be the factors influencing landslide displacement.

3.3. Decomposition of Original Data

For the raw data, the CEEMDAN method decomposition training set of landslide displacement, rainfall, and reservoir water level, with the status and trend of landslide data, namely, 96 consecutive data points for decomposition, was used. The decomposition of landslide displacement will obtain three components, the decomposition of rainfall will obtain five components, the decomposition of the reservoir water level will obtain four components, the decomposition of landslides will obtain three state variables, and the decomposition of the landslide trend will yield six components. The results breakdown is shown in Figures 6–11.

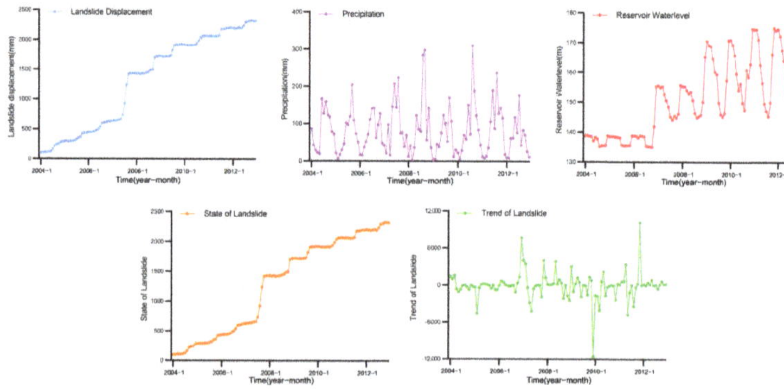

Figure 6. Original data of Baishuihe landslide.

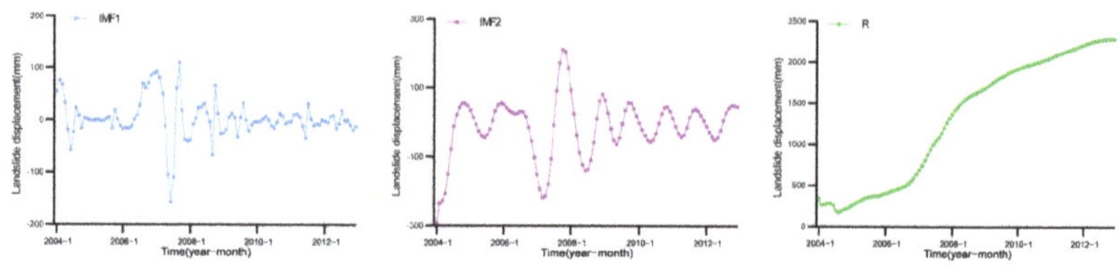

Figure 7. CEEMDAN method decomposing original landslide displacement data results.

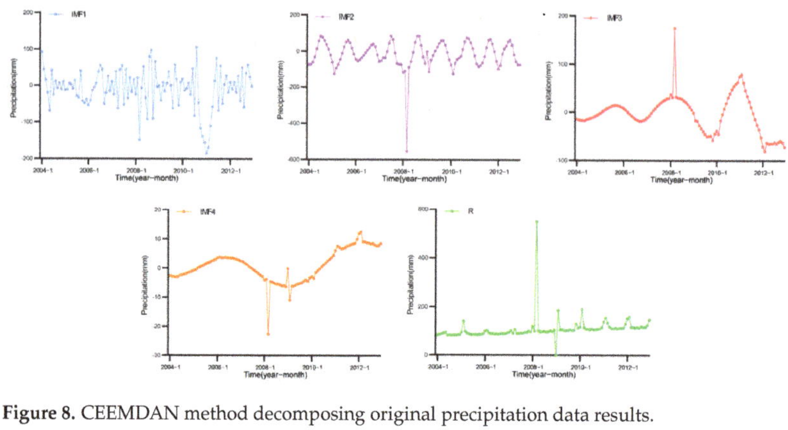

Figure 8. CEEMDAN method decomposing original precipitation data results.

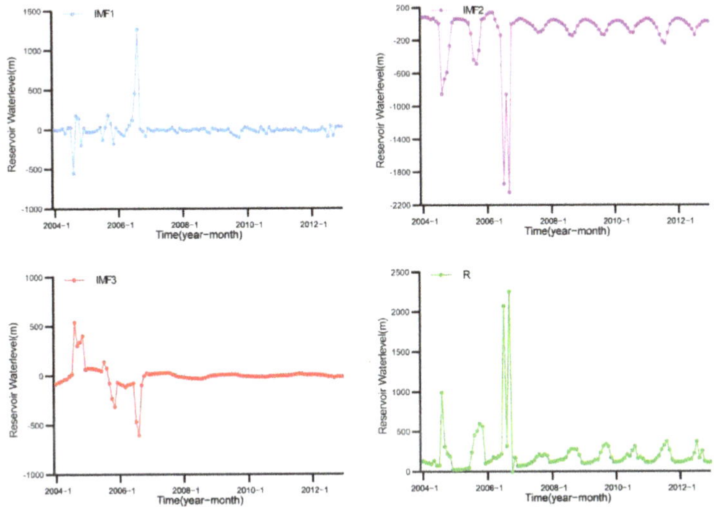

Figure 9. CEEMDAN method decomposing reservoir water level original data results.

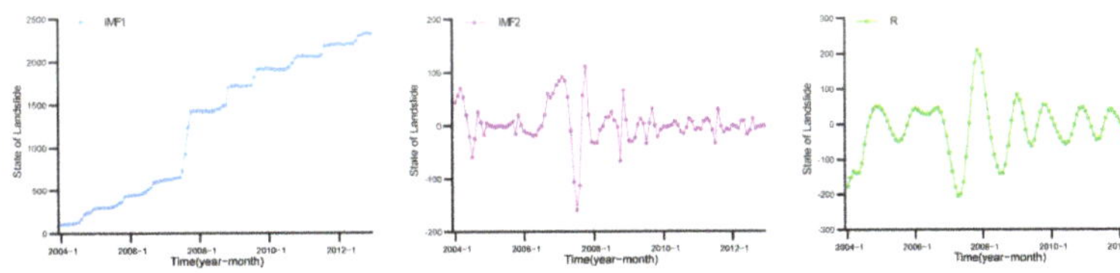

Figure 10. CEEMDAN method decomposing original state data results.

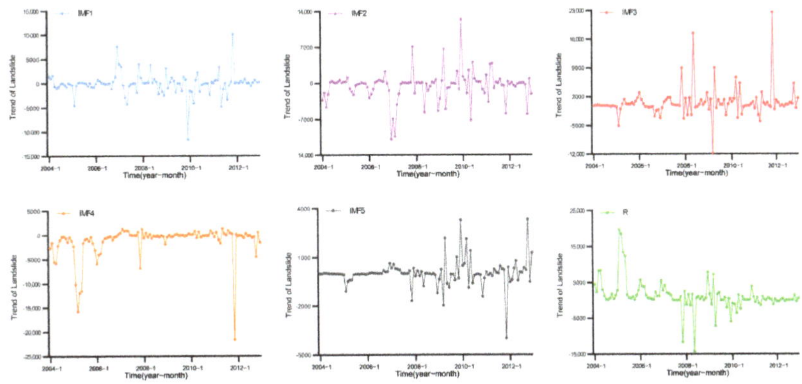

Figure 11. CEEMDAN method decomposing original trend data results.

3.4. GRA–MIC Algorithm Calculation of the Correlation

After the landslide displacement had been decomposed by the CEEMDAN method, multiple subsequences with different frequencies could be obtained. However, not all factors had an effect on landslide displacement. Using factors with less influence on landslide displacement to train the prediction on model will reduce the prediction accuracy, while using factors with greater influence will help improve the prediction performance of the model. Many studies use the MIC algorithm or GRA algorithm to calculate the correlation degree between landslide displacement and environmental factors [61,62], and both algorithms quantify the correlation degree from their own single perspective. Considering these two algorithms, this study proposes a GRA–MIC algorithm and, when combined with the GRA algorithm and the MIC algorithm, it can consider the correlation between displacement and environmental factors from two perspectives and improve the prediction accuracy. Moreover, in the Discussion section, the prediction is compared with that using the GRA algorithm or the MIC algorithm alone. The correlation calculation results are shown in Tables 1 and 2.

Table 1. GRA between landslide displacement and influencing factors.

Landslide Displacement	Influencing Factors	Influencing Factor Subsequences					
		IMF1	IMF2	IMF3	IMF4	IMF5	IMF6
IMF1	Precipitation	0.794	0.712	0.715	0.674	0.622	/
	Reservoir water level	0.791	0.782	0.712	0.623	/	/
	State of landslide	0.904	0.805	0.625	/	/	/
	Trend of landslide	0.887	0.836	0.735	0.721	0.749	0.671
IMF2	Precipitation	0.755	0.691	0.707	0.672	0.616	/
	Reservoir water level	0.755	0.738	0.689	0.622	/	/
	State of landslide	0.793	0.904	0.623	/	/	/
	Trend of landslide	0.804	0.797	0.724	0.701	0.703	0.627
R	Precipitation	0.620	0.619	0.590	0.623	0.900	/
	Reservoir water level	0.638	0.639	0.616	0.927	/	/
	State of landslide	0.605	0.603	0.989	/	/	/
	Trend of landslide	0.623	0.628	0.602	0.726	0.607	0.491

Table 2. MIC between landslide displacement and influencing factors.

Landslide Displacement	Influencing Factors	Influencing Factor Subsequences					
		IMF1	IMF2	IMF3	IMF4	IMF5	IMF6
IMF1	Precipitation	0.194	0.212	0.264	0.269	0.337	/
	Reservoir water level	0.263	0.235	0.290	0.337	/	/
	State of landslide	0.338	0.266	0.337	/	/	/
	Trend of landslide	0.304	0.291	0.247	0.273	0.329	0.337
IMF2	Precipitation	0.255	0.300	0.337	0.482	0.531	/
	Reservoir water level	0.268	0.238	0.331	0.381	/	/
	State of landslide	0.319	0.757	0.531	/	/	/
	Trend of landslide	0.179	0.304	0.306	0.303	0.400	0.512
R	Precipitation	0.309	0.370	0.954	0.852	0.913	/
	Reservoir water level	0.423	0.468	0.837	0.789	/	/
	State of landslide	0.323	0.538	0.679	/	/	/
	Trend of landslide	0.236	0.382	0.598	0.978	0.842	0.877

After obtaining the results of the correlation calculation with the GRA–MIC algorithm, it is necessary to select appropriate factors to participate in the training and prediction of the model. Selecting factors with a correlation that is too low will result in the selection of too many data that are not related to landslide displacement, which will reduce the accuracy of the landslide displacement prediction. Although the selection of factors with a high correlation is beneficial to the prediction process, there are few qualified data, which will lead to insufficient training of the model, affecting the prediction performance of the model. In this paper, data satisfying the conditional GRA > 0.65 and the conditional MIC > 0.3 were selected.

3.5. *Predicted Landslide Displacement*

According to the results in Tables 1 and 2, the input data of the IMF1 component of landslide prediction finally selected six influencing factor subseries, the input data of the IMF2 component of landslide prediction finally selected nine influencing factor subseries, and the input data of the R component of landslide prediction finally selected five influencing factor subseries. The first 96 selected data were used to train the CNN–BiLSTM model, and the last 12 data were used to test the prediction accuracy. The learning rate of the CNN–BiLSTM model was set to 0.01, the number of iterations was set to 1000, and the number of hidden stratification points was set to 100. The prediction results of the three components of landslide displacement are shown in Figure 12a–c. The final predicted landslide displacement can be obtained by adding the three components, as shown in Figure 12d. The prediction model proposed in this paper can predict the displacement of the Baishuihe landslide.

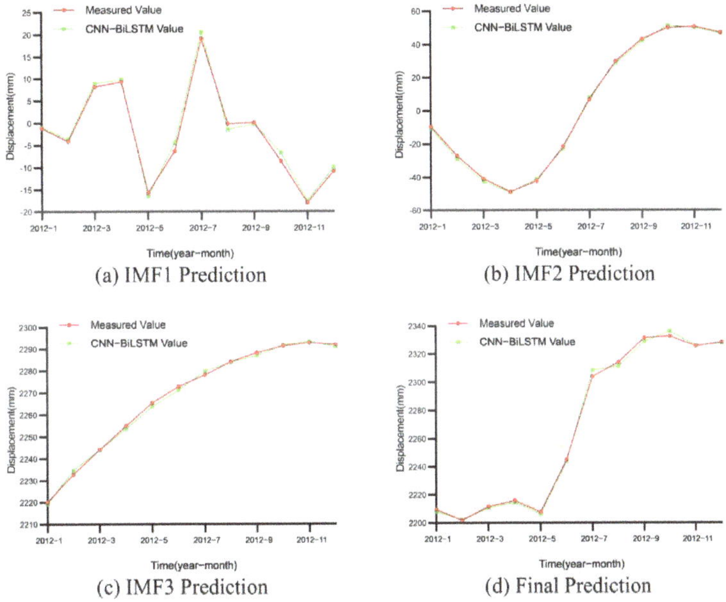

Figure 12. Prediction result of the CNN–BiLSTM model for the Baishuihe landslide displacement.

After the GRA–MIC algorithm screening, the CNN–BiLSTM model could achieve good results in both the displacement component and total displacement prediction, and the error between the final results and the actual measured data was controlled within a limited range.

4. Discussion

To better verify the performance of the proposed model, when other conditions are the same, in this paper, CNN–BiLSTM with GRA–MIC, CNN–BiLSTM with MIC, CNN–BiLSTM with GRA and CNN–BiLSTM without GRA–MIC were used to predict and compare the three components of landslide displacement. The comparison results of the CNN–BiLSTM model for each component and the total displacement are shown in Figure 13.

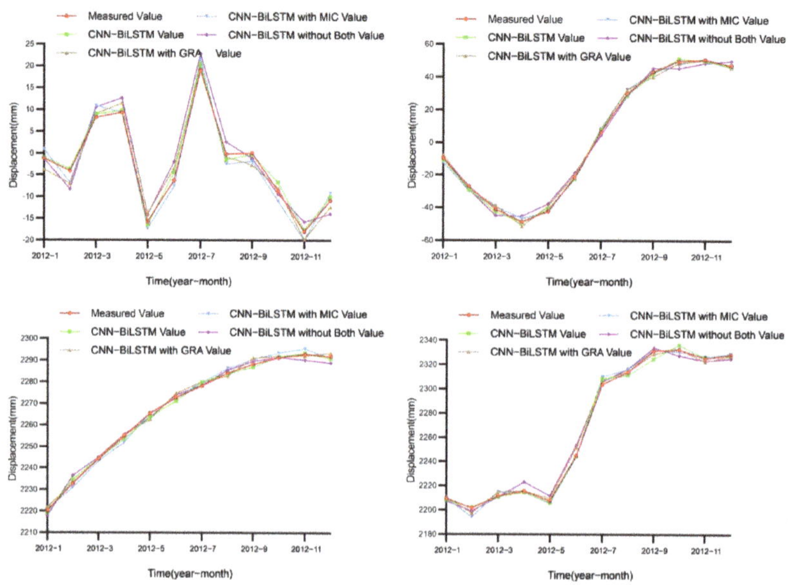

Figure 13. Prediction results of CNN–BiLSTM model with different inputs.

When the CNN–BiLSTM model was used for prediction, good prediction results could be achieved under different quantitative correlation algorithms, which reflects the excellent prediction performance of the CNN–BiLSTM model. For a better comparison, this paper uses four evaluation indicators to quantify the prediction performance, and the comparison results are shown in Table 3.

Table 3. Comparison of prediction performances of the CNN–BiLSTM model under different inputs.

Models	*MAE*	*MAPE*	*RMSE*	R^2 (%)	Minimum Error	Maximum Error	Total Error
CNN–BiLSTM	1.789	0.078	2.206	99.84	0.02	6.77	25.62
CNN–BiLSTM with GRA	2.335	0.103	2.981	99.70	0.02	6.54	28.02
CNN–BiLSTM with MIC	2.323	0.102	3.240	99.65	0.18	7.51	28.04
CNN–BiLSTM without Both	3.630	0.161	4.238	99.40	0.82	8.52	43.56

As shown in Table 3, when the GRA or MIC algorithms were used, appropriate influencing factors could be effectively selected, and the result was better than that when neither of the two algorithms were used, which reflects the role played by the GRA and MIC algorithms. When the GRA–MIC algorithm was used in the model, better influencing factors were selected from two different perspectives, and data with low correlations were removed. Compared with the GRA or MIC algorithms, the prediction results of the model were further improved. Due to the reduction in input data, the GRA–MIC algorithm not only improved the efficiency of the whole prediction process of the model, but also improved the prediction performance of the model.

In addition to comparing the prediction performance of the CNN–BiLSTM model in different correlation algorithms, this paper also used an additional seven deep learning algorithms for comparison: the CNN–RNN, CNN–LSTM, CNN–GRU, BiLSTM, RNN, LSTM, and GRU models. In the case of the GRA–MIC algorithm and other identical cases, the results comparisons of the eight models are shown in Figure 14 and Table 4.

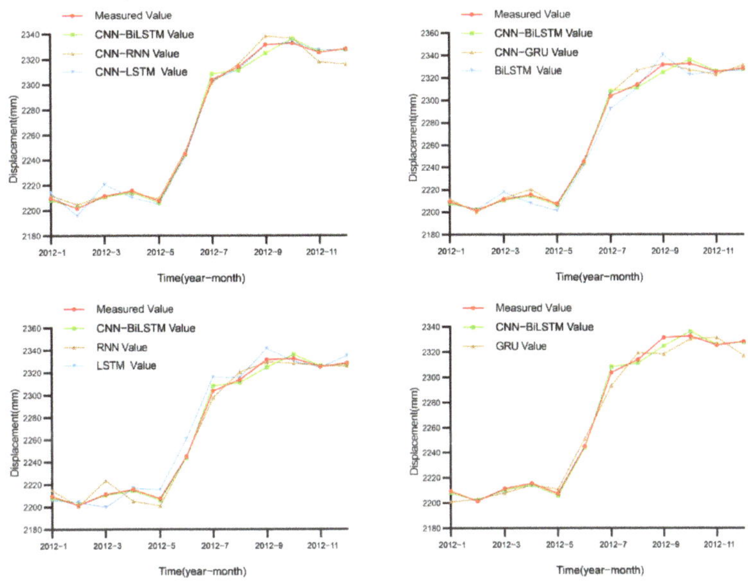

Figure 14. Comparison of the prediction effect between the CNN–BiLSTM model and other models.

Table 4. Comparison of prediction performances of different models with the same input.

Models	*MAE*	*MAPE*	*RMSE*	R^2 (%)	Minimum Error	Maximum Error	Total Error
CNN–BiLSTM	1.789	0.078	2.206	99.84	0.02	6.77	25.62
CNN–RNN	3.841	0.167	5.018	99.17	0.31	12.28	46.09
CNN–LSTM	3.063	0.137	4.012	99.47	0.23	9.36	36.76
CNN–GRU	3.302	0.144	4.578	99.31	0.64	12.69	39.62
BiLSTM	5.018	0.220	6.300	98.70	0.36	11.24	60.19
RNN	5.442	0.239	7.274	98.26	0.07	11.93	58.11
LSTM	4.888	0.215	7.013	98.38	0.74	15.79	77.49
GRU	6.076	0.266	7.203	98.29	1.21	13.37	72.91

Table 4 shows that because of the complexity and uncertainty of the landslide, a suitable time series data classification of the CNN model was adopted to forecast the displacement characteristics of the future and then build other models to forecast the concrete values; the effective reduction of the single model for complex data fitting ability was insufficient, and increasing the CNN part model could obtain a better effect. Compared with RNN models, traditional LSTM and GRU models have better prediction performances because the internal gate structures of the LSTM and GRU models adjust the input data flow and solve the problems of gradient disappearance and gradient explosion. Because of the similar structures, the prediction performances of the LSTM and GRU models are similar. Since the BiLSTM model adopts a bidirectional LSTM module, it can more fully train data and extract periodic information from environmental data in the training process. Compared with the traditional LSTM model, it improves the efficiency of data use and the accuracy of prediction.

5. Conclusions

The effective analysis and utilization of landslide displacement and influencing factor data is particularly important to improve the accuracy of landslide displacement prediction and ensure early warnings of landslides. Additionally, it provides a geological theoretical basis for the policymaking of land use management. Due to the problem of random displacement being ignored in time series analysis, the accuracy of the time series analysis method is limited when it is used in rainfall landslide displacement prediction. In this paper, a rainfall landslide displacement prediction method based on the time-frequency analysis method was proposed. The CEEMDAN method was used to decompose landslide displacement data into multiple subseries with different frequencies, two new concepts that evaluate the state of the landslide and the trend of the landslide were proposed, and the GRA–MIC joint association method was used to select the main influencing factors of each subseries. Then, CNN–BiLSTM, a fusion model based on deep learning, was used to train and predict landslide displacement. The model combines the CNN model with the BiLSTM model so that the model can more fully extract the features of landslide displacement data to provide a more effective method to use landslide displacement data. The prediction of landslide displacement showed that the fusion model combining CNN and BiLSTM was more effective than the single model in predicting the landslide displacement of Baishuihe, and the GRA–MIC joint association method was better than the single method in selecting influencing factors. This paper provides a research basis for landslide early warning based on landslide displacement.

Author Contributions: Conceptualization, Z.L., Y.J. and X.S.; methodology, Z.L., Y.J. and X.S.; formal analysis, Z.L.; software, Z.L.; validation, Z.L., Y.J. and X.S.; resources, Y.J. and X.S.; writing—original draft preparation, Z.L. and X.S.; writing—review and editing, Z.L., Y.J. and X.S.; funding acquisition, Y.J. and X.S. All authors have read and agreed to the published version of the manuscript.

Funding: This research was funded by the National Natural Science Foundation of China (62161007, 62061010), Department of Science and Technology of Guangxi Zhuang Autonomous Region (AA20302022, AA19254029, AB21196041, AB22035074, AD22080061), Guilin Science and Technology Project (20210222-1).

Institutional Review Board Statement: Not applicable.

Informed Consent Statement: Not applicable.

Data Availability Statement: The datasets analyzed in the current study are available from the corresponding author on reasonable request.

Conflicts of Interest: The authors declare no conflict of interest.

References

1. Mohanty, U.C.; Prasad, K.B.; Mohapatra, M.; Sarat, C.S. Guest editorial to the special issue of natural hazards: Climate change and coastal vulnerability. *Nat. Hazards* **2020**, *102*, 553–555. [CrossRef]
2. Mahdi, P.; Amiya, G.; Hamid, R.P.; Fatemeh, R.; Saro, L. Spatial prediction of landslide susceptibility using hybrid support vector regression (SVR) and the adaptive neuro-fuzzy inference system (ANFIS) with various metaheuristic algorithms. *Sci. Total Environ.* **2020**, *741*, 139937.
3. Suk, W.K.; Kun, W.C.; Minseok, K.; Filippo, C.; Byoungkoo, C.; Jung, I.S. Effect of antecedent rainfall conditions and their variations on shallow landslide-triggering rainfall thresholds in South Korea. *Landslide* **2021**, *18*, 569–582.
4. Minu, T.A.; Neelima, S.; Maria, A.B.; Biswajeet, P.; Binh, T.P.; Samuele, S. Using Field-Based Monitoring to Enhance the Performance of Rainfall Thresholds for Landslide Warning. *Water* **2020**, *12*, 3453.
5. Rahat, K.; Suhail, Y.; Abdul, H.; Uddin, M.I. Exploring a Design of Landslide Monitoring System. *Complexity* **2021**, *2021*, 5552417.
6. Leijin, L.; Feng, H.; Hongjiang, L. The use of remote sensing satellite using deep learning in emergency monitoring of high-level landslides disaster in Jinsha River. *J. Supercomput.* **2021**, *77*, 8728–8744.
7. Rodriguez-Caballero, E.; Rodriguez-Lozano, B.; Segura-Tejada, R.; Blanco-Sacristán, J.; Cantón, Y. Landslides on dry badlands: UAV images to identify the drivers controlling their unexpected occurrence on vegetated hillslopes. *J. Arid. Environ.* **2021**, *187*, 104434. [CrossRef]
8. Moritz, G.; John, S.; Kurosch, T. Internet of Things Geosensor Network for Cost-Effective Landslide Early Warning Systems. *Sensors* **2021**, *21*, 2609.

9. Pietro, M.; Mariano, D.N.; Luigi, G.; Massimo, R.; Chester, S.; Mariagiulia, A.C.; Diego, D.M. Landslide Awareness System (LAwS) to Increase the Resilience and Safety of Transport Infrastructure: The Case Study of Pan-American Highway (Cuenca–Ecuador). *Remote Sens.* **2021**, *13*, 1564.
10. Ascanio, R.; Samuele, S.; Vanessa, C.; Antonio, M.; Angela, G.; Nicola, C. Definition of 3D rainfall thresholds to increase operative landslide early warning system performances. *Landslide* **2020**, *18*, 1045–1057.
11. National Bureau of Statistics of the People's Republic of China. In *China Statistical Yearbook*; China Statistics Press: Beijing, China, 2021.
12. Won, Y.L.; Seon, K.P.; Hyo, H.S. The optimal rainfall thresholds and probabilistic rainfall conditions for a landslide early warning system for Chuncheon, Republic of Korea. *Landslide* **2021**, *18*, 1721–1739.
13. Wang, C.; Zhao, Y.; Bai, L.; Guo, W.; Meng, Q. Landslide Displacement Prediction Method Based on GA-Elman Model. *Appl. Sci.* **2021**, *11*, 11030. [CrossRef]
14. Liu, Y.; Xu, C.; Huang, B.; Ren, X.; Liu, C.; Hu, B.; Chen, Z. Landslide displacement prediction based on multi-source data fusion and sensitivity states. *Eng. Geol.* **2020**, *271*, 105608. [CrossRef]
15. Li, S.H.; Wu, L.; Chen, J.J.; Huang, R. Multiple data-driven approach for predicting landslide deformation. *Landslide* **2020**, *17*, 709–718. [CrossRef]
16. Wu, L.Z.; Li, S.H.; Huang, R.Q.; Xu, Q. A new grey prediction model and its application to predicting landslide displacement. *Appl. Soft Comput.* **2020**, *95*, 106543. [CrossRef]
17. Jiang, Y.; Luo, H.; Xu, Q.; Lu, Z.; Liao, L.; Li, H.; Hao, L. A Graph Convolutional Incorporating GRU Network for Landslide Displacement Forecasting Based on Spatiotemporal Analysis of GNSS Observations. *Remote Sens.* **2022**, *14*, 1016. [CrossRef]
18. Lian, C.; Zhu, L.; Zeng, Z.; Su, Y.; Yao, W.; Tang, H. Constructing prediction intervals for landslide displacement using bootstrapping random vector functional link networks selective ensemble with neural networks switched. *Neurocomputing* **2018**, *291*, 1–10. [CrossRef]
19. Long, J.; Li, C.; Liu, Y.; Feng, P.; Zuo, Q. A multi-feature fusion transfer learning method for displacement prediction of rainfall reservoir-induced landslide with step-like deformation characteristics. *Eng. Geol.* **2022**, *297*, 106494. [CrossRef]
20. Xie, P.; Zhou, A.; Chai, B. The Application of Long Short-Term Memory(LSTM) Method on Displacement Prediction of Multifactor-Induced Landslides. *IEEE Access.* **2019**, *7*, 54305–54311. [CrossRef]
21. Han, H.; Shi, B.; Zhang, L. Prediction of landslide sharp increase displacement by SVM with considering hysteresis of groundwater change. *Eng. Geol.* **2021**, *280*, 105876. [CrossRef]
22. Li, S.; Wu, N. A new grey prediction model and its application in landslide displacement prediction. *Chaos Solitons Fractals* **2021**, *147*, 110969. [CrossRef]
23. Deng, L.; Smith, A.; Dixon, N.; Yuan, H. Machine learning prediction of landslide deformation behaviour using acoustic emission and rainfall measurements. *Eng. Geol.* **2021**, *293*, 106315. [CrossRef]
24. Hu, X.; Wu, S.; Zhang, G.; Zheng, W.; Liu, C.; He, C.; Liu, Z.; Guo, X.; Zhang, H. Landslide displacement prediction using kinematics-based random forests method: A case study in Jinping Reservoir Area, China. *Eng. Geol.* **2021**, *283*, 105975. [CrossRef]
25. Wang, R.; Zhang, K.; Wang, W.; Meng, Y.; Yang, L.; Huan, H. Hydrodynamic landslide displacement prediction using combined extreme learning machine and random search support vector regression model. *Eur. J. Environ. Civ. Eng.* **2020**, *2020*, 2345–2357. [CrossRef]
26. Zhang, Y.; Tang, J.; He, Z.; Tan, J.; Li, C. A novel displacement prediction method using gated recurrent unit model with time series analysis in the Erdaohe landslide. *Nat. Hazards* **2021**, *105*, 783–813. [CrossRef]
27. Yang, B.; Yin, K.; Lacasse, S.; Liu, Z. Time series analysis and long short-term memory neural network to predict landslide displacement. *Landslides* **2019**, *16*, 677–694. [CrossRef]
28. Miao, F.; Wu, Y.; Xie, Y.; Li, Y. Prediction of landslide displacement with step-like behavior based on multialgorithm optimization and a support vector regression model. *Landslide* **2017**, *15*, 475–488. [CrossRef]
29. Liu, Z.; Guo, D.; Lacasse, S.; Li, J.; Yang, B.; Choi, J. Algorithms for intelligent prediction of landslide displacements. *J. Zhejiang Univ. Sci. A* **2020**, *21*, 412–429. [CrossRef]
30. Wang, Y.; Tang, H.; Huang, J.; Wen, T.; Ma, J.; Zhang, J. A comparative study of different machine learning methods for reservoir landslide displacement prediction. *Eng. Geol.* **2022**, *298*, 106544. [CrossRef]
31. Lin, Z.; Sun, X.; Ji, Y. Landslide Displacement Prediction Model Using Time Series Analysis Method and Modified LSTM Model. *Electronics* **2022**, *11*, 1519. [CrossRef]
32. Lin, Z.; Sun, X.; Ji, Y. Landslide Displacement Prediction based on Time Series Analysis and Double-BiLSTM Model. *Int. J. Environ. Res. Public Health* **2022**, *19*, 2077. [CrossRef]
33. Guo, Z.; Chen, L.; Gui, L.; Du, J.; Do, H.M. Landslide displacement prediction based on variational mode decomposition and WA-GWO-BP model. *Landslides* **2019**, *17*, 567–583. [CrossRef]
34. Liu, Q.; Lu, G.; Dong, J. Prediction of landslide displacement with step-like curve using variational mode decomposition and periodic neural network. *Bull. Eng. Geol. Environ.* **2021**, *80*, 3783–3799. [CrossRef]
35. Xu, S.; Niu, R. Displacement prediction of Baijiabao landslide based on empirical modedecomposition and long short-term memory neural network in Three Gorgesarea, China. *Comput. Geosci.* **2018**, *111*, 87–96. [CrossRef]
36. Cai, Z.; Xu, W.; Meng, Y.; Chong, S.; Wang, R. Prediction of landslide displacement based on GA-LSSVM with multiple factors. *Bull. Eng. Geol. Environ.* **2015**, *75*, 637–646. [CrossRef]

37. Zhou, C.; Yin, K.; Ying, C.; Emanuele, I.; Bayes, A.; Filippo, C. Displacement prediction of step-likelandslide by applying a novel kernel extreme learning machine method. *Landslides* **2018**, *15*, 2211–2225. [CrossRef]
38. Huang, F.; Yin, K.; Zhang, G.; Gui, L.; Yang, B.; Liu, L. Landslide displacement prediction using discrete wavelet transform and extreme learning machine based on chaos theory. *Environ. Earth Sci.* **2016**, *75*, 1376. [CrossRef]
39. Yi, W.; Chuannuo, X.; Yu, W.; Cheng, X. A comprehensive diagnosis method of rolling bearing fault based on CEEMDAN-DFA-improved wavelet threshold function and QPSO-MPE-SVM. *Entropy* **2021**, *23*, 1142.
40. Peterson, O.J.; Siaw, F.; Anokye, M.A.; Samuei, A.; Emmaniel, N.G.; Prof, D.A.; George, T. COVID-19 as Information Transmitter to Global Equity Markets: Evidence from CEEMDAN-Based Transfer Entropy Approach. *Math. Probl. Eng.* **2021**, *2021*, 8258778.
41. Cem, E.; Mustafa, T. Wind speed estimation using novelty hybrid adaptive estimation model based on decomposition and deep learning methods (ICEEMDAN-CNN). *Energy* **2022**, *249*, 123785.
42. Vishnu, P.M.; Anil, K.B.; Srinivasa, R.S. Grey Relational Analysis-Based Objective Function Optimization for Predictive Torque Control of Induction Machine. *IEEE Trans. Ind. Appl.* **2020**, *57*, 835–844.
43. Arash, K.; Ali, P.; Ali, S.; Tao, M. Optimization of a novel photovoltaic thermal module in series with a solar collector using Taguchi based grey relational analysis. *Sol. Energy* **2021**, *215*, 492–507.
44. Senthilkumar, S.; Karthick, A.; Madavan, R.; Moshi AA, M.; Bharathi, S.S.; Saroja, S.; Dhanalakshmi, C.S. Optimization of Transformer Oil blended with Natural Ester Oils using Taguchi-based Grey Relational Analysis. *Fuel* **2021**, *288*, 119629. [CrossRef]
45. Reshef, D.N.; Reshef, Y.; Grossman, S.R.; Lander, E.S.; Finucane, H.K.; McVean, G.; Turnbaugh, P.J.; Mitzenmacher, M.; Sabeti, P.C. Detecting novel associations in large data sets. *Science* **2011**, *334*, 1518–1524. [CrossRef]
46. Yalan, J.; Chaoshun, L.; Zhixin, Y.; Yujie, Z.; Xianbo, W. Remaining Useful Life Estimation Combining Two-Step Maximal Information Coefficient and Temporal Convolutional Network With Attention Mechanism. *IEEE Access.* **2021**, *9*, 16323–16336.
47. Young, R.C.; Rhee, M.K. Face Video Retrieval Based on the Deep CNN With RBF Loss. *IEEE Trans. Image Process.* **2021**, *30*, 1015–1029.
48. Zian, L.; Ji, Y.; Liang, W.; Sun, X. Landslide Displacement Prediction Based on Time-Frequency Analysis and LMD-BiLSTM Model. *Mathematics* **2022**, *10*, 2203.
49. Yujie, F.; Jian, L.; Yang, L.; Suge, W.; Deyu, L.; Xiaoli, L. Multiple Perspective Attention Based on Double BiLSTM for Aspect and Sentiment Pair Extract. *Neurocomputing* **2021**, *438*, 302–311.
50. Zhenzhu, M.; Jinxin, Z.; Yating, H.; Christophe, A. Temporal Prediction of Landslide-Generated Waves Using a Theoretical–Statistical Combined Method. *J. Mar. Sci. Eng.* **2023**, *11*, 1151.
51. Du, H.; Song, D.; Chen, Z.; Shu, H.; Guo, Z. Prediction model oriented for landslide displacement with step-like curve by applying ensemble empirical mode decomposition and the PSO-ELM method. *J. Clean. Prod.* **2020**, *270*, 122248. [CrossRef]
52. Zhen, G.; Bin, Y.; Mengyan, H.; Wensi, W.; Yu, J.; Fang, Z. A novel hybrid method for flight departure delay prediction using Random Forest Regression and Maximal Information Coefficient. *Aerosp. Sci. Technol.* **2021**, *116*, 106822.
53. Benyang, Z.; Wei, L.; Li, X.; Shiyu, L.; Yang, Z.; Huang, Y.; Zhou, M. A CNN-based FBG demodulation method adopting the GAF-assisted ascending dimension of complicated signal. *Opt. Commun.* **2021**, *499*, 127296.
54. Ningtao, C.; Dajing, C.; Bin, L.; Jing, F.; Hongyang, W. A biosensing method for the direct serological detection of liver diseases by integrating a SERS-based sensor and a CNN classifier. *Biosens. Bioelectron.* **2021**, *186*, 113246.
55. Yu, C.; Fang, R.; Liang, T.; Sha, Z.; Li, S.; Yi, Y.; Zhou, W.; Song, H. Stock Price Forecast Based on CNN-BiLSTM-ECA Model. *Sci. Program.* **2021**, *2021*, 2446543.
56. Mohammed, A.B.; Muhammad, Y.N.; Imad, E.A.; Shaharin, A.S. BiLSTM Network-Based Approach for Solar Irradiance Forecasting in Continental Climate Zones. *Energies* **2022**, *15*, 2226.
57. Devi, K.; Paulraj, D.; Muthusenthil, B. Deep Learning Based Security Model for Cloud based Task Scheduling. *KSII Trans. Internet Inf. Syst.* **2020**, *14*, 3663–3679.
58. Lian, C.; Zeng, Z.; Wang, X.; Yao, W.; Su, Y.; Tang, H. Landslide displacement interval prediction using lower upper bound estimation method with pre-trained random vector functional link network initialization. *Neural Netw.* **2020**, *130*, 286–296. [CrossRef]
59. Xing, Y.; Yue, J.; Chen, C.; Qin, Y.; Hu, J. A hybrid prediction model of landslide displacement with risk-averse adaptation. *Comput. Geosci.* **2020**, *141*, 104527. [CrossRef]
60. Zhou, C.; Yin, K.; Cao, Y.; Ahmed, B. Application of time series analysis and PSO–SVM model in predicting the Bazimen landslide in the Three Gorges Reservoir, China. *Eng. Geol.* **2016**, *204*, 108–120. [CrossRef]
61. Huang, F.; Huang, J.; Jiang, S.; Zhou, C. Landslide displacement prediction based on multivariate chaotic model and extreme learning machine. *Eng. Geol.* **2017**, *218*, 173–186. [CrossRef]
62. Wang, Y.; Tang, H.; Wen, T.; Ma, J. A hybrid intelligent approach for constructing landslide displacement prediction intervals. *Appl. Soft Comput.* **2019**, *81*, 105506. [CrossRef]

sustainability

MDPI

Article

Spatio-Temporal Variation of Habitat Quality for Bird Species in China Caused by Land Use Change during 1995–2015

Bingkui Qiu [1], Jinjiang Yao [2,*], Siyu Han [3] and Zhe Zhu [4,*]

1 Department of Tourism Management, Jinzhong University, Jinzhong 030619, China
2 Science and Technology Innovation and Public Management Research Center of Shanghai, Fudan University, Shanghai 200433, China
3 College of Public Administration, Huazhong University of Science and Technology, Wuhan 430079, China
4 School of Management, Wuhan Institute of Technology, Wuhan 430205, China
* Correspondence: jinjiang.yao11@alumni.imperial.ac.uk (J.Y.); zhuzhe@wit.edu.cn (Z.Z.)

Abstract: The analysis of land use change (LUC) has become an important criterion for evaluating the impact of human activities on the natural environment. Habitat loss and degradation caused by LUC are the main threats to biodiversity worldwide. Research on the impact of long-term, wide-scope, and fine-scale LUC on bird habitats is currently limited due to a lack of adequate data. In this study, conducted in China, 9 km grid units were sampled randomly between 1995 and 2015. Logistic regression was used to calculate the probability that each unit grid contained suitable habitat (hereinafter, abbreviated as PGSH) for 981 bird species and analyze the spatial-temporal characteristics of PGSH accordingly. The results showed that: (1) The habitat quality of 84 bird species deteriorated, but for 582 bird species, habitat quality improved. (2) There is an inverted U-shaped relationship between the intensity of LUC and the PGSH. The LUC intensity threshold is approximately 67.21%. (3) Based on the counterfactual scenario analysis, the construction of the Three North Shelterbelt has increased the PGSH for all bird species from 20.76% before restoration to 21.38% after restoration. Within the LUC grid representing the transformation of farmland back to forests, the average PGSH for all birds increased from 73.97% to 75.04%. These results may provide a reference for measuring the impacts of LUC on bird species, enabling the protection of bird species and habitats that need it most.

Keywords: land use change; habitats quality; counterfactual analysis; China

Citation: Qiu, B.; Yao, J.; Han, S.; Zhu, Z. Spatio-Temporal Variation of Habitat Quality for Bird Species in China Caused by Land Use Change during 1995–2015. *Sustainability* **2022**, *14*, 10078. https://doi.org/10.3390/su141610078

Academic Editor: Fabio Carlucci

Received: 13 July 2022
Accepted: 11 August 2022
Published: 15 August 2022

1. Introduction

Humans and other living creatures depend on our natural environment for survival. Natural landscapes have undergone a long-term transformation, largely as a result of increasing human populations and their associated activities; consequently, land use has changed greatly over time. Unreasonable utilization of land resources has caused a series of major global problems, such as environmental pollution, vegetation destruction, land degradation, species extinction, and resource scarcity [1,2].

The analysis of LUC has become an important criterion for evaluating the impact of human activities on the natural environment [3]. The driving factors of land use change are complex. They are not only affected by natural factors, such as climate, land slope, and drought [4–6], but also affected by social and economic factors, of which population growth, economic development, urbanization process, and improvement of traffic conditions are of particular concern [7,8]. Habitat loss and degradation caused by the transformation of land use are the main threats to biodiversity worldwide [9–11]. Habitats provide important resources for all living organisms, such as sufficient food resources, suitable breeding sites, protection from natural enemies, and challenging climatic conditions. Among wildlife species, birds are highly sensitive to habitat changes, and can therefore act as indicators of habitat changes [12,13]. With the advancement of urbanization, habitat reduction and habitat fragmentation caused by economic development and human activities are having increasing impacts on bird communities [14–17], which also impacts species composition [18–20], predation behaviors [21], and migration paths. Studies have found that the number of birds in North America has decreased by around 29% since 1970, equivalent to nearly 3 billion birds, with habitat loss being the main reason for this steep decline [2].

With growing economies and urbanization, the trend of land use change in developing countries represented by China is more obvious than that in Western developed countries [22]. In the last few decades, great changes have taken place in the land use pattern of developing countries. Large areas of undeveloped land around urban centers, such as cultivated land, forests, and wetlands, have been urbanized, which is a rapidly expanding trend in China. It is the large availability of undeveloped land that has supported the urbanization process and brought about great economic achievements since China's reform. However, the drastic changes in land use have affected avian habitat, which poses a huge threat to bird communities. There is a close relationship between the living conditions of birds and their habitats. The conditions of the habitats will affect all stages of the life stage of birds [23]. The food sources, activity sites, and breeding sites that are indispensable for the survival of birds depend on the habitat environment. However, land use may directly reduce the land types that birds mainly depend on, such as forests, wetlands, and swamps, resulting in the fragmentation and loss of habitats and further affecting the species distribution pattern [24], accelerating species extinction [25] and reducing bird biodiversity [26]. This issue has attracted extensive attention of scholars. In 1999, the first review paper on birds' habitat selection in China was published, which is a phased summary of avian habitat research in China and the prospects for future research [27]. The China Biodiversity Red List, released in 2015, showed that habitat degradation and loss caused by deforestation, alternative planting of an economic forest, and wetland reclamation is the key factor affecting avian survival, accounting for 80.8% of all factors [28]. In recent years, due to the acceleration of urbanization and the increase of land use intensity, the overwintering habitat of Red Crowned Crane in Northern Jiangsu Province, located in the eastern province, has gradually decreased, posing a serious threat to the survival of the Red Crowned Crane population [29]. The decrease of birds in Hainan Island, in the southern tropics of China, is mainly due to urban development [30]. Coastal wetland areas in the subtropical region of Xiamen have decreased, and many wetland birds that rely on coastal wetlands for survival and reproduction have lost important habitat [31]. The Yellow River Wetland Nature Reserve, located in the temperate zone in China, has experienced a massive reduction, and the natural reed marshes and tidal flats have been reclaimed into fish ponds, lotus ponds, and rice fields. As a result, the overall area of avian habitat has reduced by 20,000 hectares, and is continually decreasing, leading to the wintering waterfowl in this area being sharply reduced (https://www.sohu.com/a/151012215_351301, accessed on 1 July 2018). These studies and reports highlight the substantial decrease of avian habitat caused by the transformation of land use and the threat that this poses to the survival of many bird species.

Observational changes in bird distribution can help inform on the extinction risks of birds [32]. However, published bird distribution data in China are province-scale based and lack spatial details of avian distribution, which hinders further research [33]. Therefore, a large amount of bird information is collected by professional birdwatchers through field surveys [34], bibliometrics [35], GPS tracking [36], citizen science [37,38], and other methods, which have become the main methods of fine-scale research on avian distribution. However, there are still very few data sources that provide such information on a national scale. China Bird Watching Database [39] and China Biodiversity Observation Network-Birds are two rare national bird observation databases, but both the number of observations and the selection of observation sample areas are far less abundant than EBird. EBird is a bird sighting record database, managed by the Cornell Lab of Ornithology in the US. It is the largest, most comprehensive, and most popular civilian science project related to biodiversity in the world [40]. The EBird Basic Dataset released by EBD_relApr-2019 has more than 600 million observation records, with each record detailing 45 observational attributes, including species name, observation time (including year, month, day, and hour), and observation location represented by longitude and latitude [41]. Therefore, the spatial cover of the data, based on longitude and latitude, and spatial superposition with land use data, can show the temporal and spatial relationship of bird distribution with LUC. Inspired by EBird, BirdReport has been developed for use in China, essentially the Chinese version of EBird.

Using observational and land use data from EBird and BirdReport, we hope to quantitatively answer the following research questions: How much impact does land use change have on bird habitat quality in China? What is the spatial and temporal pattern of this impact? To the best of the author's knowledge, this study is the first exploration to use fine broad-scale data on the distribution and habitats of bird species in China. It can help to inform on which bird species are most threatened by LUC, which will allow for corresponding measures to protect their habitats.

2. Study Area and Data Source

This study was conducted in China from 1995 to 2015. This period was an important stage of China's economic development. China's market economic system was set up and the economy developed rapidly shortly after 1995. However, 20 years later, China's economic growth slowed, and land and space development were restricted [42]. Land use transformation was therefore most prevalent during this period, which made it most appropriate to study the impact of LUC on avian habitats during this period. Data for land use in 1995, 2000, 2005, 2010, and 2015 were selected from the Institute of Geography affiliated with the Chinese Academy of Sciences to investigate spatial and temporal changes of land use. The spatial resolution of these data was 30 m, including six first-class levels: cultivate, forest, grass, water, build-up, and non-use land. This dataset was the most accurate land use data available in China. The accuracy and practicability of the classification have been demonstrated in the literature [43].

The avian observation data originated from EBird (https://ebird.org/home, accessed on 1 June 2019) and BirdReport (www.birdreport.cn, accessed on 1 June 2019). Each dataset contained the attributes longitude and latitude, bird name, and year. According to the observation year of land use data, we extracted records from 1995, 2000, 2005, 2010, and 2015. There were 128,543 records of 1022 species of birds. Records with less than 10 observations were eliminated as they did not meet the required threshold (10) for logistic regression analysis, which left us with 981 birds' species for analysis. The spatial distribution of land use and avian observation sites is shown in Figure 1.

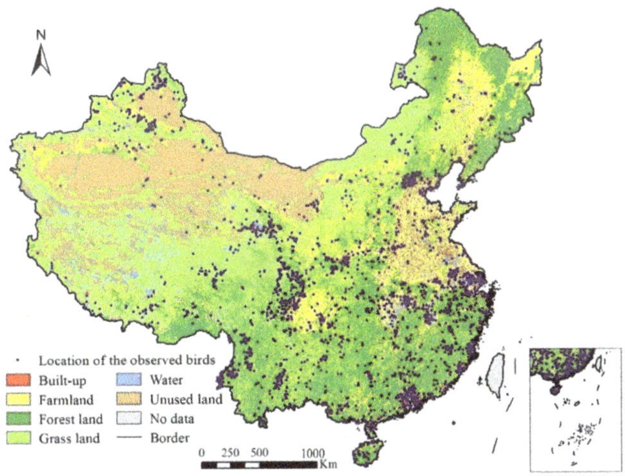

Figure 1. Spatial distribution of land use (2015) and sampled avian observation sites (to improve visualization, the resolution of the land use grid was set to 1 km, and the bird distribution was based on 10,000 randomly selected bird observation points).

3. Methodology

Exploring the impact of land use change (LUC) on bird habitat requires an analysis of habitat characteristics, including the structure of the land type, and the preference of spatial proximity. However, what scope is used to calculate the composition and proportion of land use types around each bird observation point, and how should the probability of research units suitable as habitat for specific birds be determined? After answering the above two questions, we can calculate the probability that each unit grid contained suitable habitat (PGSH) over time, then analyze its spatial distribution and spatial-temporal evolution, and detect the impact of LUC policy on bird habitat change. In general, we followed the framework of the methods shown in Figure 2. The details of the methods involved are stated in turn below.

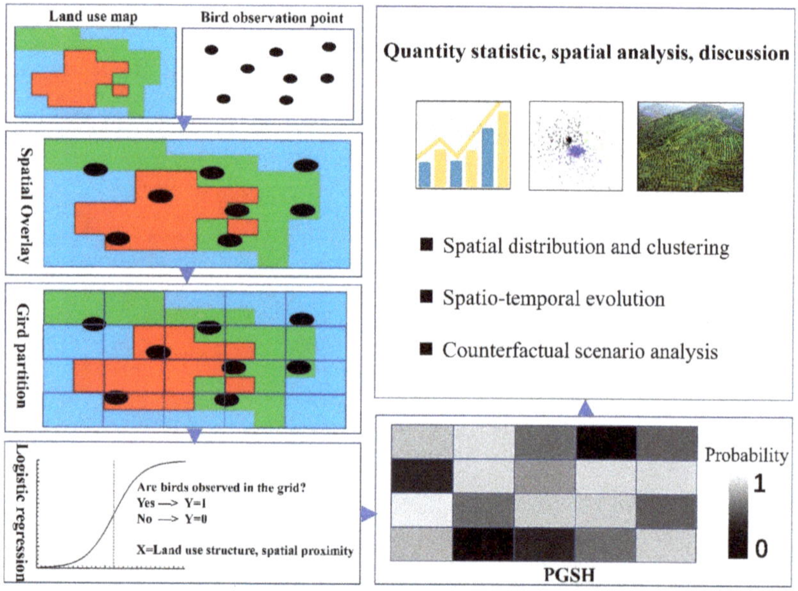

Figure 2. Methods framework diagram.

Use grids to divide basic analysis units. Since we need to make statistics on the composition and proportion of land types near each bird observation point, it is necessary to determine a statistical range for each observation point. However, there is no authoritative data to demonstrate the specific activity range of birds, and the collected observation points are not regularly distributed in space, Therefore, we used Thiessen polygon, a method proposed by Dutch climatologist A. H. Thiessen to calculate the average rainfall according to the rainfall of discrete meteorological stations [44]. The rainfall intensity of a unique weather station included in this polygon represents the rainfall intensity in this polygon area. In our case, that is, each polygon represents the statistical range of the habitat of the observation bird, and a total of 119,753 polygons were divided with a radius of approximately 9 km, corresponding to the average polygon area as the grid width, and the grid range as the statistical range. Note that 9 km is not the average radius of activity of the birds, it refers to the statistical range determined in the context of the current distribution of bird observation points.

Use logistic regression to calculate the PGSH: We collected the land use characteristics of the grid where the observation points for where the bird has appeared and has not appeared in 1995, 2000, 2005, 2010, and 2015, respectively. The composition and proportion of various land types of each bird habitat were calculated, along with the distance from cities and water as spatial proximity. The land use structure feature reflects the preference of different birds for the land use composition of the habitat. For example, wader birds inhabit wetlands, while woodpeckers are associated with forests. As the densest agglomerations of human activity, cities may have adverse effects on the migration and habitat of birds, while water sources can provide water and other resources for birds. The urban area is directly characterized by the construction polygon extracted from LUC, and the waters are replaced by rivers, canals, and lakes. Therefore, for each sampled grid, the following record can be used:

$$C_b = \left(B, R_{cultivate}, R_{forest}, R_{grass}, R_{water}, R_{built-up}, R_{nonuse}, Dis_{city}, Dis_{river}, Dis_{lake}\right) \quad (1)$$

where the value of B is 0 or 1; if a bird is observed in the selected grid then B is 1, otherwise B is 0. $R_{cultivate}, R_{forest}, R_{grass}, R_{water}, R_{built-up}, R_{nonuse}$ are the proportion of land type: cultivated land, forest land, grassland, waterbody, and built-up land within the statistical scope. $Dis_{city}, Dis_{river}, Dis_{lake}$ are distances from the grid center to the nearest city, river, and lake, respectively. When we counted the above-mentioned characteristic variables in each grid for five years—1995, 2000, 2005, 2010, and 2015—the probability of each grid being suitable for habitat could be calculated according to the following:

$$P_{im} = \frac{e^{y_m}}{1 + e^{y_m}} \quad (2)$$

where P is PGSH for bird habitat i at grid m. y_m can be calculated based on each variable's value and corresponding weight at grid m. P is in a range of 0–1. The closer P is to 1, the higher the probability of it being suitable for habitat. For the five sampled years between 1995 and 2015, we determined threatened bird habitat by more than three consecutive periods of decreased PGSH, and when the PGSH increased for more than three consecutive periods, it was categorized as continuous improvement.

Use spatial autocorrelation to detect the spatial distribution characteristic. Global Moran's I can measure spatial autocorrelation based on element locations and element values [45]. Given a set of elements and related attributes, this index evaluates whether the expressed pattern is a clustering pattern, a discrete pattern, or a random pattern. Z scores and p values were used to evaluate the significance of the index. The value of Global Moran's I falls in the interval from −1.0 to +1.0. When the value is positive, it means that there is a spatial agglomeration of elements, and the larger the value is, the more obvious the agglomeration. Conversely, when the value is negative, it means that there is spatial diffusion of elements, and the smaller the value is, the more obvious the diffusion is. When the Global Moran's I value is 0, it means a random distribution of elements. Global Moran's I can only reflect the global distribution characteristics of elements but cannot detect the local clustering of elements. Local Moran's I gives a set of elements (input element class) and an analysis field (input field), which can identify the spatial clustering of elements with high or low values [46]. In this study, Moran's I and local Moran's I were used to detect the spatial distribution clustering characteristics of the probability of bird habitat suitable for grid-scale. We used tools in ArcGIS10.2 to realize the calculation of Global Moran's I, and the cartographic display of Local Moran's I.

Use counterfactual analysis to evaluate the effect of land use policies. Counterfactual reasoning refers to the negation and representation of a fact that has occurred in the past, to construct a hypothesis of possibility [47]. A counterfactual approach is appropriate for answering fundamental questions, such as what would have happened if there had been no intervention, or if there had been different policy systems. In the counterfactual analysis, an unobserved case (called a counter fact) is designed to be compared with the actual case to illustrate the important factors that explain the impact of the policy. In this article, if we examine the impact of changes in a certain land type A on the habitat of birds, we will examine the following scenarios. From A to other land types and other land types to A, we compared the changes in the PGSH in the factual scenarios and in the hypothetical unhappened scenarios.

4. Results and Analysis

4.1. Land Use Change in 1995–2015

The land use transition matrix represented by area from 1995 to 2015 is shown in Table 1. In terms of area proportion change (Figure 3), the proportion of cultivated land (CL), forest land (FL), grassland (GL), water (WL), built-up land (BL), and unused land (UL) changed from 18.47%, 23.97%, 31.48%, 2.75%, 1.79%, and 21.54% in 1995 to 18.89%, 23.76%, 27.92%, 3.01%, 2.93%, and 23.49%, respectively, in 2015. The largest decline was found in GL, which decreased by 3.56 percentage points. BL and UL increased by 1.14 and 1.95 percentage points, respectively, with small changes in the other three land use types. However, the relative variation rate was 2.27%, −0.88%, −11.31%, 9.45%, 63.69%, and 9.05%, respectively.

Table 1. Land use transition matrix from 1995 to 2015 in China (km^2).

Land Use Type	CL	FL	GL	WL	BL	UL
CL	1,543,309.64	56,402.96	55,947.74	18,442.05	88,532.43	10,224.89
FL	86,234.84	2,016,160.76	125,749.32	8406.80	33,374.31	30,836.77
GL	111,111.40	176,691.31	2,151,035.08	43,172.08	13,472.24	527,014.15
WL	12,957.20	4594.05	20,414.69	180,167.35	4207.66	41,929.24
BL	25,952.83	2084.32	2254.03	4406.07	136,487.09	942.86
UL	33,822.70	24,570.33	324,941.50	34,439.86	5527.71	1,644,181.76

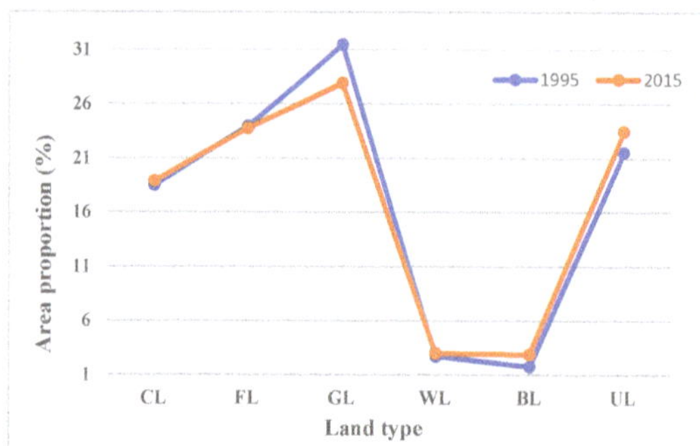

Figure 3. Area proportion change of six land type during the 1995–2015 period.

4.2. Quantitative Changes of Bird Habitat Suitability

The number of bird species at risk between 1995 and 2015 was 84 (84/981). Among them, four bird species, white-winged magpie, limestone leaf warbler, rusty-flanked tree-creeper, and rusty-fronted barwing, were found to be of particular conservation concern, because, in 2015, the average PGSH for these species across China was only 0.7%, 3.9%, 7.1%, and 7.3% (the average PGSH of all birds was about 48.6%), respectively. If no vigorous conservation measures are adopted to protect them, they risk potential extinction in the relatively near future. Habitat suitability for 582 bird species (582/981) continued to improve, which far exceeds the number of threatened birds. Figure 4 presents 20 species of birds, showing low PGSH (average PGSH less than 10% in 2015) but constant improvement. The IUCN Red List of Endangered Birds lists a total of 86 endangered bird species and 83 other supplementary rare birds, of which 6 species are threatened: Hainan partridge, yellow-bellied tragopan, Chinese monal, great bustard, spotted greenshank, and fairy pitta. The habitats of 18 bird species on the Red List have been improved continuously.

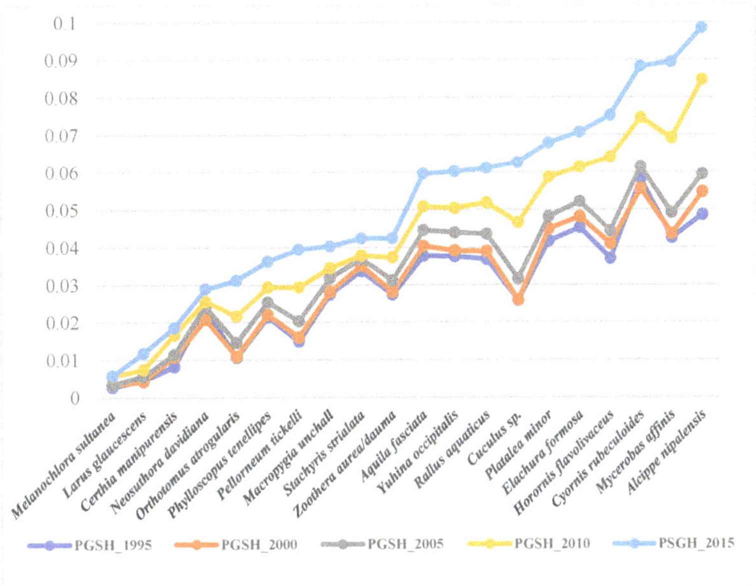

Figure 4. Twenty species of birds with low PGSH but constant improvement (PGSH change trend).

4.3. The Spatial Distribution of Suitability for Bird Habitats

The average PGSH for all 981 species of birds analyzed in each 9 km grid unit is shown in Figure 5a. The higher the grid value, the more important it is for maintaining bird species diversity. The Global Morans' I index is 0.938, which takes on a typically obvious spatial agglomeration feature. Detected by local clusters, high aggregation areas are southern China and northeast China, which are crucial forest areas of China that play an important role in maintaining bird habitats. An interesting finding was that high aggregation areas and low aggregation areas were split by the Chinese population distribution line "Hu-Line" (Figure 5b). High-value areas were mainly gathered on the right side of the line, if 0.8 is the threshold, then the right proportion is 78.84%. If the threshold is 0.9, the corresponding number is 86.60%. Therefore, bird-friendly areas overlap with the higher human population density side of the Hu-Lin, but human activities pose a huge challenge to the protection of birds.

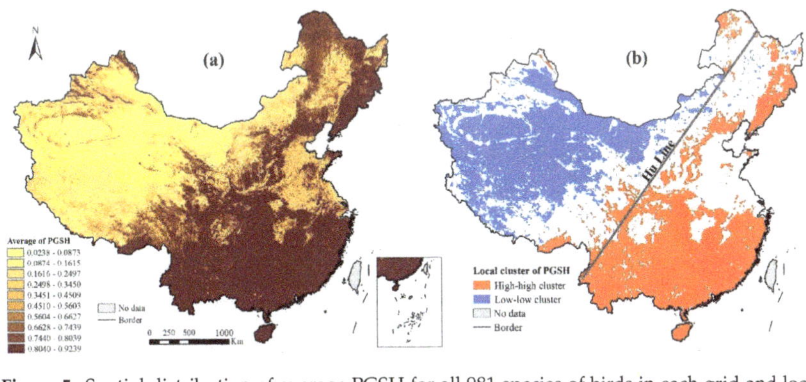

Figure 5. Spatial distribution of average PGSH for all 981 species of birds in each grid and local cluster. ((**a**): Average PGSH for all 981 species of birds analyzed in each grid unit; (**b**): Local cluster of PGSH, note that the right side of Hu-Line accounts for about 43.8% of the national area, containing 95% of the total population).

From the four time periods formed by five years as an interval, the number of grids continuously reduced in terms of average PGSH, which was 7238, accounting for 6.27% of the total number of grids, while the number of grids that continuously improved was 18,498, accounting for 16.02%. The spatial distribution of the average PGSH reduction and increases are shown in Figure 6. The area of improvement was substantially higher than that of deterioration. This discovery will lead us to re-examine the relationship between LUC and bird habitat changes. The continuously deteriorating areas were mainly located in areas containing three different land use types, namely, the forest areas in the northeast, the deserts and non-use lands in Xinjiang and Tibet in the west of China, and the grasslands in Inner Mongolia in the north. The areas of continuous improvement were more widely distributed, among which the Qinghai-Tibet Plateau was the most concentrated area of improvement, but we were surprised by the improved agglomeration area formed in the Yangtze River Delta region, having the most developed economy in China.

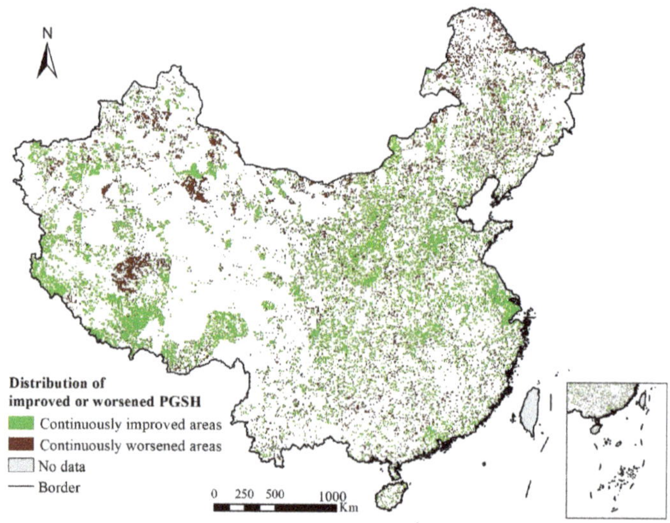

Figure 6. Spatial distribution of continuously improved/reduced grids with PGSH.

Integrating the four periods from 1995 to 2015 to analyze the spatial-temporal changes in the PGSH in each grid, we found the lowest overall PGSH but the most noticeable growth in the Qinghai-Tibet Plateau, which formed a gathering area. However, much of the grassland in this area is becoming bare and unused and is experiencing rising temperatures, thus the environment is becoming more suitable for highland birds. PGSH in north China, which has declined significantly on account of the transition of large areas of forest to non-forestry lands, should also be closely monitored, because of its importance for bird species in China. Furthermore, we found that some PGSH with a high level of urbanization had significant growth, such as Shanghai and Jiangsu in China's Yangtze River Delta region, which is one of the most developed regions in China, where LUC driven by human construction is prevalent. However, there are also many areas where LUC is significant while PGSH is on the decline.

5. Discussion

5.1. The Relationship between the Intensity of LUC and PGSH

The intensity of LUC (LandCR, equal to the changed land area/total area) in each grid had a correlation coefficient of 0.038 **, with the change of PGSH in 2015 indicating a significant positive relationship between the higher LUC and the higher PGSH improvement. However, if LUC and PGSH are always in a linear relationship, it means that the more drastic the land use change, the more favorable the improvement of the bird habitat, which is obviously not in line with the cognition. Several studies have also found that the impact of land-use change on birds is complex and nonlinear [48–50]. We speculate that there is a threshold value for the degree of LUC to PGSH. Before this threshold value, a certain degree of LUC is beneficial to PGSH; however, exceeding this threshold will seriously disturb the living environment of birds, which is somewhat similar to the theory of Environmental Kuznets Curve [51]. Therefore, in order to determine this threshold, we added the square term of LUC (LandCR2) to the independent variable to build a new regression model. The result is shown in Table 2, which shows that the coefficient of LandCR was positive, while the coefficient of LandCR2 was negative, indicating that LUC and PGSH have an inverted U-shaped relationship. We identified the threshold as being approximately 0.6721, meaning that when the LUC is less than 0.6721, a higher LUC can promote the increase of PGSH, but beyond this value, it will reduce the PGSH value. Of all the grids in which LUC occurred in China, the PGSH of 90,752 (93.33%) grids was less than 0.6721, while the 6489 grids larger than 0.6721 accounted for 6.67%.

Table 2. The regression result of LandCR and LandCR2.

Variable	Coefficient
LandCR	0.082 ***
LandCR2	−0.061 ***

Dependent variable: PGSH; *** denotes significant at the $p < 0.01$ level.

5.2. *Influence of Two Land Use Policies on PGSH*

Cities are the areas with the highest concentration of human activities, which can impact bird populations living in these areas. Because China experienced an unprecedented increase of urbanization from 1995 to 2015, we are concerned about the negative impacts that this human-led expansion has had on bird habitats. Fortunately, forests serve as the main habitat for most bird species in China, which are less disturbed by human activity than cities. During this study period, there were two major artificial expansion projects of forest land in China: returning farmland to forest, Chinese term "tuigenghuanlin" (mainly sloping and desertified farmland with severe soil erosion and low yield), and the construction of the Three-North Shelterbelt Forest (mainly to alleviate the impact of sandstorms in northern China). We question what changes these developments have made to suitable bird habitat in these areas based on the counterfactual method to compare the PGSH under the situation of returning farmland to forest and the construction of the Three-North Shelterbelt, did not happen (hypothetical state) and actual state in these grids.

Returning farmland to forest: although the total amount of forest land changed little during the study, it may have changed spatially. Because forests play a vital role in the maintenance of bird habitat, we evaluated the impact of the "returning farmland to forest" policy on birds. Since the pilot project was implemented in 1999, the area returned from farmland to forest between 2000 and 2015 has been approximately 46,082 km^2. If fragmented areas with an area of less than 10,000 m^2 are removed, the remaining area is 43,934 km^2, accounting for around 1.93% of the total area of forest land in 2015. This forest land was distributed across 49,859 grids. The counterfactual analysis found that the policy of returning farmland to forest had no obvious benefits for improved bird habitat. This may be related to the unreasonable selection of tree species, planting site, and disturbance of nutrient cycle [52,53]. In the grid that implemented this policy, the probability of all birds inhabiting increased by only 1.07 percentage points, from an average of 73.97% to 75.04%. Even the inhabiting probability of six species of birds, namely spotted warbler, light-tailed warbler, brown-crested cuckoo falcon, Emei flycatcher warbler, wren, and unidentified falcon, had decreased.

Three-North Shelterbelt: the forest area within the Three-North Shelterbelt increased by 43,811 km^2 during the study period and the average PGSH of all birds in the grid where forest increase occurred changed from 0.2076 before restoration to 0.2138 after restoration.

It is believed that the Three-North Shelterbelt not only plays a direct role in improving land desertification but also improves the quality of bird habitat. This may be related to the important role of the corridors of ecological network in maintaining biodiversity, which has been greatly improved since the implementation of this policy [54,55]. We found that the habitat of 667 species of birds has been improved, but the habitat quality for 312 species deteriorated. The black-backed swallowtail had the highest improvement degree, increasing from 0.2877 to 0.368, while the PGSH of brown-winged snow finches, giant-billed sand finches, white-winged woodpeckers, and Mongolian sand finches, decreased by more than 10 percentage points.

Although returning farmland to forest and the construction of the Three-North Shelterbelt have increased the area of forest land to a certain extent, the areas scattered in each grid are small, accounting for 1.14% and 8.6748%, respectively. Therefore, it could be concluded that birds prefer large and agglomerated areas over small, fragmented ones.

6. Conclusions

Using multi-temporal land use data and the national bird observation database in China, this study systematically analyzed the impact of LUCs on 981 species of birds from 1995 to 2015. We used logistic regression to calculate the PGSH on all grid cells for each species. Overall, we found that the number of birds whose habitat quality continued to improve (582) was significantly higher than the number of birds under constant threat (84). Interestingly, the distribution of PGSH coincides with the boundary line of China's human population (Hu-line), with a clear divide between high PGSH in the east and low PGSH in the west. Within a certain range, PGSH was generally higher in the region with high human activity, but when urbanization intensity exceeds 67.21%, the continued increase of human activity would likely threaten bird habitats. China's policy of returning farmland to forests and the Three-North Shelterbelt project increased the area of green space, but the impact on PGSH was limited, with an average increase of less than 2%.

Although we studied the spatial and temporal changes of bird PGSH within the multi-data source over a relatively long period and across a broad research range, it provided the potential for comparative analysis of impacts of LUCs on different bird PGSH. However, because the data depend on citizen contributions to EBird and BirdReport, there may be bias for locations and observed species for specific contributors, thus these findings may have limitations for national extrapolation. In addition, the factors affecting the distribution of bird habitats are complex; for example, feed condition, presence of freshwater, climate, and temperature are important factors to consider. Simply considering land use and spatial proximity may lead to a certain degree of bias in the results. Thirdly, we found that there is an inverted U-shaped relationship between LUC and PGSH, just like the environmental Kuznets Curve; however, we did not give too much explanation for this phenomenon, which requires solid econometric statistics and discussion, which is beyond the scope of this paper. Although this paper has the above shortcomings, we believe that this research provides a useful attempt at analyzing substantial (two large datasets), large-scale (China) data. The results provide a useful reference for identifying bird species and habitats that require most conservation attention in the face of continued land use transformation.

Author Contributions: Conceptualization, B.Q. and J.Y.; methodology, Z.Z.; software, S.H.; validation, J.Y., Z.Z., and B.Q.; formal analysis, Z.Z.; investigation, J.Y.; resources, S.H.; data curation, S.H.; writing—original draft preparation, B.Q.; writing—review and editing, J.Y.; visualization, S.H.; supervision, Z.Z. All authors have read and agreed to the published version of the manuscript.

Funding: This research received no external funding.

Informed Consent Statement: Not applicable.

Data Availability Statement: The data presented in this study are available on request from the corresponding author.

Conflicts of Interest: The authors declare no conflict of interest.

References

1. Lambin, E.F.; Meyfroidt, P. Global LUC, economic globalization, and the looming land scarcity. *Proc. Natl. Acad. Sci. USA* **2011**, *108*, 3465–3472. [CrossRef]
2. Rosenberg, K.V.; Dokter, A.M.; Blancher, P.J.; Sauer, J.R.; Smith, A.C.; Smith, P.A.; Marra, P.P. Decline of the North American avifauna. *Science* **2019**, *366*, 120–124. [CrossRef] [PubMed]
3. Sterling, S.M.; Ducharne, A.; Polcher, J. The impact of global land-cover change on the terrestrial water cycle. *Nat. Clim. Chang.* **2013**, *3*, 385. [CrossRef]
4. Wu, H.; Lin, A.; Xing, X.; Song, D.; Li, Y. Identifying core driving factors of urban land use change from global land cover products and POI data using the random forest method. *Int. J. Appl. Earth Obs. Geoinf.* **2021**, *103*, 102475. [CrossRef]
5. Liu, Y.; He, Q.; Tan, R.; Liu, Y.; Yin, C. Modeling different urban growth patterns based on the evolution of urban form: A case study from Huangpi, central China. *Appl. Geogr.* **2016**, *66*, 109–118. [CrossRef]
6. Pielke Sr, R.A. Land use and climate change. *Science* **2005**, *310*, 1625–1626. [CrossRef] [PubMed]
7. Zhang, D.; Liu, X.; Wu, X.; Yao, Y.; Wu, X.; Chen, Y. Multiple intra-urban land use simulations and driving factors analysis: A case study in Huicheng, China. *GISci. Remote Sens.* **2019**, *56*, 282–308. [CrossRef]
8. Du, X.; Jin, X.; Yang, X.; Yang, X.; Zhou, Y. Spatial pattern of land use change and its driving force in Jiangsu Province. *Int. J. Environ. Res. Public Health* **2014**, *11*, 3215–3232. [CrossRef] [PubMed]
9. Sala, O.E.; Stuart Chapin, F.I.I.I.; Armesto, J.J.; Berlow, E.; Bloomfield, J.; Dirzo, R.; Huber-Sanwald, E.; Huenneke, L.F.; Wall, D.H. Global biodiversity scenarios for the year 2100. *Science* **2000**, *287*, 1770–1774. [CrossRef] [PubMed]
10. Northrup, J.M.; Rivers, J.W.; Yang, Z.; Betts, M.G. Synergistic effects of climate and land-use change influence broad-scale avian population declines. *Glob. Chang. Biol.* **2019**, *25*, 1561–1575. [CrossRef] [PubMed]
11. Powers, R.P.; Jetz, W. Global habitat loss and extinction risk of terrestrial vertebrates under future land-use-change scenarios. *Nat. Clim. Chang.* **2019**, *9*, 323. [CrossRef]
12. Mistry, J.; Berardi, A.; Simpson, M. Birds as indicators of wetland status and change in the North Rupununi, Guyana. *Biodivers. Conserv.* **2008**, *17*, 2383–2409. [CrossRef]
13. Ali, A.A.; Idris, N.H.; Ishak, M.H.I. The influence of land use land cover changes on urban bird communities. *Int. Arch. Photogramm. Remote Sens. Spat. Inf. Sci.* **2019**. [CrossRef]
14. Regos, A.; Imbeau, L.; Desrochers, M.; Leduc, A.; Robert, M.; Jobin, B.; Brotons, L.; Drapeau, P. Hindcasting the impacts of land-use changes on bird communities with species distribution models of Bird Atlas data. *Ecol. Appl.* **2018**, *28*, 1867–1883. [CrossRef]
15. Scholtz, R.; Polo, J.A.; Fuhlendorf, S.D.; Duckworth, G.D. Land cover dynamics influence distribution of breeding birds in the Great Plains, USA. *Biol. Conserv.* **2017**, *209*, 323–331. [CrossRef]
16. Newbold, T.; Scharlemann, J.P.; Butchart, S.H.; Şekercioğlu, Ç.H.; Alkemade, R.; Booth, H.; Purves, D.W. Ecological traits affect the response of tropical forest bird species to land-use intensity. *Proc. R. Soc. B Biol. Sci.* **2013**, *280*, 20122131. [CrossRef]
17. Sohl, T.L. The relative impacts of climate and land-use change on conterminous United States bird species from 2001 to 2075. *PLoS ONE* **2014**, *9*, e112251.
18. Chiawo, D.O.; Kombe, W.N.; Craig, A.J. Bird responses to LUC: Guild diversity in a Kenyan coastal forest and adjoining habitats. *Emu-Austral Ornithol.* **2018**, *118*, 281–292. [CrossRef]
19. Chapman, P.M.; Tobias, J.A.; Edwards, D.P.; Davies, R.G. Contrasting impacts of land-use change on phylogenetic and functional diversity of tropical forest birds. *J. Appl. Ecol.* **2018**, *55*, 1604–1614. [CrossRef]
20. Waltert, M.; Mardiastuti, A.; Mühlenberg, M. Effects of land use on bird species richness in Sulawesi, Indonesia. *Conserv. Biol.* **2004**, *18*, 1339–1346. [CrossRef]
21. Okada, S.; Lindenmayer, D.B.; Wood, J.T. Does LUC influence predation of bird nests? *Austral Ecol.* **2019**, *44*, 768–776. [CrossRef]
22. Zhang, Y.; Wang, H.; Xie, P.; Rao, Y.; He, Q. Revisiting spatiotemporal changes in global urban expansion during 1995 to 2015. *Complexity* **2020**, *2020*, 6139158. [CrossRef]
23. Wang, C.; He, J.; Liu, D.; Yu, X.; Shi, Q. Impact of land use change on bird habitat connectivity: A case study in Ezhou City. *Acta Ecol. Sin.* **2022**, *42*, 4197–4208. (In Chinese with English Abstract)
24. Cintra, R.; Magnusson, W.E.; Albernaz, A. Spatial and temporal changes in bird assemblages in forest fragments in an eastern Amazonian savannah. *Ecol. Evol.* **2013**, *3*, 3249–3262. [CrossRef]
25. Crooks, K.R. Relative sensitivities of mammalian carnivores to habitat fragmentation. *Conserv. Biol.* **2002**, *16*, 488–502. [CrossRef]
26. Fahrig, L. Effects of habitat fragmentation on biodiversity. *Annu. Rev. Ecol. Evol. Syst.* **2003**, *34*, 487–515. [CrossRef]
27. Zhang, Z.; Zheng, M. *Advances in the Study of Bird Habitat Selection*; China Forestry Press: Beijing, China, 1999.
28. Ministry of Environmental Protection, the People's Republic of China. China Biodiversity Red List. 2015. Available online: http://www.mee.gov.cn/gzfw/spk/hbbvideo/Videos/201605/t20160531_352442.shtml (accessed on 1 July 2020).
29. Liu, L.; Liu, H.Y.; Li, Y.F.; Wang, J.; Xie, F.F. Dynamic changes in population size and habitat distribution of wintering red-crowned crane in northern Jiangsu Province. *Acta Ecol. Sin.* **2018**, *38*, 926–933, (In Chinese with English Abstract).
30. Xu, Y.; Lin, S.; He, J.; Xin, Y.; Zhang, L.; Jiang, H.; Li, Y. Tropical birds are declining in the Hainan Island of China. *Biol. Conserv.* **2017**, *210*, 9–18. [CrossRef]
31. Fang, W.Z.; Chen, Z.H.; Chen, X.L.; Lin, Q.X. Avian community diversity during winter in Xiamen wetlands. *Mar. Sci.* **2007**, *31*, 10–16. (In Chinese with English Abstract)
32. Jiang, Z.; Ma, K. Status quo, challenges and strategy in conservation biology. *Biodivers. Sci.* **2009**, *17*, 107–116.

33. Dai, S.; Feng, D.; Chan, K.; Gong, P.; Xu, B. A spatialized digital database for all bird species in China. *Sci. China Life Sci.* **2019**, *62*, 661–667. [CrossRef]
34. He, F.Q.; Fellowes, J.R.; Chan, B.P.; Lau, M.W.; Lin, J.S.; Shing, L.K. An update on the distribution of the 'Endangered' White-eared Night Heron Gorsachius magnificus in China. *Bird Conserv. Int.* **2007**, *17*, 93–101. [CrossRef]
35. Chen, J.; Zhang, B.; Ma, K.; Jiang, Z. Bibliometric analysis of status quo of conservation biology in China. *Biodivers. Sci.* **2009**, *17*, 423–429.
36. Si, Y.; Xin, Q.; Prins, H.H.; De Boer, W.F.; Gong, P. Improving the quantification of waterfowl migration with remote sensing and bird tracking. *Sci. Bull.* **2015**, *60*, 1984–1993. [CrossRef]
37. Bonney, R.; Cooper, C.B.; Dickinson, J.; Kelling, S.; Phillips, T.; Rosenberg, K.V.; Shirk, J. Citizen science: A developing tool for expanding science knowledge and scientific literacy. *BioScience* **2009**, *59*, 977–984. [CrossRef]
38. Supp, S.R.; Sorte, F.A.L.; Cormier, T.A.; Lim, M.C.; Powers, D.R.; Wethington, S.M.; Graham, C.H. Citizen-science data provides new insight into annual and seasonal variation in migration patterns. *Ecosphere* **2015**, *6*, 1–19. [CrossRef]
39. Li, X.; Liang, L.; Gong, P.; Liu, Y.; Liang, F. Bird watching in China reveals bird distribution changes. *Chin. Sci. Bull.* **2013**, *58*, 649–656. [CrossRef]
40. Sullivan, B.L.; Wood, C.L.; Iliff, M.J.; Bonney, R.E.; Fink, D.; Kelling, S. eBird: A citizen-based bird observation network in the biological sciences. *Biol. Conserv.* **2009**, *142*, 2282–2292. [CrossRef]
41. EBird Basic Dataset (EBD). *Version: EBD_relApr-2019*; Cornell Lab of Ornithology: Ithaca, NY, USA, 2019.
42. He, Q.; Song, Y.; Liu, Y.; Yin, C. Diffusion or coalescence? Urban growth pattern and change in 363 Chinese cities from 1995 to 2015. *Sustain. Cities Soc.* **2017**, *35*, 729–739. [CrossRef]
43. Lai, L.; Huang, X.; Yang, H.; Chuai, X.; Zhang, M.; Zhong, T.; Chen, Z.; Chen, Y.; Wang, X.; Thompson, J.R. Carbon emissions from land-use change and management in China between 1990 and 2010. *Sci. Adv.* **2016**, *2*, e1601063. [CrossRef]
44. Thiessen, A.H. Precipitation averages for large areas. *Mon. Weather. Rev.* **1911**, *39*, 1082–1089. [CrossRef]
45. Fu, W.; Jiang, P.; Zhou, G.; Zhao, K. Using Moran's I and GIS to study the spatial pattern of forest litter carbon density in a subtropical region of southeastern China. *Biogeosciences* **2014**, *11*, 2401–2409. [CrossRef]
46. Anselin, L. Local indicators of spatial association—LISA. *Geogr. Anal.* **1995**, *27*, 93–115. [CrossRef]
47. Bennett, J. Event causation: The counterfactual analysis. *Philos. Perspect.* **1987**, *1*, 367–386. [CrossRef]
48. Butler, S.; Boccaccio, L.; Gregory, R.; Vorisek, P.; Norris, K. Quantifying the impact of land-use change to European farmland bird populations. *Agric. Ecosyst. Environ.* **2010**, *137*, 348–357. [CrossRef]
49. De Lima, R.F.; Dallimer, M.; Atkinson, P.; Barlow, J. Biodiversity and land-use change: Understanding the complex responses of an endemic-rich bird assemblage. *Divers. Distrib.* **2013**, *19*, 411–422. [CrossRef]
50. Acevedo, M.; Restrepo, C. Land-cover and land-use change and its contribution to the large-scale organization of Puerto Rico's bird assemblages. *Divers. Distrib.* **2008**, *14*, 114–122. [CrossRef]
51. Dinda, S. Environmental Kuznets curve hypothesis: A survey. *Ecol. Econ.* **2004**, *49*, 431–455. [CrossRef]
52. Trac, C.; Schmidt, A.; Harrell, S.; Hinckley, T. Environmental reviews and case studies: Is the returning farmland to forest program a success? Three case studies from Sichuan. *Environ. Pract.* **2013**, *15*, 350–366. [CrossRef]
53. Li, Z.; Sun, X.; Huang, Z.; Zhang, X.; Wang, Z.; Li, S.; Zhai, B. Changes in nutrient balance, environmental effects, and green development after returning farmland to forests: A case study in Ningxia, China. *Sci. Total Environ.* **2020**, *735*, 139370. [CrossRef]
54. Mu, H.; Li, X.; Ma, H.; Du, X.; Huang, J.; Su, W.; Li, B. Evaluation of the policy-driven ecological network in the Three-North Shelterbelt region of China. *Landsc. Urban Plan.* **2022**, *218*, 104305. [CrossRef]
55. Wang, H.; He, M.; Ran, N.; Xie, D.; Wang, Q.; Teng, M.; Wang, P. China's key forestry ecological development programs: Implementation, environmental impact and challenges. *Forests* **2021**, *12*, 101. [CrossRef]

sustainability

MDPI

Article

The Effect of Land Tenure Institutional Factors on Small Landholders' Sustainable Land Management Investment: Evidence from the Highlands of Ethiopia

Shewakena Aytenfisu Abab [1,*], Feyera Senbeta [2] and Tamirat Tefera Negash [1]

1 Center for Environment and Development, College of Development Study, Addis Ababa University, Addis Ababa P.O. Box 1176, Ethiopia; tamirat.tefera@aau.edu.et
2 Department of Biological Sciences, Faculty of Sciences, Botswana University of Agriculture and Natural Resources, Gaborone 0027, Botswana; fwakjira@buan.ac.bw
* Correspondence: shewakena.aytenfisu@aau.edu.et

Abstract: Sustainable Land Management (SLM) is one of the key policy responses being implemented to curb land degradation in the highlands of Ethiopia. However, there is scant evidence to what extent Land Tenure Institutional Factors (LTIFs) influence small landholders' on-farm investment in SLM. The overall objective of this study is, therefore, to understand the extent to which LTIFs influence on-farm SLM investment in the highlands of Ethiopia through unbundling tenure security (de jure, de facto, and perceived) across a bundle of rights. Survey data were collected between April and May 2021 from 2296 smallholder households and 6692 parcels of 19 highland woredas (districts) in three regional states (Amhara, Oromia, and SNNP) in Ethiopia. A probit regression model was used to estimate the average marginal effects of LTIFs quantitatively and supported by an in-depth qualitative analysis. The results revealed that 10 out of 16 LTIF-related variables have significantly influenced households' on-farm investment in SLM with average marginal effect ranging from a minimum of 3% (tree tenure security risks) to a maximum of 14% (possession of land certificates), at 95% confidence interval, compared to a mean probability of 45%. The results also revealed that some of the households' socio-economic and demographic factors and parcel-specific variables have significantly influenced on-farm SLM investment. These imply two policy issues. Firstly, it strengthens the notion that security of tenure may be a necessary condition, but not a sufficient, factor to incentivize smallholders' on-farm SLM investment. Secondly, an in-depth analysis of the security of tenure categories across a bundle of rights is necessary to help formulate context-specific SLM policy and strategy incentivizing smallholders' on-farm SLM investment.

Keywords: land degradation; bundle of rights; security of tenure; SLM; investment; Ethiopia

Citation: Abab, S.A.; Senbeta, F.; Negash, T.T. The Effect of Land Tenure Institutional Factors on Small Landholders' Sustainable Land Management Investment: Evidence from the Highlands of Ethiopia. *Sustainability* **2023**, *15*, 9150. https://doi.org/10.3390/su15129150

Academic Editors: Lu Zhang, Bing Kuang and Bohan Yang

Received: 6 April 2023
Revised: 17 May 2023
Accepted: 2 June 2023
Published: 6 June 2023

1. Introduction

Policy makers, practitioners, and researchers are becoming more conscious of the importance of clear, secure, and inclusive access to and control over land resources because of increased competition for land resources and mounting climate change uncertainties. Under the United Nations, land tenure indicators are adopted as a fundamental element of measuring the global sustainable development goals (SDGs). For instance, SDG 1.4.2 aims to measure the "proportion of the total adult population with secure tenure rights to land including (a) with legally recognized documentation, and (b) who perceive their rights to land as secure, disaggregated by sex and tenure type (%)" [1]. In addition, SDG 5.a.1 stresses women's land tenure, SDG 2, Target 2.3.1 and 2.3.2 address smallholder farmers; and Target 2.4.1 also focuses on agricultural land [2]. Land tenure also influences land use and is thus key to achieving SDG 14 (b) and SDG 15 on the sustainable use of land and natural resources. Likewise, land tenure is also vital as it is often considered a driver of conflict if managed poorly, yet it is a source of resilience if managed properly, and hence affects SDG 16, promoting peace and inclusive societies and institutions.

Land tenure importance is also manifested by households' land use decisions at the local level, indicating the need for physical capital to spur economic growth and land governance systems [3]. Land tenure security is an important development agenda for strengthening land governance systems, thereby improving social stability, spurring economic growth, and promoting the environmental sustainability of citizens, communities, and business firms [4,5]. The assumption is that recognizing land and resource rights will benefit the rights holders by 'unlocking' capital through access to credit or by enabling full utilization of production factors, reducing uncertainty, providing opportunities and empowerment, and incentivizing the sustainable use of natural resources [6].

Tseng et al. [6] and Robinson et al. [7] also identified two dimensions of tenure issues with a strong potential to influence land-based investment decisions, including the type of rights landholders have and the security of those rights. The bundle of rights includes access, use, management, exclusion, alienation, and the rights to be compensated during compulsory expropriation [7,8]. In contrast, security is understood as a perception by right holders that rights are recognized and protected [9]. Land rights are secure when a person perceives them to be stable and predictable over a reasonable period and protected from expropriation or arbitrary change [10]. This is consistent with the SDG 1.4.2a secured tenure rights definition.

Other scholars [7,11,12] distinguished the categories of land tenure security as (1) de jure/legal, (2) de facto/contextual, and (3) perceived/socio-psychological tenure security. This category of security is associated with a given tenure system such as freehold, lease hold, or customary and the myriad social, economic, political, and environmental factors that condition the de facto performance of such an arrangement [13]. This arrangement may be formal, informal, or applied through customary institutions that can be a major hinder or enable sustainable land management or development [14]. According to Masuda et al. [5,13], Holland et al. [12], and Robinson et al. [7], as societies grow and land pressure increases, there becomes a need for clear and transparent processes that assign and enforce rights among various parties and spell out the rules for how rights can be accessed, transferred, terminated, or gained. Locke [10] even argued the primary function of government is to secure and protect such property rights. This means that sustained land tenure security most likely comes with the state-recognized backing of land rights [6]. However, such institutional genesis is a long-term process that needs to grow within the existing socio-economic and political system.

Land tenure institutions are, therefore, fundamentally important in enhancing land-based investment and promoting the efficient allocation of economic resources [15–17]. A relationship between a rights holder and a subject parcel of landholding depends on the characteristics of the bundle of rights that qualify its usefulness in economic exchange and influence economic behavior on investments [15,18,19] and the financing of these investments [20]. The governance of these relationships is mainly administered by land tenure institutions. The inefficiency and ineffectiveness of those institutions affect the quality of tenure security. Uncertainty about tenure rights also creates insecurities about land tenure and frequently leads to poor uses of limited resources as these influence the practices, abilities, and choices of landholders in line with the adoption, sustainability, effectiveness, and efficiency of their investment [4,6].

Despite a notable increase in rigorous systematic reviews in recent years, much of the evidence on land tenure remains linked to tenure security achieved through land titling and its implication on environment and development outcomes. For instance, Tseng et al. [6] recently reviewed about 117 studies to understand the causal effect of land tenure security interventions such as land titling and formalization on human well-being or environmental outcomes, of which two-thirds of the studies reported positive links. Likewise, Lawry et al. [21] undertook a similar systematic review and found that land tenure recognition positively affected productivity and income gains substantially through perceived tenure security and investment. However, these reviews show the existing body of land tenure literature focus on a single category of tenure security and its implication on socio-economic and environmental outcomes. This approach hinders an in-depth yet broader understanding of LTIF effects along categories of land tenure security and the context-specific bundle of rights on development and environment.

Considering the land tenure theory and existing evidence base, the link between LTIFs, such as the three categories of land tenure security and investment in SLM, appears inconclusive, at least in the Ethiopian context. This is because, firstly, landholders have only perpetual usufruct rights and could not be used as collateral to access formal credit until recently, or land exchange or sale was forbidden [22]. This implies that SLM investment made in the rural landscapes, specifically at the farm level by smallholder households, cannot be attributed to either greater access to credit or enhanced functions of the land market as land sales are ruled out by law. Secondly, perceived tenure security and de facto tenure security vary in a range of transferability of legally (de jure) recognized bundle of rights such as risks related to inheritance, gift, lease/rentals, conservation, tree tenure, expropriation and compensation, and land redistribution. Thirdly, most previous studies rely on a small sample size of cross-sectional data targeted to a specific watershed and biomes and looked at secure tenure without any categorization, thus limiting more rigorous and in-depth analysis of factors influencing on-farm investment in SLM among small landholders. Specifically, this hinders the full understanding of the LTIFs linked to the country's SLM policy implementation effectiveness in guiding context-specific small landholders' on-farm SLM investment.

Historically, Ethiopia's highland agriculture is dominated by small landholders' farmland tenure model and characterized by fragmentations. Land degradation is one of the major environmental and development challenges compounded by climate change risks. SLM is also considered one of the key policy responses being implemented using watershed as a unit of planning and watershed users' cooperative societies as governance structures. However, at the landscape level, the landholding types are a mosaic of communal, private, and state/public lands associated with certain socio-ecological systems and highly dominated by smallholder land tenure. The interaction and relations of these diverse tenure types affect the land use practice and on-farm SLM investment of small landholders and its sustainability in the study areas.

For instance, legally recognized rights might be represented by registering those rights and provisioning land certificates as a de facto protection of those land rights. However, the impact of land certificates on tenure security differs by how perceived tenure security is measured [23]. Where the perceived tenure security dimension is specifically measured along risks related to the bundle of rights such as inheritance, land redistribution, expropriation and compensation, land transfer through land rent/sharecropping, conservation, tree tenure security, as well as credit transactions. This bundle of rights is an important set of land rights recognized in the existing legal framework of the country. Hence, understanding these dynamics of land tenure institutional factors through the lens of categories of tenure security across the bundle of rights is vital to design pragmatic SLM policy and context-specific implementation strategy.

Therefore, the objective of this study is to understand the extent to which LTIFs influence the probability of households' on-farm SLM investment in the highlands of Ethiopia; through the lens of the three categories of security of tenure (de jure, de facto, and perceived) across the bundle of rights in the context of the existing legal framework. To this end, this study employed household and parcel-level survey data accessed from the USAID publicly available data repository and addresses the limitations and contributes to the existing body of evidence.

2. Materials and Methods

2.1. Sample Size and Data

This study was conducted in three highland regional states of Ethiopia (Amhara, Oromia, South Nation Nationalities and People (SNNP)), 19 woreda, and 183 kebeles. This study used the 2021 survey data collected for the follow-on impact evaluation (IE) study of the USAID-funded land administration programs, namely the Ethiopia Land Tenure Administration Program (ELTAP, run between 2005 and 2008) and Ethiopia Land Administration Program (ELAP, run between 2008 and 2013). This is because the 2021 follow-on survey data were found as the best publicly available recent data in the sector, with a trove of survey data that help to respond to the research objective of the current study. The impact evaluation studies of USAID examined the impact and limitations of the land certification intervention on rural land users over a 15 years' time horizon.

As part of this panel dataset, data were previously collected in three rounds, namely, 2008 as a baseline, 2015 as an end line, and 2021 as a follow-on impact evaluation study [24,25]. In all waves of data collection, the researchers collected data using a head-of-the-household survey and a wife survey that was applied to the head of the household (male or female) and their spouse or wives in the case of polygamous households. Unlike the two previous surveys, the 2021 survey excluded households from Tigray and the 12 kebeles in Amhara because of the conflict and security issues, while the ELAP targeted households were also excluded to reduce selection bias on the results since they were targeted by the land administration programs with higher potential for agricultural investment [24], Figure 1.

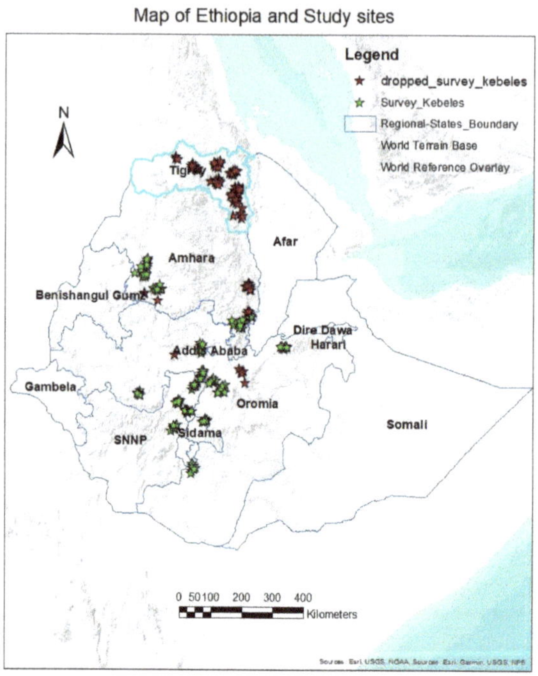

Figure 1. Map of Ethiopia and study survey sites. Compiled by the Author, 2023. Red shows the study sites excluded from the 2021 survey due to security issues in Tigray and Amhara regional states as well as ELAP-supported woredas excluded to avoid selection bias. Source: USAID [26], [Data set]. https://www.land-links.org/ accessed on 21 December 2022.

In terms of sample size, systematic stratified sampling method was employed. Firstly, six program woredas were selected from the four program participating regions. Secondly, within each woreda, a stratified systematic selection of kebeles was made based on distance from capital/road (3 categories or clusters identified, i.e., with 5 KMs—near, 10 KMs—medium, and above 10 KMs—far). Hence, 8 treatment and 3 control groups per kebeles, and for the control group, 3 kebeles were randomly chosen per woreda. Thirdly, within selected treatment and control kebeles, the selection of households was made using stratified random sampling proportional to the number of male- and female-headed households in the kebeles, which includes 15 per treatment and 10 per control kebeles. Accordingly, the 2008 survey covered 3600 households across 284 kebeles in the four regions (Amhara, Oromia, SNNP, and Tigray). The survey yielded 2754 wives in 2643 male-headed households and 698 women in female-headed households [25]. Likewise, the 2015 survey also collected data for 3412 wives in 3412 male-headed households and 914 women in female-headed households. On the other hand, the 2021 survey collected data from the same households from April to May 2021 who had been interviewed in 2008 and 2015. However, 3 percent household attrition is observed for several reasons, including household change of place, death, separation or dissolution of household, and illness, among others. The survey in 2021 includes 2306 households, of which ten households were dropped from the sample because their information on land certification status was incomplete, meaning the final sample size was only 2296 [24].

While the original plan was to use the three waves of the panel data, after a thorough review of the panel datasets, the authors decided to use the 2021 survey data only because of several important limitations to the design and instruments among the three waves. Firstly, despite the same household survey module employed, baseline data were not collected at the parcel level, which reduces the study's ability to assess parcel-level SLM activities under the current study rigorously. Secondly, the 2021 follow-on survey contains detail data both at the household and parcel level, including (a) the socio-economic and demographic issues, (b) land tenure and land certification status, (c) engagement of households in land transactions such as land rentals/sharecropping, inheritance, gift, and credit, (d) land dispute incidents, (e) level of awareness on land rights, (f) perception in land tenure security and related risks, (g) land use quality, (h) soil and water conservation investment and productive assets building, among others.

Thirdly, over the past decade and a half, the difference between the treatment and control groups in terms of land tenure security improving interventions have been closed, such as land registration and certification. Meaning, most households in the control groups received treatment overtime. Fourthly, the discrepancies in the resolution or presence of certain variables across the baseline, end line, and follow-on datasets mean direct implications on the sample size and the ability to fully utilize certain finer resolutions between baseline and end line datasets compared to the follow-on under the current study. Finally, in terms of methodology, both the baseline and end line impact evaluation studies used a Difference in Difference (DiD) econometrics model and analysis, while the follow-on evaluation also included Continuous Treatment (CT) analysis additionally and compared the results of the two. Hence, the current study, while employing the follow-on survey data only, departs in methodology as well to better understand the role of LTIFs and their average marginal effect on the dependent variable. To perform the statistical analysis, the authors employed STAT version 14 software and ArcMap for mapping and visualizing survey kebeles spatial distribution in the country, Figure 1.

2.2. Empirical Model Specification

This study employed a probit regression model to test the hypothesis that the three categories of tenure security may have different effects on the probability of households' on-farm SLM investment across the bundle of rights. This approach also helps us to understand better the average marginal effect of the LTIFs influencing the probability of households' any on-farm SWC practice as proven SLM investment. The model looked at the fitted probability of the dependent variable due to the influence of the set of explanatory variables presented in Equation (7) below, meaning the outcome variable is determined or predicted as a non-linear model that forces the probability function to fall between 0 and 1 based on communitive density functions of independent variables derived from the robust standard errors distribution.

Using this model, the authors are able to estimate the probability of households' on-farm SLM investment, accounting for the households' demographic and socio-economic variables, quality or characteristics of the parcel of land, and the three categories of land tenure security variables across the bundle of rights. Hence, the authors estimate four models and compare the results of the regression analysis, i.e., Model 1 includes household and parcel variables only, Model 2 includes household, parcel, and de jure variables only, Model 3 includes household, parcel, de jure, and de facto variables only, and the final or Model 4 includes household, parcel, de jure, de facto, and perceived tenure risk variables. The probit regression equation is specified as:

$$P(Y = 1|X) = G(X\beta) = \int (2\pi)^{-5} \exp\left(-\frac{X\beta^2}{2}\right) \quad (1)$$

$$the\ G\ function\ of\ X\beta = \beta_0 + \beta_1 X_1 + \beta_2 X_2 + \beta_3 X_3 + \beta_4 X_4 \quad (2)$$

That means:

$$fitted\ probability = \hat{p}(Y = 1|X) = G(X\hat{\beta}) \quad (3)$$

where:

$$lim_{X\beta \to \infty}\, G(X\hat{\beta}) = 1$$

$$lim_{X\beta \to \infty}\, G(X\hat{\beta}) = 0$$

To estimate the $\hat{\beta}$ coefficient, we use the maximum likelihood estimation that maximizes the joint probability of the outcome variable and constructed based on the product of each observation probability of observing what we see, which can be written as follows:

$$L = \prod_{i=1}^{N} P_i^{Y_i} \ln(1 - P_i)^{(1-Y_i)} \quad (4)$$

Taking logs, we can attain the "log likelihood" as follows:

$$lnL = \sum_{i=1}^{N} Y_i \ln(P_i) + (1 - Y_i) \ln(1 - P_i) \quad (5)$$

The marginal effects depend on X, where the average marginal effect calculates each individual observation's marginal effect and then takes the mean, which is the derivative of G with respect to $X\beta$ constructed as:

$$\frac{\partial P(Y = 1|X)}{\partial X_1} = \beta_1 G'(X\beta) \quad (6)$$

where $G'(X\beta)$ will change as X changes, which allows for diminishing returns or a non-linear relationship; therefore, the final probit empirical model is constructed as:

$$Y_{ih} = \beta_0 + \beta_1 X_1 + \beta_2 X_2 + \beta_3 X_3 + \beta_4 X_4 + \ldots \beta_{24} X_{24} + \varepsilon_{ih} \quad (7)$$

where Y is a dummy outcome or dependent variable which represents a small landholder's investment in any SWC practice by a household h, and X is the set of explanatory variables related to households, parcels, and categories of the three secure tenure rights across the bundle of rights as recognized in the existing legal framework. More specifically, Y = dependent variable (investment in any SWC); β_0 = constant term; X_1 is age; X_2 is sex; X_3 is highest school grade; X_4 is marital status; X_5 is land area; X_6 represents time to walk to parcel; X_7 is walking distance to parcel; X_8 is water erosion risk, X_9 is usufruct rights-de jure; X_{10} is transfer rights (rent/sharecropping)-de jure, X_{11} is bequest-de jure; X_{12} is collateral rights-de jure; X_{13} is decision on what to grow/invest-de facto; X_{13} represents decision on the use of the produce-de facto; X_{14} represents decision on transfer, i.e., who decides on the transfer (rent/sharecropping-OUT) to others-de facto; X_{15} represents credit obtained-de facto; X_{16} represents possession of First Level Landholding Certificate-de facto; X_{17} is possession of Second Level Landholding Certificate-de facto; X_{18} represents bequest-perceived; X_{19} is transfer to others (rent-out/sharecropping)-perceived; X_{20} represents credit transaction-perceived; X_{21} is perceived conservation risk-perceived risk, X_{22} is tree tenure risk-perceived risk; X_{23} represents land redistribution risk-perceived risk; X_{24} represents enter in to any business transaction risk-perceived risk; ε_{ih} is the error term of explanatory variables.

This study also employed qualitative data collected from focus group discussions among kebele administrative officials of the study sites and legal and administrative document reviews which complement and substantiate the quantitative analysis.

3. Results

3.1. Characteristics of this Study's Kebeles (Villages)

To better understand the overall characteristics of this study's kebeles, focus group discussions were conducted that covered the estimated number of populations, mean livelihood of residents, land use, land scarcity, migration (in and out), and services, including road, market, mobile phone network coverage, financial institutions, transportation, and religious institutions.

The surveyed kebeles are spatially distributed in the three highland regions with 1500 m above mean sea level and are characterized by high population density. The mean number of households in the study kebeles was 1072, with a standard deviation of 885 households and 5.8 average persons per household. There is migration in and out of kebeles, with slightly more than half of kebeles reporting net out-migration. This net out-migration may likely increase household labor productivity and improve land use efficiency by freeing some land for the land rental market. Agriculture is the predominant land use system on small private landholdings and livelihood sources, while 14 percent of kebeles have no remaining bush or forest land. About 83 percent have fewer than 25 percent of kebele land area left as bush/forest land. In addition to agriculture, pity trade, and casual labor are the primary means of livelihood for kebele residents. About 72 percent of kebele main roads are all-weather roads, meaning they are accessible year-round, while 77 percent of kebeles have a large weekly market that exchanges goods and services locally.

About 92 percent of the kebeles had access to mobile phone network coverage, which facilitated information flow, thereby reducing the cost of information and, by extension, services. However, only 4% of kebeles had access to a bank service within their vicinity, but 39% had access to a microfinance institution's financial service. This shows kebeles have limited access to formal financial services. Thus, about 80 percent of the total credit service is provided to rural households by financial cooperatives and Microfinance Institutions (MFIs). Moreover, only 10 percent of the kebele authorities reported that there was a project-based SWC intervention between 2016 and 2021. This also shows that this study's kebeles receive limited project-based extension services related to SLM that may hinder on-farm investment among smallholder households. On average, survey kebeles were about 23 km by road to the nearest major urban center, which implied that kebele residents face some barriers to accessing services outside their kebele. For instance, one-fourth of kebeles do not have passable roads year-round and face substantial costs for public transportation, estimated to be 10 percent of the daily household per capita expenditures, meaning landholders who reside in remote areas have limited time and resources to travel to woreda offices to access land tenure related services and may sometimes involve opportunity costs for leaving their on-farm investments during their travel.

On the other hand, despite being a place of worship, churches and mosques are used as important avenues to disseminate information about administrative and community development extension services. According to the FGD, there were an average of 4.36 churches and 2.56 mosques in the survey kebeles. Those community-based religious institutions facilitate various awareness-raising meetings and serve as information disseminating points, including SWC campaigns and programs.

3.2. Descriptive Statistics of Surveyed Households

Table 1 summarizes detailed demographic and socio-economic characteristics of the survey respondents, parcel characteristics, land tenure institutional factors (LTIFs), description of the variables, means, frequencies, and standard deviations. As explained earlier in Section 2.1, the datasets also collected a range of land tenure-related information. The authors grouped the LTIFs into three categories of security of tenure: de jure/legal security, de facto/contextual security, and perceived security, as also used by Asaaga et al. [11].

Table 1. Descriptive and summary statistics of variables.

Variables	Description of Variables	Expected Sign	Mean/Ferq.	Std. Dev.
	Dependent Variable			
Invested in any soil and water conservation practices	Households invested in any SWC in their landholding, dummy (1 = yes, 0 = otherwise)	±	0.45	0.50
Stone bund	Length of constructed stone bunds (in meters), continuous	±	10.93	46.62
Soil bund	Length of constructed soil bund (in meters), continuous	±	27.75	75.16
Water retention structure	Number of on-farm water retention structures (ponds, retention ditches) constructed, continuous	±	0.087	0.004
Trees planted per hectare	Number of trees planted, continuous	±	105.54	819.75
Perennials planted per hectare	Number of perennials crops planted, continuous	±	162.49	734.79
	Independent variables			
	Household demographic and socio-economic variables			
Age	Age of the household head, continuous	±	55.17	14.26
Gender	Gender of the household head, dummy (1 = man, 2 = woman)	±	1.22	0.41
Education	Highest level of education completed, categorical Illiterate Read-only Read & write Grade 4 complete Grade 8 complete Grades 10–12 complete Above grade 12	±	 51.14 2.55 11.77 19.42 8.24 4.94 1.95	
Marital status	Marital status of the household head, categorical: 1 = Unmarried/Never married, 2 = Married, 3 = Divorced, 4 Widowed, 5 = cohabiting, 6 = preferred not to respond	±	 0.02 0.77 0.03 0.18 0.00 0.00	

Table 1. *Cont.*

Variables	Description of Variables	Expected Sign	Mean/Ferq.	Std. Dev.
Means of land originally acquired	How was land originally acquired? categorical Inherited Official land redistribution Gift Others	±	 39.24 37.51 8.64 14.61	
Holding Size	Area of land possessed in hectares, continuous	±	1.68	1.78
	Parcel variables			
Time travel to parcel	Time to walk to parcel one way (in minutes), continuous	±	15.3	24.79
Parcel distance	Walking distance to parcel one way (in meters), continuous	±	1460.26	2314.94
Water erosion risk	Parcels located on sloping lands with soil erosion risk, dummy (1 = yes, 0 = otherwise)	±	0.389	0.488
	Legal tenure security			
Usufruct rights	land laws allow to use of the parcel, dummy (1 = yes, 0 = otherwise)	±	0.976	0.154
Transfer rights (rent/sharecropping)	land laws allow making a contract (rent/sharecropping) dummy (1 = yes, 0 = otherwise)	±	0.965	0.185
Bequeath or inherit rights	land laws allow to bequest it to hires, dummy (1 = yes, 0 = otherwise)	±	0.946	0.226
Collateral rights	land laws allow to use of land as collateral, dummy (1 = yes, 0 = otherwise)	±	0.801	0.398
	De facto/contextual security of tenure			
What to grow?	Who decides on what crop (s) to grow, Continuous (1 = Husband, 2 = Wife, 3 = Husband and Wife, 4 = Children, 5 = Family, 6 = Single Household Head, 8 = Household Head and children, 97 = Other)	±	15.17 0.51 59.80 2.22 5.52 13.90 2.61 0.26	
on the use of produce	Who decides on the use of produce from the land? Continuous (1 = Husband, 2 = Wife, 3 = Husband and Wife, 4 = Children, 5 = Family, 6 = Single Household Head, 8 = Household Head and children, 97 = Other)	±	11.14 0.85 63.83 1.84 6.01 15.37 0.86 0.12	
on the transfer	Who decides on the transfer of land use rights (rent/sharecropping-out) to others? Continuous (1 = Husband, 2 = Wife, 3 = Husband and Wife, 4 = Children, 5 = the whole family, 6 = Single Household Head, 8 = Household Head and children, 97 = Other)	±	11.23 0.63 63.90 1.78 5.42 16.42 0.34 0.29	
Credit obtained	HH obtained credit (formal or informal) during the last 2 years, dummy (1 = yes, 0 = otherwise)	±	0.058	0.233

Table 1. *Cont.*

Variables	Description of Variables	Expected Sign	Mean/Ferq.	Std. Dev.
	Perceived tenure security risks			
Conservation risk	HH head fully convinced to benefit from SWC measures they may undertake., categorical (1 = Strongly Agree, 2 = Agree, 3 Disagree, 4 Strongly Disagree)	±	52.69 42.53 3.44 1.34	
Tree tenure risk	HH head fully convinced not to benefit from trees planted, categorical (1 = Strongly Agree, 2 = Agree, 3 Disagree, 4 Strongly Disagree)	±	9.59 11.73 35.06 43.62	
Land redistribution risk	HH believes that redistribution of land is likely to take place in their kebele in 5 years, categorical (1 = Strongly believe, 2 = Believe, 3 = Don't believe, 4 Strongly don't believe)	±	3.48 6.98 34.13 55.45	
Rent out risk	HH feels that renting out is a risky business, categorical (1 = Strongly Agree, 2 = Agree, 3 Disagree, 4 Strongly Disagree)	±	30.19 39.72 23.62 6.46	
Business transaction risk	HH will feel more secure entering any sort of business transaction involving credit with a farmer who has a Land Certificate than who does not have, categorical (1 = Strongly Agree, 2 = Agree, 3 Disagree, 4 Strongly Disagree)	±	53.98 39.01 6.47 0.53	

Based on the survey result, 78 percent of the respondents were male-headed households, while the remaining 22 percent were female-headed households with an average household size of 5.3. In addition, 76 and 18 percent of household heads were married and widower/ed, respectively, while 3 percent of household heads were divorced. For the entire sample, the average age of the household heads in the study area was 55, indicating that most of the household heads were active and productive. Moreover, the survey result shows that the majority (51 percent) of the respondents were illiterate, while about 47 percent of the respondents can read and write information about their land use rights, restrictions, and responsibilities.

The survey also collected data on the land parcel's biophysical characteristics or quality. Most households in the study areas were characterized as smallholders, with an average of 1.59 hectares and 3.1 parcels per landholding, which indicates a slightly higher than the national average landholding size, i.e., 1.22 hectares but with lower fragmentation [26]. This may have implications on households' on-farm SLM investment. It is also noteworthy that a land holding may consist of one or more parcels within a kebele, which is the lowest administrative and land registration unit. In terms of land use type and proportion, households reported that about 80, 7, 10, and 3 percent of their landholding area was used for annual crops, perennial crops, grazing land, and woodlots, respectively. This indicates that most land uses were dedicated to food crop production and little for conservation.

The average walking distance from home to the farm/parcel of land was 1.5 KMs which takes 15 min. This may have an implication on small landholders' on-farm SLM investment that saves time and increases labor efficiency. The survey also collected data on whether households are in areas where land use policy mandates soil and water conservation (SWC) investment due to the topographic nature (slope gradient) and soil erosion prevalence of their parcels of landholdings. Accordingly, about 39 percent of the households reported that they had at least one or more parcels located on sloping lands with high exposure to soil erosion. As a result, the survey also revealed that two-thirds of the small landholder households had been required by the woreda/kebele administration to implement SWC measures that the land use regulation mandated investment in SWC.

Regarding investment in SLM, such as practicing any SWC, about 45 percent of the households constructed and/or maintained any SWC such as stone bund, soil bund, water retention structures, planted trees, and perennials crops on their parcels of landholdings. For instance, the average length of stone bund constructed per parcel of landholding using the household's own resources was 11 m, with a maximum of 600 m on average per parcel area of 0.38 hectares in the past year. Likewise, the average length of soil bund constructed by the household's own resources on the same parcel was 28 m with a maximum of 800 m, showing households employed at least two or more complementary physical SWC practices on their farm. The survey results also revealed that one in ten parcels of landholding had on-farm water retention structures, such as ponds, that were constructed using the household's own resources. Moreover, the average number of trees and perennials crops planted by households (using their own resources) was about 106 and 163, respectively. This also shows households were practicing/complementing the physical SWC with long-term biological measures.

Regarding legal security or de jure tenure security, the survey measured whether households know what type of land rights are recognized under the existing land laws, including usufruct, transfer (rent/sharecropping), bequest/inheritance, and collateral. According to the survey results, households reported that they know their land rights are recognized in the land law, including 98 percent to usufruct, 97 percent to transfer, including rent and sharecropping, 95 to bequest, and 80 to collateralize their rights. This indicates respondents were aware of what type of tenure rights are recognized and secured in the land laws, meaning their rights are legally recognized and protected by the land laws.

Regarding de facto tenure security, the survey collected data on the decision-making power of the households on the crops to grow, the use of the produce, the transfer of their land parcels, actual credit obtained between 2019 and 2021, and whether they received land certificates (either FLLC or SLLC). Accordingly, about 60, 64, and 64 percent of decisions on the crop to grow, the use of the produce, and the transfer rights were made by both husband and wife jointly, respectively. Whereas 15, 11, and 11 percent of decisions were made by husband only, respectively. Whereas less than 1 percent of the decisions were accounted for or made by the wife only on the mentioned variables, meaning women have less decision-making power on their land rights matters within their household or joint holding. On the other hand, only six percent of household heads responded that they obtained credit over the past two years, meaning small landholders are still credit constrained, which may limit their on-farm SLM investment capacities. Moreover, in the 2021 survey, about 32 and 58 percent of respondents received FLLC and SLLC, respectively. About 90 percent of the respondent household heads are categorized as having "any certificate" in the study areas.

Perceived tenure security was also measured in terms of the right to bequeath, anticipated land redistribution within 5 years, and participation in credit transactions. The risk dimensions of perceived tenure security were also measured in terms of conservation security risks, tree tenure security risks, land redistribution risks, land rental risks, and participation in any sort of transaction involving credit if it were with a farmer who has a land certificate of possession over their land than that a farmer who does not have a land certificate. Accordingly, about 39 percent of households perceived that the inheritance right was secured after land certification, while one-fourth of the households expected land redistribution within five years. This means about 61 percent of respondent household heads feel their inheritance rights are insecure, while about 76 percent feel that they are secured from further land redistribution in the coming five years, meaning more needs to be conducted in terms of removing such perceptions. Moreover, the survey revealed that 83 percent of households feel more secure in credit transactions with land certificate holders, meaning they can lend or borrow money from anyone with a land certificate. This shows the issuance of landholding certificates strengthens the legacy informal credit market and leverages the creditworthiness of small landholders among their communities.

Regarding the risks of perceived tenure security-related variables, about 96 percent of respondents either strongly agree or agree that they are fully convinced that they will stand to benefit in the future from whatever SWC measures they may undertake on their land at present. This indicates that respondent household heads perceive no risk of losing the benefits of their present investment in the future as their land tenure is secured. On the other hand, about 79 percent of respondents either strongly disagree or disagree that they are fully convinced that they will not stand to benefit in the future from trees that they may plant on their land at present. This means one-fifth of respondent households perceive tree tenure insecurity that likely disincentives on-farm investment such as agroforestry which is one of the proven on-farm SLM practices.

The survey results also revealed that about 70 percent of respondent households either strongly agree or agree that they feel renting out their land for money or on a sharecropping basis, even for one cropping season, is a risky business that they should avoid unless they have no other options of overcoming their difficulties. Small landholder households perceive land rental as risky, even for one cropping season. This may hinder the emergence of the land rental market in the study areas, even for short-term contracting, which may lead to land use inefficiency. Contrary to this, the survey results showed that about 93 percent of respondent household heads would either strongly agree or agree on they would feel more secure entering any sort of business transaction involving credit if it were with a farmer who has a landholding certificate of possession over their land than that a farmer who does not have a land certificate. This implied that landholding certificates facilitate credit markets among landholders who possessed landholding certificates. However, in the past two years, only six percent of respondent household heads borrowed money using their landholding certificates as collateral from financial institutions or informal lenders.

3.3. *Estimates of the Parameters of the Probit Regression Model*

As explained in the methods section, the authors estimate four models for the outcome variable. Summary estimates of the probit regression models results of the probability of households' on-farm SLM investment and the average marginal effects of the explanatory variables are presented in Table 2. The probit regression model 4, the best out-fitted model among the four estimated models, results indicated that among the 25 hypothesized explanatory variables, 16 variables were found to influence the small landholder households' on-farm investment significantly in SLM. Out of the total 25 hypothesized explanatory variables, 16 variables (two-thirds of the variables) are related to LTIFs. The results of regression Model 4 revealed that 10 out of 16 LTIF-related variables have significantly influenced the small landholder households' on-farm investment in SLM but in different directions. From the results, this study's regression model, i.e., Model 4, has outperformed by 14.06 percent compared with the baseline model. The likelihood ratio Chi-square of 713.56 with a p-value of 0.000 indicates that the research model is statistically significant.

Those variables with positive average marginal effects include marital status, FLLC, SLLC, perceived land rental risks, tree tenure security risk, water erosion risk, and decision on land rental. Whereas gender, age, education, means of original land acquisition, land area, credit obtained, perceived conservation security risks, the decision on the use of the produce, and the laws recognize bequest, have negative average marginal effects on the small landholder households' on-farm investment in SLM in the study areas. The average marginal effects of each parameter and their implications are presented and analyzed as follows.

Gender—the results revealed that the gender of the household head significantly and negatively influenced the on-farm investment in SLM. The survey result revealed that a household headed by a woman has a 10 percent reduced probability of investing in any SWC, compared to a mean probability of 45 percent with a 95 percent confidence interval. There is a significant and negative gender differential effect on investing in the on-farm SLM between a man and a female-headed household.

Table 2. Estimates of the probit regression model summary of average marginal effects on the probability of on-farm SLM investment of households. Source: calculated by the author based on the survey data obtained from the USAID data repository, 2022.

Categories of Variables	Independent Variables	Model 1		Model 2		Model 3		Model 4	
		Coefficient (Robust Std. Errors)	dy/dx	Coefficient (Robust Std. Errors)	dy/dx	Coefficient (Robust Std. Errors)	dy/dx	Coefficient (Robust Std. Errors)	dy/dx
HH demographic and socio-economic variables	Sex	−0.4311 (0.1036)	−0.0937 **	−0.5118 (0.1087)	−0.1106 **	−0.5371 (0.1115)	−0.1159 **	−0.4539 (0.1128)	−0.0958 **
	Age	−0.0054 (0.0015)	−0.0011 **	−0.0059 (0.0016)	−0.0013 **	−0.0082 (0.0017)	−0.0017 **	−0.0081 (0.0017)	−0.0017 **
	Education	−0.1115 (0.0132)	−0.0242 **	−0.1088 (0.0136)	−0.0238 **	−0.1036 (0.0145)	−0.0223 **	−0.1068 (0.0145)	−0.0225 **
	Marital status	0.1347 (0.0552)	0.0293	0.1803 (0.0587)	0.0405 **	0.2249 (0.0615)	0.0485 **	0.1965 (0.0620)	0.0415 **
	Acquisition	−0.0267 (0.0091)	−0.0058 **	−0.0308 (0.0098)	−0.0068 **	−0.0329 (0.0106)	−0.0071 **	−0.0320 (0.0106)	−0.0067 **
	Land area	−0.0574 (0.0124)	−0.0124 **	−0.0619 (0.0130)	−0.0132 **	−0.0645 (0.0138)	−0.0139 **	−0.0605 (0.0135)	−0.0127 **
Parcel specific characteristics	Time	0.0048 (0.0016)	0.0010 **	0.0055 (0.0018)	0.0012 **	0.0075 (0.0028)	0.0016	0.0066 (0.0027)	0.0014
	distance	−0.0000 (0.0000)	−0.0000	−0.0000 0.0000	−0.0000	−0.0000 (0.0000)	−0.0000	−0.0000 (0.0000)	−0.0000
	Water erosion	0.6969 (0.0364)	0.1516 **	0.7009 (0.0377)	0.1525 **	0.7214 (0.0396)	0.1557 **	0.7478 (0.0403)	0.1579 **
De jure/legal tenure security	usufruct			−0.4372 (0.1209)	−0.0957 **	−0.2480 (0.1392)	−0.0535	−0.3023 (0.1476)	−0.0638
	Rent			0.2338 (0.1234)	0.0511	0.1396 (0.1318)	0.0301	0.1300 (0.1374)	0.0274
	bequest			−0.2532 (0.0931)	−0.0554	−0.3715 (0.0969)	−0.0802 **	−0.3947 (0.0989)	−0.0833 **
	collateral			0.1008 (0.0524)	0.0220	0.0129 (0.0560)	0.0027	−0.0094 (0.0581)	−0.0019
De facto/actual tenure security	Decision on crop					−0.0736 (0.0419)	−0.0158	−0.07324 (0.0427)	−0.0154
	Decision on use					−0.2593 (0.0636)	−0.0559 **	−0.2545 (0.0640)	−0.0537 **
	Decision on rent					0.2816 (0.0563)	0.0607 **	0.2674 (0.0560)	0.0564 **
	FLLC					0.6298 (0.0767)	0.1359 **	0.6803 (0.0794)	0.1437 **
	SLLC					0.3526 (0.0460)	0.0761 **	0.3588 (0.0466)	0.0758 **
	Credit					−0.4178 0.0963	−0.0901 **	−0.3849 (0.0966)	−0.0813 **
Perceived tenure security risks	redistribution							0.0223 (0.0266)	0.0047
	Inheritance							0.0458 (0.0296)	0.0096
	Rent out							0.0880 (0.0265)	0.0185 **
	collateral							0.0593 (0.0357)	0.0125
	conservation							−0.2653 (0.0348)	−0.0560 **
	Tree tenure							0.1403 (0.0207)	0.0296 **
Constant		−0.4657	0.1437 **	−0.0657	0.1972	−0.0507	0.1990	−1.0748	0.2967 **

Note: n = 6692 Wald Chi^2(25) = 713.56, Prob > Chi^2 = 0.0000; Pseudo R^2 = 0.1406; $p < 0.05$ **; Robust standard errors are given in parentheses. The average marginal effect (dy/dx) is calculated at the mean for continuous and discrete change from 0 to 1 for dummy variables.

Age—the age of a household head negatively and significantly influenced investment in the on-farm SLM with an average marginal effect of 0.2 percent with a 95 confidence interval compared to a mean age of 55. Meaning every one-year increase in the age of the household head leads to a 0.2 percent decrease in the probability of on-farm investment in SLM. This may relate to the decrease in household labor within the household and inability of a household to conduct farm management as households aging.

Education—the results revealed that educational attainment negatively and significantly affected the small landholder households' on-farm investment in SLM in the study areas with an average marginal effect of 2.3 percent with a 95 confidence interval. This shows that when the educational attainment of the household head increases by one grade level, the probability of investment in the on-farm SLM decreases by 2.3 percent.

Marital Status—the results of the current study revealed that the marital status of the head of the household positively and significantly affected the small landholders' on-farm investment in SLM with an average marginal effect of 5 percent with a 95 confidence interval. This shows that households headed by married couples have a 5 percent higher on-farm SLM investment probability than households headed by unmarried individuals or widowers.

Means of land acquisition—access to land through administrative allocation is becoming impossible due to a shortage of land caused by the increasing population. Access to land determines the on-farm investment of households. In this regard, the results unfolded that means of original land acquisition negatively and significantly influenced the probability of households' on-farm SLM investment with an average marginal effect of 0.7 percent with a 95 percent confidence interval, compared to a mean of 45 percent. This shows small landholder households who originally acquired their landholdings currently under their possession other than administrative land redistribution or allocations have a 0.7 percent reduced probability of on-farm SLM investment incentives. Given that the last administrative land redistribution was conducted 30 years ago and about 40 percent of the land was acquired through inheritances in the study areas, SLM policy and strategy need to consider this factor.

Land area—the survey results revealed that land area is also found to influence the probability of households' on-farm investment negatively and significantly in SLM in the study areas with an average marginal effect of 1.4 percent with a 95 confidence interval. Meaning every one-unit increase in the land held by the household head leads to a 1.4 percent decrease in the probability of investment in the on-farm SLM.

Time taken and distance from homestead to parcel—the survey result revealed that both time taken and distance to parcel are also found to effect the probability of households' on-farm investment in SLM positively and negatively, respectively, but insignificantly, with an average marginal effect of 0.14 and 0.00 percent, respectively. Meaning every one-minute increase in the travel time from home to the parcel leads to a 0.14 percent increase in the probability of investment in on-farm SLM. In contrast, every 100 m increase in distance from home to the parcels leads to a 0.01 percent decrease in the probability of households' on-farm SLM investment, compared to a mean distance of 1.5 KMs. This may have an insignificant effect since most of the parcels possessed by the landholders are found reachable in 15 min, with an average walking distance of 1.5 KMs.

Water erosion risk– the results also revealed that water erosion risk is found to affect the probability of households' on-farm SLM investment positively and significantly. The results revealed that households who held a parcel of landholding located on sloping lands with soil erosion risk from water had a 16 percent increased incentive to invest in SLM technologies with a 95 percent confidence interval, compared to a mean probability of 45 percent. This implies the higher the water erosion risk, the better probability of incentives to invest in on-farm SLM practices.

Possession of landholding certificates—land registration and certification is one of the mechanisms sought for improving tenure security in Ethiopia to incentivize long-term land-based investment such as on-farm SLM and climate-smart agriculture. Under the current study, the survey result revealed that possessions of either FLLC or SLLC were found to influence the probability of households' on-farm SLM investment positively and significantly in the study areas. The results indicated that landholding certification increases the probability of investing in on-farm SLM with an average marginal effect of 14 and 8 percent for FLLC and SLLC, respectively, at a 95 percent confidence interval and compared to a mean probability of 45 percent. This shows that small landholders who possessed either FLLC or SLLC for their parcels have a 14 and 8 percent increased probability of on-farm SLM investment than those households without either FLLC or SLLC for their parcels, respectively. However, from these data, it is less clear why SLLC has a lower impact than FLLC on incentivizing households' on-farm SLM investment.

Credit—the survey results also revealed that credit access significantly but negatively affected the probability of households' on-farm investment in SLM with an average marginal effect of 8 percent at a 97 confidence interval, compared to a mean probability of 45 percent. Meaning households without credit have an 8 percent reduced probability of investing in on-farm SLM.

Conservation security risk—perceived conservation risk is found to negatively influence the probability of households' on-farm investment in SLM significantly. The results revealed that those households who are fully convinced or believe in the future benefit from an SLM investment have a 6 percent increased probability of investment incentives in SLM at a 95 confidence interval, compared to a mean probability of 45 percent.

Tree tenure security risk—in another measure of tree tenure security risk, the results revealed that those households who are fully convinced that they will not stand to benefit in the future from trees have a 3 percent reduced investment probability of on-farm SLM. Meaning households who foresee a tree tenure insecurity risk will likely be disincentivized to invest in on-farm tree planting at a 95 confidence interval, compared to a mean probability of 45 percent.

Regarding de jure tenure security, household heads who know their bequest land rights are recognized and protected by the land laws have an 8 percent increased probability of on-farm SLM investment at a 95 confidence interval, compared to a mean probability of 45 percent. Meaning those households who were aware of their bequest rights recognized before the laws were better off investing in on-farm SLM.

4. Discussion

This section discusses the results of the current study by comparing them with previous studies on factors that influence smallholder households' SLM investment. The role of land tenure institutions, be it formal, customary, or informal, in sustainable land use and resource management has paramount importance. This is because the way land tenure institutions are organized and enforced can greatly influence how communities and landholders use land resources and whether durable sustainability on-farm SLM investments are being made. Regardless of the forms of tenure rights, their recognition and protection are also critical factors for sustainable land use and resource management. For instance, secure private land use rights, without enforced land use planning which regulates land use zoning and other environmental management measures, may result in adverse environmental outcomes. Based on the findings of the current study, this section particularly discusses the effect of LTIFs represented by the three categories of land tenure security and their corresponding bundles of land tenure rights on the probability of households' on-farm investment in SLM in the study areas.

4.1. Whether De Jure Land Tenure Security Influences Households' On-Farm Investment in SLM

The current study revealed that legal recognition of land use rights of households in the study areas provides de jure tenure security, such as the right to bequest one's landholding to heirs, significantly affecting the probability of households' on-farm investment in SLM. This is consistent with Boone [27], who found legal empowerment of the poor through property rights reform in Sub-Saharan Africa incentivizes land-based investment.

In their recent systematic review, Tseng et al. [6] also found similar and strong support for strengthening land tenure security largely led to positive human well-being and environmental outcomes, particularly through formalization, land use planning, and land policy reform. Aggarwal et al. [28] conducted an assessment in 23 countries and found that governments are increasingly giving legal recognition to community forest rights but fewer legal protection and more barrier to using those rights.

Accordingly, in rural Ethiopia, individual land rights are generally recognized under the federal and regional land administration and use proclamations. The 1995 Constitution of the country enshrined the ownership of land to the state. The state body is almost always implicated as a duty holder as the entity with the power to arrest and adjudicate. Ethiopian nationals can have individual usufruct rights in that peasants and pastoralists can obtain land for cultivation and grazing purposes free of charge for an indefinite time. Proclamation 456/2005 of the federal democratic republic of Ethiopia also recognizes acquiring of individual landholding rights through allocation, redistribution, settlement programs, donation, and/or inheritance free of charge. However, neither collateralization of landholding rights to access credit nor acquiring land through sales or any other exchange are ruled out by the existing legal framework. Those recognized rights by law are exclusive but not absolute because landholders' tenure rights are generally bounded by limits on externalities, such as preventing soil and water pollution. This indicates that the existing legal framework recognizes and provides protection of small landholders' rights clearly and implies there is a de jure tenure security except for collateral and land sale. However, local conditions determine which of these bundles of rights are protected in practice. For instance, forest tenure rights held by individuals are recognized in the existing legal framework, e.g., Proc. No. 456/2005 Art 2/11 and Proc. No 1065/2018, Art 2/6 of the forest proclamation. In addition, communal forest tenure rights are recognized in the same proclamations, Art 2/12 and Art 2/7, with adequate duration and scope, respectively.

The econometric results revealed that those households who were aware of their usufruct rights, transfer rights through rent/sharecropping, and bequeath/inheritance rights recognized by the existing laws have a 5 to 8 percent better probability of investing in on-farm SLM. This implies that legal literacy or awareness of what bundles of tenure rights are recognized and protected by the existing land and forest laws makes a significant difference in on-farm SLM investment among small landholders in the study areas. This is consistent with what Vu H. and Goto D. [29] found in Vietnam that awareness about land tenure security towards agricultural land tenure rights increases sustainable land-based investment. However, the results also revealed that landholders knew that the land laws did not recognize collateralization of land rights, hence insignificantly influencing small landholders' on-farm SLM investment. Meaning collateralization of land rights was not an option for small landholders to access credit and finance on-farm SLM investment until recently. However, since 2019/20, there has been a policy change in land use rights as collateral to borrow mainly from financial institutions. Hence, de jure tenure security significantly influences the probability of households' investment in the on-farm SLM in the context of inheritance than usufruct, rental, and collateral bundles of tenure rights. Therefore, the provisions of succession in the land law need to be clear and strengthened.

4.2. Whether De Facto Land Tenure Security Influences Landholders' On-Farm Investment in SLM

De facto tenure security is also measured across the bundles of tenure rights as recognized in the existing legal framework, practiced by the smallholder households, and protected by the state or local governments in the study areas. Overall, the econometric model findings show that the de facto tenure security set was found to influence the smallholder households' on-farm investment in SLM significantly. In their systematic review, Tseng et al. [6] found that changes from de jure to de facto tenure security demonstrated by the formalization of land rights lead to better environmental outcomes. Regardless of countries' specific legal systems, legal documentation of rights refers to the recording and publication of information on the nature and location of land, rights, and rights holders [30].

In a formal system or statutory context, land titling is sought as one of the mechanisms that provide rights holders with a secure tenure right and incentivizes them to use land efficiently by investing in land conservation and improvement [5,18]. Since early 2000, the government of Ethiopia has launched one of the biggest two-stage land registration and certification programs in Africa with the aim to improve land tenure security in the highlands of Ethiopia and incentivize long-term land-based investment such as SLM practices and curb land degradation [31].

As well documented in the existing literature, land degradation is one of the major environmental and development challenges in the highlands of Ethiopia that reduces agricultural production, increases food insecurity, and disrupts sustainable ecosystem functions [32–34]. Guided by Ethiopia's Sustainable Investment Framework (ESIF) for SLM, the government of Ethiopia embarked on a national SLM flagship program in 2010. ESIF presumed that the removal of the key barrier of insecure land tenure is believed to be one of the way-outs to greater adoption of SLM practices and reduces further land degradation [35]. Component two of the ESIF recommends the improvement of the land administration and certification system. Under ESIF, the combination of participatory and integrated watershed management and secure land tenure rights is expected to lead to increased adoption of SLM practices, reducing land degradation, increasing carbon sequestration, and delivering more resilient and sustainable livelihoods.

The econometric model of the current study revealed that possession of either FLLC or SLLC was found to positively influence the probability of households' on-farm investment in SLM significantly. This is consistent with Adere et al. [34] findings in southern Ethiopia that land certification has a positive but heterogeneous impact on different SWC techniques among farmers with different risk preferences in that the effect is stronger for more risk-averse farmers in Ethiopia. Deininger et al. [36] also found consistent evidence of the impact of land certification on tenure security, investment, and land market participation in Ethiopia. Gebremedhin et al. [37] also found that land tenure security contributes to land conservation by influencing SWC actions in watersheds and enhancing household willingness to invest in high-cost and long-term conservation practices in Ethiopia's Tigray regional state. Likewise, Frank [38] also found, in some Sub-Sharan Africa countries, the clarity and recognition of land tenure rights through land registration and certification of small landholders and communities incentivized land managers to engage in higher value and more productive land use practices. Mugagga [39] also found predominantly land tenure secure communities through communal land certification invested in longer-term soil conservation measures in Uganda.

However, it is worth noting that having a certificate does not necessarily fully secure or causes a person to believe that there is an absolute guarantee. Meaning the impact of land certification on tenure security differs by how perceived tenure security is measured. In addition, the type of tenure security risks matters the intensity and adoption of SLM practices. This leads us to the discussion on the third category of land tenure security, i.e., perceived tenure security risks related to the bundle of rights recognized in the existing land laws. Before turning into the perceived tenure security discussion, it is important to highlight another de facto tenure security bundle of rights, i.e., collateralization of land use rights and its effect on the smallholder households' on-farm investment in SLM.

The econometric analysis revealed that credit access significantly but negatively affected smallholder households' investment in SLM. This result is in line with the work of Mulwa et al. [40], who found that access to credit allowed households to adopt SWC activities that helped them to invest more in agricultural inputs in Malawi. Abeje et al. [33] also found access to credit has a positive effect on adopting a higher number of SLM practices in Ethiopia. Similarly, Asaaga et al. [11] found that access to credit plays a critical mediating role in the relationship between tenure security and SLM investment in Ghana. This implied that in the absence of access to credit, small landholders may still find it difficult to invest in on-farm resource-intensive SLM investments such as SWC measures.

Contrarily, the econometric analysis shows that decision-making on the crop to grow was found to influence negatively but insignificantly the probability of households' investment in on-farm SLM. Contrarily, the decision on the use of the produce was found to negatively influence the probability of households' on-farm SLM investment significantly. This shows that although landholders have legally secured joint tenure rights under the current legal framework and documented joint title, there is a de facto tenure insecurity within intra-household. This is consistent with the results found by Feyertag et al. [41] in that women are more likely to feel threatened by internal sources of insecurity within the family or community. Chigbu et al. [42] raised the alarm concerning the failure to understand female differentials in land tenure access and security could lead to engendering policies that benefit only a section of communities rather than all women within the community. Meaning context specific inter and intra-household de facto tenure security, such as decisions on what to grow on-farm and the use of the produce, must be considered in SLM policy.

On the other hand, the decision on the transfer of rights through rent and/or sharecropping was found to positively influence the probability of households' on-farm investment in SLM significantly. This indicates that smallholder households with joint landholding rights recognized through joint land certificates should decide on the transfer of their joint landholding rights in the form of land rent/sharecropping. This is consistent with the legally recognized requirements in that the parties need to agree and provide their consent jointly to enter a land rental/sharecropping contract arrangement. This implies that joint landholding rights holders in the study areas have secured de facto tenure security that facilitates on-farm investment in SLM. This is consistence with what Ghebru and Girmachew [23] found in Ethiopia that the value-added direct and spillover effects of SLLC favor the supply side of the land rental market, the likelihood of renting/sharecropping in land is significantly enhanced even for non-beneficiary households who reside in or around land certification treated program woredas. Hence, it can be inferred that households relate their on-farm SLM investment with de facto tenure security significantly but specific to bundles of rights and contexts. Therefore, this is another strong evidence of the need to make SLM policy context specific.

4.3. Whether Perceived Land Tenure Security Risks Influence Housholds' On-Farm Investment in SLM

Coming to the perceived tenure security of small landholder households, the econometric analysis of the current study shows that this category of tenure security was found to influence the probability of households' on-farm investment in SLM significantly but in different directions. For instance, regarding perceived conservation security risk, the econometric analysis shows that perceived conservation security risk is found to negatively influence the probability of households' on-farm investment in SLM significantly. This is in line with Gebremedhin et al. [37], who found that investment in stone terraces was positively influenced by factors associated with long-term investment perspectives, such as the capacity to invest and land tenure security in the Tigray region of Ethiopia. On the other hand, Ghebru and Girmachew [23] found that while SLLC has a positive effect in reducing private land tenure risks, this intervention negatively affects men's perceived risk of private tenure security. The fact that the SLLC is predominantly implemented by issuing joint landholding certificates to heads and spouses could explain the extra sense of security married women perceive while men perceive the contrary [23].

Likewise, perceived tree tenure security risk is found to positively influence the probability of households' on-farm investment in SLM significantly. This shows that there should be clarity on the security of tree and land tenure nexus. As per the existing legal framework, in Ethiopia, land tenure rights and tree tenure rights are exclusively independent bundles of rights. As mentioned earlier, landholders have perpetual land use rights, while the forest law proclaimed individual forest ownership rights, including tree planting and use of forest and non-forest timber products. However, this might not be well understood among smallholder households, which likely negatively influences on-farm SLM investment, such as agroforestry which is one of the proven on-farm SLM technologies promoted under the ESIF. However, other previous studies in Ethiopia also found that improvement in perceived tenure security has been witnessed after the land certification program [23,24,43,44].

The econometric analysis further shows that perceived tenure security risk to enter any sort of business transaction involving credit was found to have an insignificant influence. Since this right was not legally recognized/secured before the survey data collection period, households feel there will be a credit transaction security risk, which hinders on-farm SLM investment. This is consistent with Adere et al. [34], who found in southern Ethiopia, risk preferences influence the SWC investment of households. Contrarily, Byamugisha [38] found that landholders having secure tenure rights and secure access to credit spur long-term productive investment in some sub-Saharan Africa countries [19]. Based on these findings, households relate their on-farm SLM investment with perceived tenure security.

4.4. Households' Socio-Economic and Demographics and On-Farm Investment in SLM

Regarding the demographic variables of the households, the current study found that gender, age, and education negatively influence households' on-farm investment in SLM significantly. For instance, the econometric analysis shows that the gender and age of the household head are found to influence the households' on-farm investment in SLM significantly negatively. This is consistent with earlier studies, such as by Ghebru and Girmachew [23], who found that female-headed households with SLLC are less likely to engage in investment and/or maintenance of sustainable land management practices compared to households without SLLC in Ethiopia.

Likewise, the educational level of the household head is found to negatively influence the probability of households' on-farm SLM investment significantly. Contrary to our expectation and with others on the effect of education [40,45,46] on the adoption of sustainable agricultural practices and climate adaptation measures, the current study finds that small landholder household heads with more years of schooling are less likely to invest in on-farm SLM. This may indicate that well-educated household heads tended to look for non-land-based livelihood options such as off-farm activities or prefer out-migration. More specifically, insufficient availability and productivity of land may also be among the disincentives of investment in the on-farm SLM among household heads with more school years attainment. This implies that households' educational attainment and on-farm investment in SLM should be seen carefully, with increasing pressure on land and decreasing productivity due to land degradation compounded by climate change risks.

The economic analysis further shows that the socio-economic factors such as marital status, means of original land acquisition, and area of landholding of smallholder households were found to influence significantly but in different directions. For instance, the econometric analysis reveals that land area negatively influences the on-farm investment in SLM significantly in the study areas. This is in line with Etsay et al. [47], who found a negative relationship between farm size and the adoption of indigenous conservation practices in the Tigray region of Ethiopia. However, the current result disagrees with the findings of Wondimu et al. [32] that land area has a significant positive effect on crop rotation in the Abay basin of the Oromia region of Ethiopia.

4.5. Parcel-Specific Factors and On-Farm Investment in SLM

The econometric analysis shows that distance and walking time to parcels were found to insignificantly influence SLM investment, while the exposure of the land parcels to water erosion risks was found to influence smallholder households' on-farm investment in SLM positively. This is consistent with Adimassu et al. [35,48], who found that farmers vulnerable to erosion hazards are more likely to invest in different land management practices, but investments were highly variable across their production domain. Wondimu et al. [32] also found that the perception of erosion hazard has a positive and significant effect on the adoption of soil bund SLM practice in the Abay basin of the Oromia regional state in Ethiopia. As land is household heads' ultimate resource for their livelihood, parcels which were exposed to water erosion were more likely to receive on-farm investment in SLM, thereby reducing land degradation and improving their productivity. Similarly, Abeje et al. [33] found that the parcel level factors influence the SLM investment, including slope gradient, fertility status, area, and distance to and from home.

5. Conclusions

This study assessed factors affecting households' on-farm SLM investment in 19 highland woredas of three regions (Amhara, Oromia, SNNP) in Ethiopia, where land degradation is considered a daunting environmental and development challenge. This study considered selected household and parcel-level variables and land tenure institutional factors. This was achieved by employing a probit regression model that estimated the average marginal effect of the explanatory variables on the outcome variable quantitatively, i.e., the probability of a household head invested in any on-farm SLM practices. While much of the existing land tenure literature recognizes the need to ensure the security of tenure in broader terms, this may hinder an in-depth yet broader perspective of understanding the effect of LTIFs along categories of land tenure security across bundles of rights and hence limits context specific SLM policy and implementation strategy. Hence, the current study empirically tested this approach and demonstrated that the approach might be replicated in other countries and contribute to the broader body of evidence.

The findings of this study revealed that households' on-farm investment in SLM is affected by several demographic and socio-economic factors, parcel-specific variables, and LTIFs. Particularly, the LTIFs were also found to jointly influence the probability of households' on-farm investment in SLM significantly but differently across the different categories of tenure security and bundles of rights. These results demonstrated that unbundling the categories of land tenure security across the bundles of rights and understanding their specific influence on households' on-farm SLM investment are important aspects of designing context-specific SLM policy and implementation strategy.

These results have three important implications. Firstly, while the household and parcel level variables are very important to consider when designing SLM investment policy at the household level, the LTIFs are equally important to consider across the bundle of rights. Meaning categories of secure tenure rights must be seen in perspective along with other influencing factors. Secondly, while securing tenure through land certification incentivizes the on-farm investment in SLM, land policies and regulatory frameworks should also consider the issues of access to credit to small landholders that create the capacity to invest in durable and intensive on-farm SLM investment. In the absence of access to credit, small landholders with secure tenure rights may still find it difficult to invest in an on-farm resource-intensive SLM investment. Hence, the SLM policy needs to strengthen access to credit for smallholder households across the country. Finally, regardless of the forms of tenure rights, their recognition (de jure) and enforcement (de facto) tenure security combined with the regulatory functions of land tenure institutions, such as the enforcement of land use regulations, are also critical factors for sustainable land use and resource management. Secure private land use rights, without enforced local level land use planning which regulates land use zoning and other environmental management measures, may bear little on-farm investment in SLM that could not balance the exploitation of land resources and may result in adverse environmental outcomes. This will likely affect the sustainability of SLM investment at the landscape level, including communal landholdings.

However, this study did not undertake an in-depth assessment of local-level land use plan implementation where available and its implication on households' on-farm SLM investment. Future research should focus on the impact of local-level land use plan implementation compliance by smallholder households and its implication on their on-farm investment endeavors in SLM.

Author Contributions: Conceptualization, S.A.A., F.S. and T.T.N.; methodology, S.A.A.; software, S.A.A.; validation, S.A.A., F.S. and T.T.N.; formal analysis, S.A.A.; investigation, S.A.A.; resources, S.A.A.; data curation, S.A.A.; writing—original draft preparation, S.A.A.; writing—review and editing, S.A.A., F.S. and T.T.N.; visualization, S.A.A.; supervision, F.S. and T.T.N.; project administration, S.A.A., F.S. and T.T.N.; funding acquisition, S.A.A. All authors have read and agreed to the published version of the manuscript.

Funding: The research was supported by the Partnerships for Enhanced Engagement in Research (PEER) Program, financed by the U.S. Agency for International Development (USAID) and administered by the U.S. National Academy of Sciences (NAS) under cooperative agreement AID-OAA-A-11-00012. The authors highly acknowledge and appreciate this generous financial support, without which this research could not have been realized.

Institutional Review Board Statement: Not applicable.

Informed Consent Statement: Informed consent was obtained from all subjects involved in the study.

Data Availability Statement: https://www.land-links.org/ accessed on 21 December 2022 and the Landesa research team.

Acknowledgments: Heartfelt thanks to Kelly Robbins, NAS, who administers the PEER grant funds, for her unconditional support of our research work. We also want to extend our appreciation and thanks to Krista Jacobs, Research, Evaluation and Learning at Landesa, and her research team for sharing the 2021 follow-on impact study survey datasets, without which this study could not have succeeded. Many thanks to Caleb Stevens for his unreserved support in connecting us with the Landesa research team and accessing survey datasets.

Conflicts of Interest: The authors declare no conflict of interest.

References

1. Chigbu, U.E.; Chilombo, A.; Lee, C.; Mabakeng, M.R.; Alexander, L.; Simataa, N.V.; Siukuta, M.; Ricardo, P. Tenure-Restoration Nexus: A Pertinent Area of Concern for Land Degradation Neutrality. *Curr. Opin. Environ. Sustain.* **2022**, *57*, 101200. [CrossRef]
2. Huck, W. Goal 5 Achieve Gender Equality and Empower All Women and Girls. *Sustain. Dev. Goals* **2022**, 213–250. [CrossRef]
3. Hettig, E.; Lay, J.; Sipangule, K. Drivers of Households' Land-Use Decisions: A Critical Review of Micro-Level Studies in Tropical Regions. *Land* **2016**, *5*, 32. [CrossRef]
4. Castro, B.; Kuntz, C. Land Tenure Insecurity and Climate Adaptation: Socio-Environmental Realities in Colombia and Implications for Integrated Environmental Rights and Participatory Policy. In *Land Tenure Security and Sustainable Development*; Springer: Berlin/Heidelberg, Germany, 2022; pp. 177–199. [CrossRef]
5. Masuda, Y.J.; Kelly, A.C.; Robinson, B.E.; Holland, M.B.; Bedford, C.; Childress, M.; Game, E.T.; Ginsburg, C.; Hilhorst, T.; Lawry, S.W.; et al. How Do Practitioners Characterize Land Tenure Security? *Conserv. Sci. Pract.* **2020**, *2*, e186. [CrossRef]
6. Tseng, T.W.J.; Robinson, B.E.; Bellemare, M.F.; BenYishay, A.; Blackman, A.; Boucher, T.; Childress, M.; Holland, M.B.; Kroeger, T.; Linkow, B.; et al. Influence of Land Tenure Interventions on Human Well-Being and Environmental Outcomes. *Nat. Sustain.* **2021**, *4*, 242–251. [CrossRef]
7. Robinson, B.E.; Masuda, Y.J.; Kelly, A.; Holland, M.B.; Bedford, C.; Childress, M.; Fletschner, D.; Game, E.T.; Ginsburg, C.; Hilhorst, T.; et al. *Incorporating Land Tenure Security into Conservation*; Wiley-Blackwell: Hoboken, NJ, USA, 2018. [CrossRef]
8. Cole, D.H.; Ostrom, E. *Property in Land and Other Resources*; Columbia University Press: New York, NY, USA, 2011.
9. Behrer, A.P.; Glaeser, E.L.; Ponzetto, G.A.M.; Shleifer, A.; Banerjee, A.; Bebchuk, L.; Chen, K.; Hart, O.; Kaplow, L.; Minton, R.; et al. *Securing Property Rights*; Harvard University Press: Cambridge, MA, USA, 2019.
10. Locke, A. *Shaping Policy for Development Property Rights and Development Briefing: Property Rights and Economic Growth*; Overseas Development Institute: London, UK, 2013.
11. Asaaga, F.A.; Hirons, M.A.; Malhi, Y. Questioning the Link between Tenure Security and Sustainable Land Management in Cocoa Landscapes in Ghana. *World Dev.* **2020**, *130*, 104913. [CrossRef]
12. Hollandd, M.; Masudaa, Y.; Robinson, B. *Land Tenure Security and Sustainable Development*; Springer: Berlin/Heidelberg, Germany, 2022.
13. Masuda, Y.J.; Robinson, B.E.; Holland, M.B.; Tseng, T.-W.J.; Frechette, A. Emerging Research Needs and Policy Priorities for Advancing Land Tenure Security and Sustainable Development. In *Land Tenure Security and Sustainable Development*; Palgrave Macmillan: London, UK, 2022; pp. 313–326. [CrossRef]
14. Mijiyawa, A.G. Determinants of Property Rights Institutions: Survey of Literature and New Evidence. *Econ. Gov.* **2013**, *14*, 127–183. [CrossRef]
15. Besley, T.; Ghatak, M. Property Rights and Economic Development. In *Handbook of Development Economics*; Elsevier: Amsterdam, The Netherlands, 2010; Volume 5, pp. 4525–4595. [CrossRef]
16. Besley, T. Property Rights and Investment Incentives: Theory and Evidence from Ghana Author. *J. Political Econ.* **1995**, *103*, 903–937. [CrossRef]

17. Munger, M.C.; North, D.C. Institutions, Institutional Change and Economic Performance. *South Econ. J.* **1991**, *58*, 296–297. [CrossRef]
18. Deininger, K.; Feder, G. Land Registration, Governance, and Development: Evidence and Implications for Policy. *World Bank Res. Obs.* **2009**, *24*, 233–266. [CrossRef]
19. Byamugisha, F.F.K. Experiences and Development Impacts of Securing Land Rights at Scale in Developing Countries: Case Studies of China and Vietnam. *Land* **2021**, *10*, 176. [CrossRef]
20. Ravnborg, H.M.; Bashaasha, B.; Pedersen, R.H.; Spichiger, R.; Turinawe, A. *Land Tenure under Transition: Tenure Security, Land Institutions and Economic Activity in Uganda*; Danish Institute for International Studies: København, Denmark, 2013.
21. Lawry, S.; Samii, C.; Hall, R.; Leopold, A.; Hornby, D.; Mtero, F. The Impact of Land Property Rights Interventions on Investment and Agricultural Productivity in Developing Countries: A Systematic Review. *J. Dev. Effect* **2017**, *9*, 61–81. [CrossRef]
22. Abab, S.A.; Senbeta, F.; Negash, T.T. The Effect of Policy and Technological Innovations of Land Tenure on Small Landholders' Credit-Worthiness: Evidence from Ethiopia. *Land* **2023**, *12*, 1055. [CrossRef]
23. Ghebru, H.; Girmachew, F. *Direct and Spillover Effects of Land Rights Formalization in Africa a Case Study of the Second-Level Land Certification (SLLC) in Ethiopia*; FAO: Rome, Italy, 2020.
24. Alvarado, G.; Kieran, C.; Jacobs, K.; Beduhn, J.; Heidenrich, T.; Linkow, B. *Ethiopia Strengthening Land Tenure and Administration Program Follow-On Report: An Impact Evaluation of Long-Term Effects of Second-Level Land Certification*; USAID: Washington, DC, USA, 2022.
25. USAID. *Ethiopia Strengthening Land Tenure and Administration an Impact Evaluation of the Effects of Second-Level Landholding Certification*; USAID: Washington DC, USA, 2016.
26. Tesfaye, A.; Nigussie, A.; Ambaw, G. *Monitoring Socioeconomic Impacts of Climate-Smart Agricultural Practices at Doyogena and Basona Worena Climate-Smart Landscapes, Ethiopia*; CGIAR Research Program on Climate Change, Agriculture and Food Security: New York, NY, USA, 2021.
27. Boone, C. Legal Empowerment of the Poor through Property Rights Reform: Tensions and Trade-Offs of Land Registration and Titling in Sub-Saharan Africa. *J. Dev. Stud.* **2019**, *55*, 384–400. [CrossRef]
28. Aggarwal, S.; Larson, A.; McDermott, C.; Katila, P.; Giessen, L. Tenure Reform for Better Forestry: An Unfinished Policy Agenda. *For. Policy Econ.* **2021**, *123*, 102376. [CrossRef]
29. Vu, H.T.; Goto, D. Does Awareness about Land Tenure Security (LTS) Increase Investments in Agriculture? Evidence from Rural Households in Vietnam. *Land Use Policy* **2020**, *97*, 104721. [CrossRef]
30. Musinguzi, M.; Enemark, S. A Fit-For-Purpose Approach to Land Administration in Africa—Supporting the 2030 Global Agenda. *Int. J. Technosci. Dev. (IJTD)* **2019**, *4*, 69–89.
31. Abab, S.A.; Wakjira, F.S.; Negash, T.T. Factors Influencing the Formalization of Rural Land Transactions in Ethiopia: A Theory of Planned Behavior Approach. *Land* **2022**, *11*, 633. [CrossRef]
32. Wondimu, L.; Jema, H.; Mengistu, K.; Bezabih, E. Determinants of Adoption of Sustainable Land Management Practice Choices among Smallholder Farmers in Abay Basin of Oromia, Ethiopia. *J. Dev. Agric. Econ.* **2021**, *13*, 1–9. [CrossRef]
33. Abeje, M.T.; Tsunekawa, A.; Adgo, E.; Haregeweyn, N.; Nigussie, Z.; Ayalew, Z.; Elias, A.; Molla, D.; Berihun, D. Exploring Drivers of Livelihood Diversification and Its Effect on Adoption of Sustainable Land Management Practices in the Upper Blue Nile Basin, Ethiopia. *Sustainability* **2019**, *11*, 2991. [CrossRef]
34. Adere, T.H.; Mertens, K.; Maertens, M.; Vranken, L. The Impact of Land Certification and Risk Preferences on Investment in Soil and Water Conservation: Evidence from Southern Ethiopia. *Land Use Policy* **2022**, *123*, 106406. [CrossRef]
35. Adimassu, Z.; Langan, S.; Johnston, R. Understanding Determinants of Farmers' Investments in Sustainable Land Management Practices in Ethiopia: Review and Synthesis. *Environ. Dev. Sustain.* **2016**, *18*, 1005–1023. [CrossRef]
36. Deininger, K.; Ayalew, D.; Tekie, A. Impacts of Land Certification on Tenure Security, Investment, and Land Markets Evidence from Ethiopia. *Land Econ.* **2008**, *87*, 312–334. [CrossRef]
37. Gebremedhin, B.; Swinton, S.M. Investment in Soil Conservation in Northern Ethiopia: The Role of Land Tenure Security and Public Programs. *Agric. Econ.* **2003**, *29*, 69–84. [CrossRef]
38. Byamugisha, F. *Securing Land Tenure and Easing Access to Land*; Japan International Cooperation Agency Research Institute: Tokyo, Japan, 2016.
39. Mugagga, F. Land Tenure and Soil Conservation Practices on the Slopes of Mt Elgon National Park, Eastern Uganda. *J. Geogr. Reg. Plan.* **2013**, *6*, 255–262. [CrossRef]
40. Mulwa, C.; Marenya, P.; Rahut, D.B.; Kassie, M. Response to Climate Risks among Smallholder Farmers in Malawi: A Multivariate Probit Assessment of the Role of Information, Household Demographics, and Farm Characteristics. *Clim Risk Manag.* **2017**, *16*, 208–221. [CrossRef]
41. Feyertag, J.; Childress, M.; Langdown, I.; Locke, A.; Nizalov, D. How Does Gender Affect the Perceived Security of Land and Property Rights? Evidence from 33 Countries. *Land Use Policy* **2021**, *104*, 105299. [CrossRef]
42. Chigbu, U.E.; Paradza, G.; Dachaga, W. Differentiations in Women's Land Tenure Experiences: Implications for Women's Land Access and Tenure Security in Sub-Saharan Africa. *Land* **2019**, *8*, 22. [CrossRef]
43. Ghebru, H.; Koru, B.; Taffesse, A.S. *Household Perception and Demand for Better Protection of Land Rights in Ethiopia*; International Food Policy Research Institute: Washington, DC, USA, 2016; pp. 1–24.

44. Azadi, H.; Movahhed Moghaddam, S.; Mahmoudi, H.; Burkart, S.; Dadi Debela, D.; Teklemariam, D.; Lodin, M.; Lebailly, P. Impacts of the Land Tenure System on Sustainable Land Use in Ethiopia. In *Transitioning to Sustainable Life on Land*; MDPI: Basel, Switzerland, 2022. [CrossRef]
45. Kassie, M.; Teklewold, H.; Jaleta, M.; Marenya, P.; Erenstein, O. Understanding the Adoption of a Portfolio of Sustainable Intensification Practices in Eastern and Southern Africa. *Land Use Policy* **2015**, *42*, 400–411. [CrossRef]
46. Teshome, A.; de Graaff, J.; Ritsema, C.; Kassie, M. Farmers' Perceptions about the Influence of Land Quality, Land Fragmentation and Tenure Systems on Sustainable Land Management in the North Western Ethiopian Highlands. *Land Degrad. Dev.* **2016**, *27*, 884–898. [CrossRef]
47. Etsay, H.; Negash, T.; Aregay, M. Factors That Influence the Implementation of Sustainable Land Management Practices by Rural Households in Tigrai Region, Ethiopia. *Ecol. Process* **2019**, *8*, 14. [CrossRef]
48. Adimassu, Z.; Kessler, A.; Hengsdijk, H. Exploring Determinants of Farmers' Investments in Land Management in the Central Rift Valley of Ethiopia. *Appl. Geogr.* **2012**, *35*, 191–198. [CrossRef]

sustainability

MDPI

Article

Reassessing Resettlement-Associated Poverty Induced by Water Conservancy Projects in China: Case Study of the "Yangtze to Huai River Inter-Basin" Water Diversion Project

Yu Lu [1] and Ziheng Shangguan [2,*]

[1] School of Public Administration, Hohai University, Nanjing 211100, China; 170213050008@hhu.edu.cn
[2] School of Business, Hubei University, Wuhan 430061, China
* Correspondence: sgzh@hubu.edu.cn

Abstract: The displacement and resettlement-associated poverty caused by water conservancy projects (WCP) is a worldwide issue. Re-settlers are often impoverished for extended periods due to loss, difficult re-establishment, and insufficient compensation. Addressing poverty has become a worldwide concern, and accurate measurements of poverty remain a fundamental issue. Before 2020, the Chinese government used the absolute income method to measure re-settler poverty. However, this method reflected neither the overall income gap nor potential benefits of social development and poverty alleviation policies. Therefore, we used the relative income and multidimensional methods alongside the absolute income poverty method to measure the poverty in recently resettled households. Based on survey data from over resettled 1000 households we conclude that: (1) The remaining poor measured by the absolute poverty line were mainly caused by serious diseases, disabilities and loss of labor ability, which means they have no ability to be lifted out of poverty except through the bottom line of local governments. As a result, the absolute poverty line loses its distinction to poverty. (2) Rural re-settlers were more resilient to forced majeure because land guarantees employment and food supply, allowing households to avoid secondary livelihood destruction. (3) Income derived measurement of re-settler poverty masks the benefits of poverty alleviation and other socioeconomic aid programs. A few households showed improvements in child school attendance, child mortality, nutrition, cooking fuel, asset ownership, and social insurance following resettlement. (4) To reduce the multidimensional gap, government aid programs should focus on years of schooling (including training), nutrition, household savings, and household labor force rather than simply providing monetary assistance. At the same time, we suggest that the government adopt a variety of compensation methods, such as: sharing the benefits of water conservancy projects, industrial support and improving the bottom line guarantee.

Keywords: poverty assessment; water conservancy project; resettlement; China

Citation: Lu, Y.; Shangguan, Z. Reassessing Resettlement-Associated Poverty Induced by Water Conservancy Projects in China: Case Study of the "Yangtze to Huai River Inter-Basin" Water Diversion Project. *Sustainability* **2023**, *15*, 9477. https://doi.org/10.3390/su15129477

Academic Editors: Lu Zhang, Bing Kuang and Bohan Yang

Received: 25 April 2023
Revised: 1 June 2023
Accepted: 8 June 2023
Published: 13 June 2023

1. Introduction

The displacement and resettlement of re-settlers as a result of water conservancy projects (WCP) such as dam construction or inter-basin water transfer often lead to poverty. Because such projects affect large areas, numerous resettlement issues occur, such as those observed in association with the resettlement of 50,000 people when the Itaipu Dam, the world's largest hydropower station was built [1], and the relocation of approximately 120,000 people as a result of Aswan High Dam construction [2]. China is home to many mega hydropower projects that have led to the relocation of many millions of people, including those forced to resettle due to the construction of the Three Gorges Dam and the South-to-North Water Diversion Project [3,4]. Displacement and resettlement have huge impacts on people [5], with issues such as loss of farmland severely affects the household incomes impacting the livelihood of the dependents [6]. Furthermore, the low compensation offered in many cases means that resettled people lack sufficient capital to build new houses, restore their capacity for production, and live at pre-displacement levels [7]. Therefore, they often endure long-term poverty as a result of resettlement [8].

The poverty of WCP-induced re-settlers is an urgent problem for the government. Since most resettlements generally occur due to water conservancy projects, the responsibility to mitigate and alleviate the poverty caused by such projects lies with the government [9]; poverty alleviation includes both the duration over which a government is to perform its responsibilities and the amount of compensation offered [10]. This is especially important because long-term impoverishment can lead to social instability [11] and environmental degradation [12]. To reduce WCP-induced poverty, the World Bank provides developing countries with special financial and technical assistance for relocation projects, allowing the formulation of better resettlement policies and post-resettlement action plans [13]. Through the World Bank's consistent efforts, these developing countries have gradually formed their own local compensation standards and allocation procedures [14–16], helping resettled people to alleviate and eliminate poverty by restoring their livelihoods [17,18].

In China, the poverty of the WCP-induced re-settlers has gradually become a matter of concern for the central government. China's early WCPs failed to address the poverty and livelihood restoration problems faced by the re-settlers. Until 1985, 60% of reservoir-induced re-settlers lived in poverty [19]. To deal with the large scale of poverty induced by WCP, the State Council of China issued the first "Report on Quickly Dealing with Reservoir Resettlement Issues" in 1986. Several revisions led to the production of two important documents in 2006: (1) State Council Decree No. 471 (2006) on the land acquisition and resettlement compensation rules associated with large- and medium-scale hydraulic and hydropower projects and (2) Suggestions of the State Council No. 17 (2006) on the improvement of follow-on support for people affected by large- and medium-scale reservoirs. These decrees are aimed at preventing WCP-induced poverty by providing pre-resettlement compensation, resettlement subsidies, and follow-up support [20]. The most influential factor responsible for improvements in China's WCP poverty problem over the last seven years, Xi Jinping's "Precise (Targted) Poverty Alleviation" campaign, was aimed at lifting 70 million Chinese people above the poverty line by 2020, and although WCP-induced re-settlers have benefitted from this program [21], the poverty ratio of these people remains higher than that of the general population, and most remain in abject poverty [22,23]. Most re-settlers are poor because their former homes were located in remote rural areas, and resettlement sites are generally established in under developed regions. Additionally, lower rates of education and reliance on basic farming skills means that many struggle to transition to other livelihoods, if their land is reduced or they become landless following resettlement [24,25].

Currently, the poverty line in China is measured using the absolute income poverty method. The poverty line was initially set at a net income of 2300 RMB per year in 2011 (equivalent to US$ 1/day), and the figure is adjusted yearly according to the consumer price index (CPI) in each province [26]. However, compared with the World Bank poverty line (US$ 1.9/day), China's standards are relatively low [27]. In addition, measuring poverty from the perspective of absolute income alone cannot reflect the overall income gap or the potential benefits of any social development or poverty alleviation policies [28]. Xu et al. (2019) recommended that China adopt the relative income poverty method to accurately measure the poverty of WCP-induced re-settlers, as this method is useful for comparing the overall income gap among different resettlement groups [29]. Wang et al. (2021) argued that the multidimensional poverty method could identify the main factors causing poverty and reflect the effects of policies [30]. However, no research currently addresses the best way to measure China's WCP poverty using both methods in tandem; no comprehensive analysis of WCP-associated poverty using multiple measurement models has as yet been performed. To comprehensively understand the current poverty levels of WCP-induced re-settlers and advise governments on their justification of compensation levels and assistance programs, multiple measurement models are required to evaluate the poverty of WCP-induced re-settlers.

In this study, the "Yangtze to Huai River inter-basin" water diversion project (YtoH Diversion) was considered as a case study and multiple poverty measurement methods were used to comprehensively analyze and interpret the poverty status and causes of poverty in China. The study contributes to the literature surrounding poverty alleviation for WCP-induced re-settlers in two ways. First, a multidimensional poverty framework suitable for China's WCP-induced re-settlers was constructed based on the Global Multidimensional Poverty Index (GMPI), which was published by the United Nations Development Program (UNDP) and Oxford University. Second, this framework was used to dynamically interpret the poverty status of re-settlers in the YtoH Diversion project.

The remainder of this paper is organized as follows. Section 2, based on a brief review of the poverty measurement related literature, provides a framework for multidimensional poverty analysis. The characteristics of the case study region, research methods, and data collection are described in Section 3. The results are presented in Section 4. The main research findings, proposes specific countermeasures for poverty alleviation under China's current WCP-induced resettlement system are discussed in Section 5, and conclusions drawn in Section 6.

2. Literature Review and Multidimensional Poverty Framework

2.1. Poverty Measurement in the Literature

The most commonly used method for identifying and measuring poverty, the monetary approach, defines poverty as consumption (or income) below a certain line [31]. According to Foster's poverty theory, the poverty line can be divided into absolute and relative poverty, which reflect "subsistence" and "basic needs", respectively [32]. Monetary approaches include the income poverty method, Engel coefficient method, Martin method, and extended linear expenditure system method (ELES) [33]. However, the WCP-induced resettlement study for China uses only the income poverty method, which requires income data for quantification [34,35], and as an increasing number of people relocate from remote rural areas to suburban and urban areas, their consumption structures and employment options change, often shifting them from income-based to consumption-based poverty [30]. Therefore, some Chinese resettlement scholars, such as Wang et al., proposed the use of the Engel coefficient, Martin, and ELES methods to measure poverty status [35]. These methods fully consider all dimensions of consumption and thus more accurately measure poverty from the perspective of demand.

Recent studies have shown that the income level of WCP-induced re-settlers in China is sufficiently high to meet their minimum or basic needs and maintain them at or above the absolute poverty line following resettlement. However, owing to higher consumption at resettlement sites and reduced livelihood capital, their poverty characteristics have been observed to gradually change from consumption- to development-based [36,37]. Therefore, some scholars, including Wang and Ke (2009), began to develop a multidimensional poverty framework to measure the poverty of re-settlers [35], which was largely based on the GMPI framework developed by the UNDP and Oxford University and includes three dimensions: education, health, and assets [38]. However, Chinese scholars argue that indicators such as electricity, and improved sanitation and drinking water supply are not applicable to China's WCP-induced re-settlers. Wang et al. (2021) added four production indicators to the GMPI: farmland quantity, quality, stable employment, and labor skills [30], while Xu et al. (2019) added a psychological dimension that included two indicators: development prospects and the willingness to return [29]. Some scholars have stated that a security dimension also needs to be considered, such as social insurance or employment training [39]. Therefore, the current multidimensional poverty indicator system is considered insufficient. In addition, indicators such as farmland quality and development prospects are not easily measured in practice, and the willingness to return is particularly subjective. Therefore, it is necessary to build a more comprehensive and practical indicator system for use with WCP-induced resettlement.

2.2. *Multidimensional Poverty Framework*

Based on the GMPI and existing research, a more comprehensive and practical multidimensional poverty index system was developed that comprises four dimensions: education, health, assets, and development, with six indicators: years of schooling (YS), child school attendance (CSA), child mortality (CM), nutrition (NU), cooking fuel (CF), and asset ownership (AO), which includes household savings (HS), social insurance (SI), and the household labor force (HLF). These indicators are measured in household units.

2.2.1. Education Dimension

The education dimension includes two indicators: YS and CSA. According to the GMPI, if no household member has completed at least five years of education, a household is considered 'deprived' in terms of this dimension, while attendance deprivation is assumed if a school-age child is not attending school up to the age at which they would complete class 8 [40]. Since the per capita years of education is on average 7.7 years in rural China [41] and even lower for WCP-induced re-settlers, 7 years was taken as the threshold for YS in this study. At the same time, taking into account the 9 years of compulsory education and labor laws in China, school-age children between the ages of 7 and 16 years who do not attend school are considered deprived in terms of attendance.

2.2.2. Health Dimension

The health dimension includes two indicators: CM and NU. The GMPI considers the CM rate to be severe if a child has died within a family within the last five years, while nutritional deprivation is assumed if at least one family member is undernourished [40]. The interpretation of mortality defined by the GMPI was used in this study. In terms of NU, combined with the "Guidelines for the Prevention and Control of Overweight and Obesity in Chinese Adults" that was officially issued by mainland China, this article considers a BMI less than 18.5 as malnutrition [42].

2.2.3. Asset Dimension

The asset dimension included two indicators: CF and AO. As defined by the GMPI, residents that use dung, wood, or charcoal for cooking are considered deprived in terms of CF, while asset ownership is considered deprived if residents do not own more than one of the following: a radio, TV, telephone, bicycle, motorbike, or refrigerator, and do not own a car or truck [40]. In this study, the indicator illustration of CFI in the GMPI was used directly. In terms of AO, on the basis of the GMPI and in combination with the "Water Conservancy and Hydropower Project Resettlement Supervision and Evaluation Regulations" (SL716-2015) in China, commonly owned belongings such as a TV, refrigerator, washing machine, air conditioner, electric fan, water heater, rice cooker, pressure cooker, induction cooker, microwave oven, and telephone were considered assets and households that owned less than three were considered deprived.

2.2.4. Development Dimension

In addition to the three dimensions of the GMPI, an additional development dimension was added that included three indicators: HI, SI, and HLF. Family savings are an important financial guarantee for the subsequent development of WCP-induced re-settlers [43]. As the livelihood monitoring cycle for projects in China is six years, this article considers a resettled household deprived when its savings are less than six times that of the locals. SI provides anti-risk security for re-settlers [44]. Currently, rural re-settlers can purchase new rural social endowment and cooperative medical insurances, whereas re-settlers in cities can purchase work injury, maternity, endowment, medical, and unemployment insurance. This study considers SI to be deprived if individuals do not have insurance coverage. In addition, the HLF is the basis of subsistence [45]; thus, a household was considered deprived if its labor force was less than 50% of the household population.

3. Materials and Methods

3.1. Research Region and Sampling

The Anhui Province section of the "Yangtze-Huai River Inter-Basin" Water Diversion Project, from which the original residents were relocated to designated sites by 2016 (see Figure 1), was considered in this study. Fieldwork was conducted over four consecutive years (2017–2020), with the first survey in January 2017 considering resettlement baseline and resident data up to 2016. Surveys were given to the same households every December from 2017 to 2020. Of the 2745 relocated households, our sample comprised 1098, leading to a sampling rate of 40% (see Table 1). A stratified sampling method was used to select the participants, with 15% earning a low income, 20% a relatively low income, 30% a middle income, 20% a relatively high income, and 15% a high income. The basic properties of the samples were as follows.

- All resettled people were of official rural household registration status prior to resettlement, and were mainly engaged in agricultural production. The local government adopted a mixed resettlement model, with 79.6% resettled in urban and 20.4% in rural sites. Rural resettled people were continuously engaged in agricultural production, while urban resettled people had to find jobs in non-farming sectors.
- The rural resettlement sites were approximately 15 to 20 km away from their original home villages, while urban resettlement sites were approximately 8 to 10 km away.

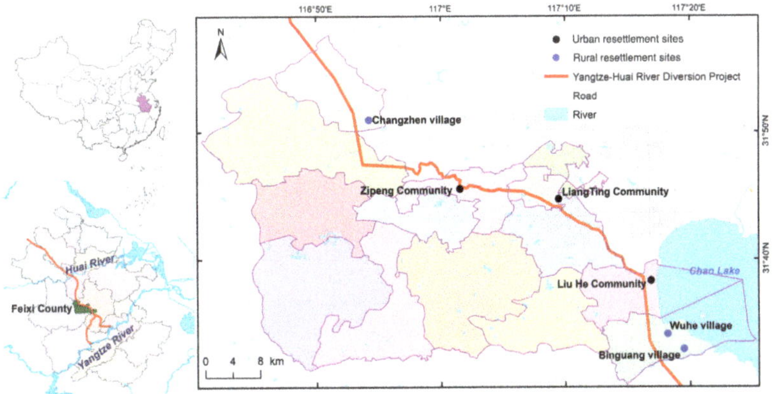

Figure 1. Resettlement sites in research sample.

Table 1. Sample selection.

Area	Location	Relocated Households	Sample Households	Geographical Coordinates
Urban	Liangting Community	158	63	117.157 E, 31.745 N
	Liuhe Community	1269	508	117.282 E, 31.638 N
	Zipeng Community	758	303	117.025 E, 31.758 N
Rural	Binguang village	404	162	117.325 E, 31.549 N
	Wuhe village	113	45	117.304 E, 31.568 N
	Changzhen village	43	17	116.902 E, 31.849 N
Total		2745	1098	

3.2. Poverty Measurement Model

The absolute and relative income poverty, and multidimensional poverty methods were used to comprehensively measure the poverty status of the resettled people. Absolute and relative income poverty were calculated using the Foster–Greer–Thorbecke (FGT) indices, and multidimensional poverty was calculated using the A–F model.

3.2.1. FGT Indices

The FGT indices are often used to analyze income poverty [46] and measure the poverty headcount ratio, poverty gap, and income inequality. The index is derived by substituting different values of parameter α into the following equation:

$$\mathrm{FGT}_{\alpha} = \frac{1}{\mathrm{N}}\sum_{\mathrm{i}=1}^{\mathrm{H}}\left(\frac{\mathrm{z}-\mathrm{y_i}}{\mathrm{z}}\right)^{\alpha} \tag{1}$$

where z is the poverty line, N is the number of people comprising an economy, H is the number of people in poverty (those with income at or below z), and y_i is the income of each individual, i. The formula reduces to the poverty headcount ratio (PHR) when $\alpha = 0$, poverty gap index (PGI) when $\alpha = 1$, and squared poverty gap index (SPGI) when $\alpha = 2$. Among these, the PHR and PGI are the most commonly used poverty indicators [47]. To facilitate comparison with the calculated multidimensional poverty result, only the PHR and PGI were calculated in this study.

The Lorentz curve equation proposed by Villasenor and Arnold was used to calculate the FGT indices [48]. This equation is expressed by the following:

$$L(1-L) = a\left(P^2 - L\right) + bL(P-1) + c(P-L) \tag{2}$$

where L is the cumulative share of income earned and P is the cumulative share of people or households from lowest to highest income (all poverty indexes in this paper are based on households). Parameters a, b, c are estimated while e, m, n, and r are obtained using the equations: $e = -(a+b+c+1)$, $m = b^2 - 4a$, $n = 2be - 4c$, $r = \sqrt{n^2 - 4me^2}$. After obtaining the quadratic Lorenz curve, the corresponding FGT indices are calculated using the formula:

$$L(P) = -\frac{1}{2}\left(bP + e + \sqrt{mP^2 + nP + e^2}\right) \tag{3}$$

Parameters s_1 and s_2 are calculated using $s_1 = (r-n)/2m$ *and* $s_2 = -(r+n)/2m$, repsectively. *PHR* and *PGI* were calculated using the following formulas:

$$PHR = -\frac{1}{2m}\left(n + r\frac{(b+2z/\mu)}{\sqrt{(b+2z/\mu)^2 - m}}\right) \tag{4}$$

$$PGI = H - \left(\frac{\mu}{z}\right)L(H) \tag{5}$$

where μ is per capita net income and z is the poverty line. The absolute poverty headcount ratio (APHR) and absolute poverty gap index (APGI) can be obtained when z denotes the absolute income poverty line. Subsequently, the relative poverty headcount ratio (RPHR) and relative poverty gap (RPGI) can be obtained when z denotes the relative income poverty line.

3.2.2. A–F Model

The A-F method, which has the advantages of being highly intuitive and suitable for policy analysis, was adopted for the assessment of multidimensional poverty [49].

Suppose there are n individuals in an economy. The poverty status of individual i is measured using m indicators. The value of the individual i for each indicator j is expressed as g_{ij}, with $g_{ij} = 1$ if indicator j of individual i is deprived, and $g_{ij} = 0$ otherwise. Setting the weight of indicator j to $w_j (0 < w_j < 1, \sum_{j=1}^{m} w_j = 1)$, the weighted score of individuals i on all m indicators can be represented by c_i, and $c_i = \sum_{j=1}^{m} w_j g_{ij}$. The critical value k $(0 < k \leq 1)$ is used to compare the degree to which individual i is deprived under m

indicators to determine their multidimensional poverty status, with individual i regarded to be in multidimensional poverty if $c_i \geq k$. No threshold is assumed when $k = 0$, otherwise, a threshold is included. Based on this method, the multidimensional poverty index (MPI) is obtained using:

$$MPI = \frac{1}{n}\sum_{i=1}^{n} c_i(k) \tag{6}$$

If $m = 1$, and $k \neq 0$, the formula indicates the censored single factor poverty index (CSFPI):

$$CSFPI = \frac{1}{n}\sum_{i=1}^{n} g_i \cdot I(c_i \geq k) \tag{7}$$

where, $I(\cdot)$ is a threshold function for which the value is 1 when $c_i \geq k$ and 0 if $c_i < k$. In this study, the value of k was set to 0.3.

The MPI can also be divided into the multidimensional poverty headcount ratio (MPHR) and the multidimensional poverty gap index (MPGI).

$$MPI = \frac{q}{n} \times \frac{1}{q}\sum_{i=1}^{n} c_i(k) = MPHR \times MPGI \tag{8}$$

where q is the number of people that have fallen into multidimensional poverty.

The ratio *CSFPI/MPHR* can be used to measure the impact of a certain indicator on multidimensional poverty, which we define as the single factor impact index (SFII):

$$SFII = \frac{CSFPI}{MPHR} \tag{9}$$

For a certain indicator i, the larger the *SFII*, the greater its impact on multidimensional poverty.

3.3. Data Processing

3.3.1. Data Used in Parameter Estimation for the Lorenz Curve

In this study, k-means cluster analysis was used to divide the per capita net income of the sample households into ten levels from low to high and the cumulative share of people at all levels was calculated as an independent variable. The cumulative share of the net income earned was obtained simultaneously as the dependent variable. Based on the above process, 10 sample points (Pi, Li) (i = 1, 2, . . . , 10) were obtained. The Lorenz curve was obtained using Equation (2) for parameter estimation. In this study, SPSS 19 was used for cluster analysis and nonlinear regression.

3.3.2. Poverty Line and per Capita Net Income

When using FGT indices to measure the absolute and relative income poverty, the poverty line z and per capita net income μ are required. The absolute income poverty line z_1, which was set at 2300 yuan by China in 2011 [50], was adjusted according to the CPI of Anhui Province. The relative income poverty line z_2 accounts for 30% of the median per capita household income of re-settlers and value μ is obtained from statistical analysis of the sampled data, as shown in Table 2.

Table 2. Poverty line and per capita net income from 2016 to 2020 (unit: RMB).

Year	z_1		z_2		μ	
	Urban	**Rural**	**Urban**	**Rural**	**Urban**	**Rural**
2016	3004	2961	3987	3713	8643	7930
2018	3104	3056	4242	4097	8386	8471
2020	3267	3224	4137	4059	8437	8202

3.3.3. Weight of Multidimensional Poverty Indicators

The indicators and definitions used in this study are presented in Table 3. The weight of each was calculated using an equal-proportion weighting method.

Table 3. Definition and weight of multidimensional poverty indicators.

Dimension	Indicator	g = 1 When Occurred, Otherwise g = 0	Weight
Education	Years of Schooling (YS)	No household member has completed 7 years of education, i.e., graduated primary school.	1/8
	Child School Attendance(CSA)	School-age children between the ages of 7 and 16 do not go to school.	1/8
Health	Child Mortality (CM)	A child has died in the family within the last 5 years.	1/8
	Nutrition (NU)	At least one family member under the age of 60 has a BMI less than 18.5.	1/8
Asset	Cooking Fuel (CF)	Uses dung, wood, or charcoal for cooking.	1/8
	Assets Ownership (AO)	Own less than three of the following assets: TV, refrigerator, washing machine, air conditioner, electric fan, water heater, rice cooker, pressure cooker, induction cooker, microwave oven, and telephone.	1/8
Development	Household Saving (HS)	Household savings of re-settlers is less than 6 times that of the locals.	1/12
	Social Insurance (SI)	Not covered by any insurance.	1/12
	Household Labour Force (HLF)	Household labour force is less than 50% of the family population.	1/12

4. Result

By comparing the calculation results for absolute income, relative income, and multidimensional poverty, the poverty status of resettled people induced by the "Yangtze to Huai River inter-basin" water diversion project, a contemporary WCP, was comprehensively measured. The parameter estimation results and curves obtained using the Lorentz curve equation is presented in Table A1 and Figure A1 in Appendix A. The calculation results for the MPI and CSPI are presented in Tables A2 and A3, respectively, of Appendix A.

4.1. *Absolute Income Poverty*

Figure 2a indicates that the APHR of urban and rural resettled people declined slightly between 2016 and 2018, indicating that, on average, the income of resettled people measured by the APHR was slightly improved, with poverty alleviation rates of 0.13% and 0.29%, respectively. However, from 2018 to 2020, the APHR indicators for both urban and rural resettlement increased, indicating a return to poverty with rates of 4.59% and 2.60%, respectively. In general, the APHR of both resettled and rural households increased over the period from 2016 (rural versus urban: 2.26% versus 1.76%) to 2020 (rural versus urban: 4.57% versus 6.22%). A similar trend can also be observed in Figure 2b, which shows that the APGI of urban re-settlers was essentially the same as that of rural re-settlers from 2016 to 2018; however, the gap widened in 2020, when the APGI of urban resettled people increased rapidly.

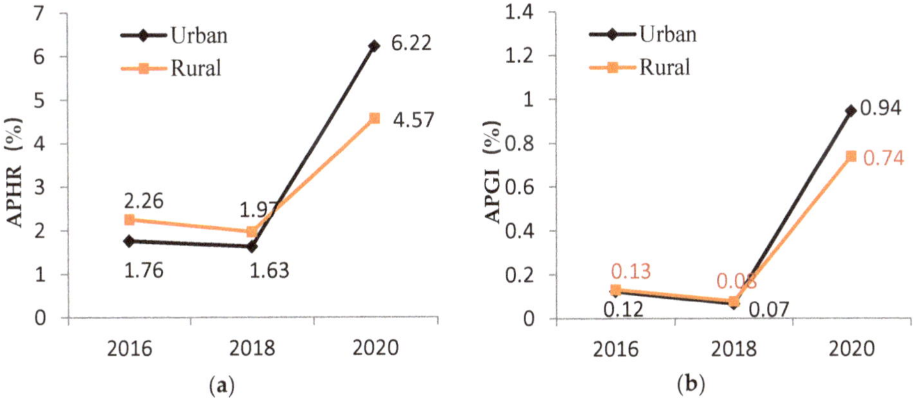

Figure 2. Intertemporal calculations showing absolute income poverty: (**a**) APHR; (**b**) APGI.

4.2. Relative Income Poverty

It is apparent from Figure 3a that the RPHR of the urban resettled increased between 2016 and 2020, indicating that the overall income gap expanded during this period, while the rural resettled show an inverted "V" shape that indicates a widening overall income gap between 2016 to 2018 followed by a narrowing from 2018 to 2020. In general, the RPHR of both resettled and rural households increased from 2016 (rural vs. urban: 8.17% vs. 6.94%) to 2020 (rural vs. urban: 9.57% vs. 13.18%). An identical pattern for the RPGI can be observed in Figure 3b.

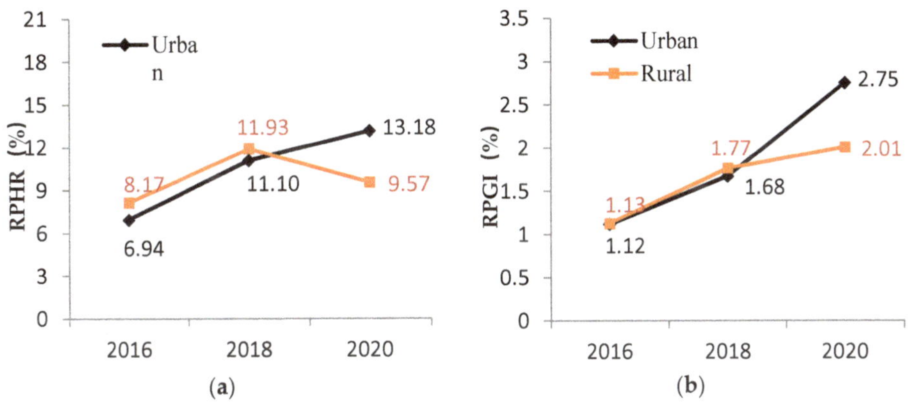

Figure 3. Intertemporal calculations of relative income poverty: (**a**) APHR; (**b**) APGI.

4.3. Multidimensional Poverty

Figure 4a indicates that the MPHR of urban resettled areas from 2016 to 2020 remained stable, with only slight fluctuations between 8.10% and 8.44%; however, the MPHR of their rural counterparts show a consistent decline from 13.34% in 2016 to 9.16% in 2020. These results indicate that the urban resettled population experienced a relatively stable MPHR, while the rural resettled population experienced a rapid reduction in MPHR. The multidimensional poverty of the resettled population thus generally improved. Similar trends can also be seen in Figure 4b, in which the MPGI is used to measure the poverty gap.

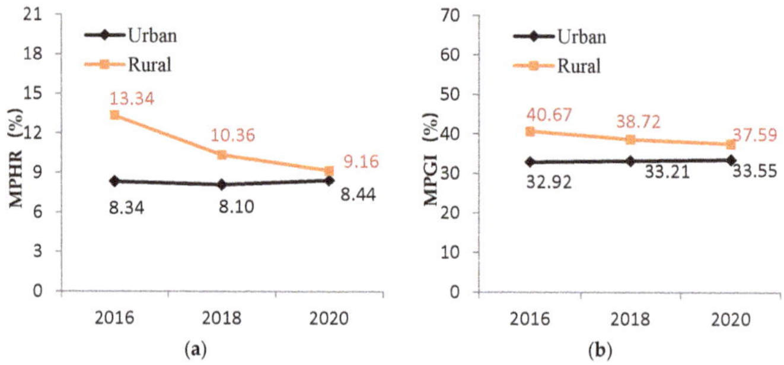

Figure 4. Intertemporal calculations of multidimensional poverty: (**a**) MPHR; (**b**) MPGI.

4.4. SFII Calculation

To identify the main factors that influence the poverty of the resettled population, the SFII was used to determine which of the key factors contribute more to multidimensional poverty. As shown in Figure 5a, YS, NU, and AO show a downward trend in urban resettled areas from 2016 to 2020, while CSA, HS, and HLF show an upward trend, and SI shows a fluctuating trend. By 2020, HS and HLF had a much greater impact on the multidimensional poverty of the urban resettled population, with SFIIs of 95% and 99%, respectively. It is therefore apparent that more attention should be devoted to HS and HLF if we are to further improve the multidimensional poverty of urban resettled populations in the future.

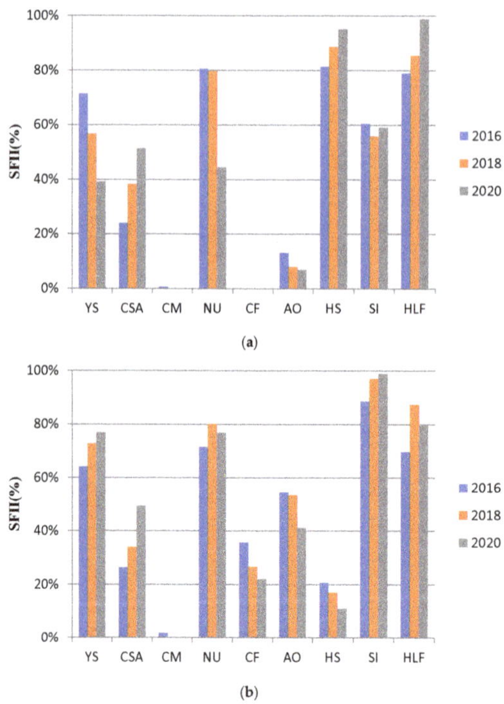

Figure 5. Intertemporal SFII Calculations: (**a**) Urban re-settlers; (**b**) Rural re-settlers. (YS = Years of Schooling; CSA = Child School Attendance; CM = Child Mortality; NU = Nutrition; CF = Cooking Fuel; AO = Assets Ownership; HS = Household Saving; SI = Social Insurance; HLF = Household Labor Force).

Figure 5b presents the SFII for rural resettlement between 2016 and 2020. CF, AO, and HS show decreasing trends, while YS, CSA, and SI show increasing trends, and NU and HLF show fluctuating trends. By 2020, YS, NU, SI, and HLF had the greatest impacts on multidimensional poverty in rural resettled areas, with SFII values of 76%, 76%, 98%, and 80%, respectively. Future alleviation efforts should therefore focus on YS, NU, SI, and HLF for these populations.

5. Discussion

Three different poverty indices were used together with the SFII to measure the poverty trends and status of the resettlement-case population. The main factors contributing to poverty within the resettled population were identified, and the consistency between the calculation results and the actual situation was discussed. The main conclusions were obtained by comparing the calculation results for absolute income poverty, relative income poverty, and multidimensional poverty.

5.1. Absolute Income Poverty Analysis

According to the current poverty line in China, z_1, the APHR index shows an overall positive result for the first two years following resettlement, with a decline in the APHR for both urban and rural re-settlers. The values of APHR obtained in the study area (1.63% for urban resettled and 1.97% for rural resettled) were slightly better than the 2.2% obtained for Anhui Province in 2018 [51]. Our survey data show that poverty in these households is largely caused by serious diseases, disabilities, and loss of labor ability [52]. Alternatively, the remaining impoverished population has no capacity to be lifted above the poverty line via the bottom-line guarantee of the local government.

However, the APHR of both urban and rural resettlements increased between 2018 and 2020, with that of urban resettlement increasing faster than that of rural resettlement. Such a high rate of re-impoverishment indicated that the resettled remained vulnerable to poverty and that the urban resettled are more sensitive to this problem. One of the main explanations for this phenomenon is the impact of the China-United States trade war, which has greatly affected employment in China's secondary and tertiary industries [53,54]. Despite the decline in agricultural sales within resettled rural areas during the trade war, the pressure of unemployment and lower income could be partially mitigated by having land as a basic guarantee. However, urban re-settlers suffer more from unemployment, resulting in a higher rate of return on poverty than those who have been resettled within the same timeframe.

5.2. Relative Income Poverty Analysis

The RPHR is influenced by the overall income gap. According to the relative income poverty line z_2, the overall income gap of the urban resettled population expanded between 2016 and 2020, whereas that of the rural resettled population expanded between 2016 and 2018 and shrank between 2018 and 2020. These trends are consistent with the change in the Lorentz curve in Figure A1; that is, the Lorentz curve of urban resettlement constantly deviated from the line of equality from 2016 to 2020, while that describing rural resettled deviated between 2016 and 2018, and approached between 2018 and 2020.

The income gap of the urban resettled expanded from 2016 to 2020, which is consistent with China's macro Gini coefficient [55]. Before 2018, the development of internet platforms such as Meituan, Alipay, JD.com, and Didi Taxi provided a large number of jobs for the tertiary industry [56], which greatly increased the income of urban re-settlers. However, the income of those engaged in secondary industries remained largely unchanged. Therefore, there was a notable increase in the income gap. However, the China-United States trade war has meant that the employment rate and income of both secondary and tertiary industries declined to varying degrees in 2019, which led to a lower increase in the income gap [57]. The income gap in the rural resettled populations also expanded from 2016 to 2018. Similar to the urban re-settlers, the main reason for this is that rural re-settlers who were engaged in tertiary industries also benefitted from the internet platforms. However, for the same reason, 68.53% of the rural resettled were forced to change their production from secondary and tertiary industries to agriculture [52]. Because the agricultural income gap is small and the income of secondary and tertiary industries has declined, the income gap of the rural resettled population narrowed between 2018 and 2020.

5.3. *Multidimensional Poverty Analysis*

The MPHR of the urban resettled remained stable from 2016 to 2020 (ranging from 8.10% to 8.44%); however, the MPHR of the rural resettled decreased substantially from 13.34% in 2016 to 9.16% in 2020. This indicates that multidimensional poverty has generally improved in the resettled population. Compared with the nationwide MPHR of 4% estimated for 2014 by Shen et al. [40], it is apparent that the deprivation associated with WCP-induced resettlement is more serious. At the same time, the much lower and more stable MPHR of the urban resettled compared to their rural counterparts indicates that the urban resettled, with official urban registration, are entitled to more material and social resources than their rural counterparts, which is consistent with the research findings of Shangguan et al. [39].

HS have been identified as one factor that contributes to the MPHR for urban resettlement. Relocating to urban areas means higher cash expenditure. The replacement of coal or wood with natural gas and the higher electricity consumption increase the cost of living. Other costs associated with transportation and improving the quality of life in urban areas further increases consumption. The most important extra cost is urban housing, which necessitates savings urban houses cost more than the resettlement compensation [58]. YS, NU, and SI have also been identified as important factors contributing to the MPHR of the rural resettled. The severe deprivation in terms of YS and NU is largely due to the relatively poor educational and material resources in rural areas [59]. The deprivation of social security occurs because rural resettlement focuses largely on short-term benefits. Since China's new rural social endowment insurance requires that insurance premiums are paid until the age of 60 to receive a pension, the resettled are more inclined to use money for their near-term living expenses. This phenomenon is consistent with the research results of Banerjee [60]. The HLF has a greater impact on the MPHR of both urban and rural re-settlers and is the basis of the migrant's livelihood, regardless of location.

6. Conclusions

This study used multiple poverty measurement methods to reassess and dynamically interpret the poverty status of China's WCP-induced re-settlers. The following conclusions were obtained: (1) Absolute poverty analysis indicates that China's current absolute poverty standards are out of date for WCP-induced re-settlers because poverty is not eliminated through the bottom-line guarantees of local government. The current absolute poverty line does not sufficiently represent the different experiences and needs of the resettled poor. (2) Through relative poverty analysis, we found that rural re-settlers are more resilient to force majeure, as witnessed during the recent pandemic. The guarantee of employment and food supply through land ownership allows re-settlers to avoid the secondary destruction of their livelihoods. (3) Comparison of the results for income poverty (both absolute and relative) with those of multidimensional poverty indicates that worsening income poverty is universal for the resettled, whereas multidimensional poverty has generally improved. Therefore, measuring poverty in terms of income alone masks the potential benefits of mitigation processes such as social development programs and poverty alleviation policies. Table A2 clearly shows that CSA, CM, NU, CF, AO, and SI for both urban and rural resettled children have all improved to varying degrees following resettlement, which is mainly due to better access to the relevant public, material, and information resources [61]. However, the higher MPHR for urban and rural resettlement indicates that multidimensional poverty could still be improved. (4) Comparison of the APGI, RPGI, and MPGI indicates a comparatively small gap between absolute and relative income poverty, whereas the gap associated with multidimensional poverty is much larger. Therefore, reducing the multidimensional poverty gap should be the focus of poverty alleviation in the later stages. Accordingly, analysis of the different included factors indicates that the YS, NU, HS, SI, and HLF are most important and should be targets for future poverty alleviation efforts. In addition, in order to improve the livelihood resilience and resist secondary disasters caused by force majeure, a stable source of income for re-settlers is also necessary. To this end, we suggest that the government adopt a variety of compensation methods, such as: sharing the benefits of water conservancy projects, industrial support and improving the bottom line guarantee. Therefore, subsequent studies should consider how to reduce multidimensional poverty of the re-settlers through the above compensation methods. At the same time, more indicators can be included in the measurement of multidimensional poverty to better reflect poverty status.

Author Contributions: Conceptualization, Z.S.; Investigation, Y.L.; Methodology, Y.L.; Software, Y.L.; Supervision, Z.S.; Validation, Z.S.; Writing—original draft, Y.L.; Writing—review and editing, Z.S. All authors have read and agreed to the published version of the manuscript.

Funding: This research was funded by the Key Program of the National Social Science Fund (grant number 19AJL016).

Institutional Review Board Statement: Not applicable.

Informed Consent Statement: Not applicable.

Data Availability Statement: Data can be requested from the first author for research purposes.

Conflicts of Interest: The authors declare no conflict of interest.

Appendix A

Table A1. Lorenz curve regression results.

Area	Year	Parameter	B	SE	95% Confidence Interval		R^2
					Lower Bound	Upper Bound	
Urban	2016	a	0.949	0.012	0.926	0.972	0.999
		b	−1.71	0.034	−1.776	−1.644	
		c	0.101	0.019	0.064	0.138	
	2018	a	0.851	0.019	0.813	0.888	0.999
		b	−1.545	0.043	−1.629	−1.461	
		c	0.158	0.022	0.115	0.201	
	2020	a	0.884	0.015	0.855	0.913	0.999
		b	−1.547	0.049	−1.643	−1.451	
		c	0.12	0.023	0.075	0.165	
Rural	2016	a	0.859	0.014	0.832	0.886	0.999
		b	−1.575	0.023	−1.619	−1.53	
		c	0.139	0.018	0.104	0.174	
	2018	a	0.809	0.01	0.789	0.828	0.999
		b	−1.416	0.029	−1.473	−1.359	
		c	0.195	0.019	0.158	0.232	
	2020	a	0.952	0.009	0.934	0.969	0.999
		b	−1.725	0.023	−1.769	−1.68	
		c	0.078	0.012	0.054	0.101	

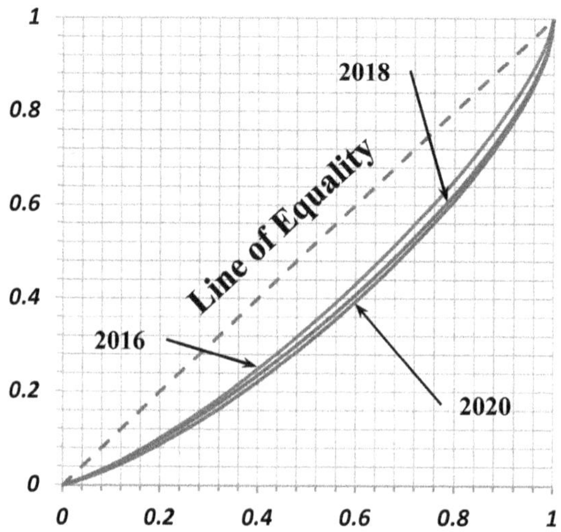

(**a**)

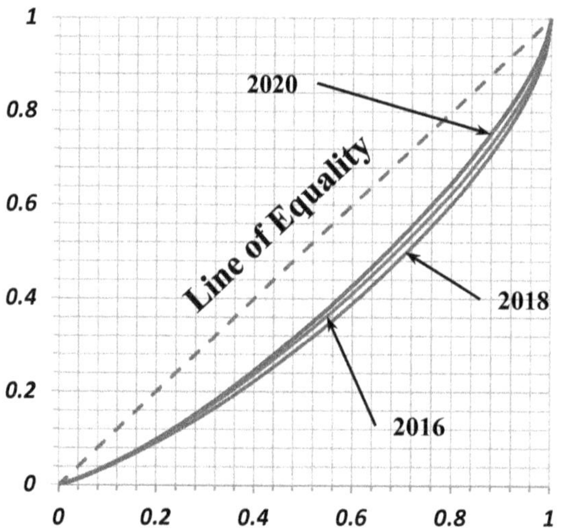

(**b**)

Figure A1. Intertemporal Lorentz curve: (**a**) Urban; (**b**) Rural.

Table A2. MPI of urban and rural re-settlers.

Area	Year	MPI	MH (%)	MPG (%)
Urban	2016	0.027	8.34	32.92
	2018	0.026	8.10	33.21
	2020	0.028	8.44	33.55
Rural	2016	0.054	13.34	40.67
	2018	0.040	10.36	38.72
	2020	0.034	9.16	37.59

Table A3. CSFPI of urban and rural re-settlers.

Index	Urban			Rural		
	2016	2018	2020	2016	2018	2020
YS	5.95	4.59	3.30	8.54	7.54	7.04
CSA	2.00	3.10	4.33	3.52	3.52	4.52
CM	0.06	0.00	0.00	0.25	0.00	0.00
NU	6.72	6.46	3.75	9.55	8.29	7.04
CF	0.00	0.00	0.00	4.77	2.76	2.01
AO	1.10	0.65	0.58	7.29	5.53	3.77
HS	6.79	7.18	8.02	2.76	1.76	1.01
SI	5.04	4.52	4.98	11.81	10.05	9.05
HLF	6.59	6.92	8.34	9.30	9.05	7.29

References

1. Pearce, F. *The Dammed: Rivers, Dams, and the Coming World Water Crisis*; The Bodley Head: London, UK, 1992; p. 155.
2. Goldsmith, E.; Hildyard, N. *The Social and Environmental Effects of Large Dams*; Volume 1: Overview; Wadebridge Ecological Centre: Cornwall, UK, 1984; p. 17.
3. Jackson, S.; Sleigh, A. Resettlement for China's Three Gorges Dam: Socio-economic impact and institutional tensions. *Communist Post-Communist Stud.* **2000**, *33*, 223–241. [CrossRef]
4. Rogers, S.; Barnett, J.; Webber, M.; Finlayson, B.; Wang, M. Governmentality and the conduct of water: China's South–North Water Transfer Project. *Trans. Inst. Br. Geogr.* **2016**, *41*, 429–441. [CrossRef]
5. Gong, Y.; Yao, K.; Zhang, R.; Liu, B.; Wang, F. Rethinking livelihood resilience after development-induced displacement and resettlement: A case study of Qianping Reservoir. *Int. J. Water Resour. Dev.* **2021**, *37*, 841–864. [CrossRef]
6. Heming, L.; Waley, P.; Rees, P. Reservoir resettlement in China: Past experience and the Three Gorges Dam. *Geogr. J.* **2001**, *167*, 195–212. [CrossRef]
7. Shangguan, Z.; Shi, G.; Wu, R.; Song, L. Analysis on the Factors Influencing the Livelihood Vulnerability of "Yangtze River to Huaihe river" Migration Based on ISM and MICMAC. *J. Stat. Inf.* **2019**, *34*, 94–100. Available online: http://tjlt-cbpt-cnki-net-s.vpn.hhu.edu.cn:8118/WKE3/WebPublication/paperDigest.aspx?paperID=8c39b535-81f5-4242-9d71-320fa3d13f7c (accessed on 9 May 2019).
8. Skeldon, R. Migration and poverty. *Asia–Pac. Popul. J.* **2003**, *17*, 67–82. [CrossRef]
9. Cernea, M.M. Compensation and benefit sharing: Why resettlement policies and practices must be reformed. *Water Sci. Eng.* **2008**, *1*, 89–120. [CrossRef]
10. Cernea, M.M. Risks, safeguards and reconstruction: A model for population displacement and resettlement. *Econ. Political Wkly.* **2000**, *35*, 3659–3678. Available online: https://www.jstor.org/stable/4409836 (accessed on 23 April 2023).
11. Wilmsen, B. Progress, problems, and prospects of dam-induced displacement and resettlement in China. *China Inf.* **2011**, *25*, 139–164. [CrossRef]
12. Choy, Y.K. Sustainable development and the social and cultural impact of a dam-induced development strategy-the Bakun experience. *Pac. Aff.* **2004**, *77*, 50–68.
13. Hay, M.; Skinner, J.; Norton, A. *Dam-Induced Displacement and Resettlement: A Literature Review*; The University of Manchester: Manchester, UK, 2019; SSRN 3538211. [CrossRef]
14. Shi, G. Comparing China's and the World Bank's resettlement policies over time: The ascent of the 'resettlement with development' paradigm. In *Challenging the Prevailing Paradigm of Displacement and Resettlement: Risks, Impoverishment, Legacies, Solutions*; Cernea, M.M., Maldonado, J.K., Eds.; Routledge: London, UK, 2018; pp. 45–56.
15. Manorom, K.; Baird, I.G.; Shoemaker, B. The World Bank, Hydropower-based Poverty Alleviation and Indigenous Peoples: On-the-Ground Realities in the Xe Bang Fai River Basin of Laos. *Forum Dev. Stud.* **2017**, *44*, 275–300. [CrossRef]
16. Singer, J.; Pham, H.T.; Hoang, H. Broadening stakeholder participation to improve outcomes for dam-forced resettlement in Vietnam. *Water Resour. Rural. Dev.* **2014**, *4*, 85–103. [CrossRef]
17. Heming, L.; Rees, P. Population displacement in the Three Gorges reservoir area of the Yangtze River, central China: Relocation policies and migrant views. *Int. J. Popul. Geogr.* **2000**, *6*, 439–462. [CrossRef]
18. Heggelund, G. Resettlement Programmes and Environmental Capacity in the Three Gorges Dam Project. *Dev. Chang.* **2006**, *37*, 179–199. [CrossRef]
19. Zhang, S. Preliminary Study on "Secondary Poverty" of Reservoir-induced migrants and Its Countermeasures. *J. Econ. Water Resour.* **1992**, *12*, 25–28. Available online: https://kns.cnki.net/kcms/detail/detail.aspx?FileName=SLJJ199204007&DbName=CJFQ1992 (accessed on 25 April 2023).
20. McDonald, B.D. From Compensation to Development: Involuntary Resettlement in the People's Republic of China. Ph.D. Thesis, The University of Melbourne, Parkville, Australia, 2006.
21. Fan, M.; Li, Y.; Li, W. Solving one problem by creating a bigger one: The consequences of ecological resettlement for grassland restoration and poverty alleviation in Northwestern China. *Land Use Policy* **2015**, *42*, 124–130. [CrossRef]

22. Luo, Y. From De-Collectivity to Involution: Social Logic of the Dispersed Backward-Settled Re locations' Poverty. *J. Anhui Norm. Univ.* **2021**, *49*, 94–102. [CrossRef]
23. Zhang, J.; Zhao, N.; Du, W. Study on Relative Poverty Governance and the Mechanism of Poverty Alleviation through Employment for Rservoir Migrants. *Soc. Secur. Stud.* **2021**, *4*, 97–104. Available online: https://kns.cnki.net/kcms/detail/detail.aspx?FileName=SHBY202104010&DbName=CJFQ2021 (accessed on 3 August 2021).
24. Shen, J.; Shi, G. Economic Analysis on the Poverty Reasons for Reservoir Resettlement. *Issues Agric. Econ.* **2008**, *12*, 43–46. Available online: https://kns.cnki.net/kcms/detail/detail.aspx?FileName=NJWT200812011&DbName=CJFQ2008 (accessed on 23 December 2008).
25. Shen, J.; Shi, G. Poverty Analysis of Involuntary Migrants in China. *Gansu Soc. Sci.* **2003**, *5*, 114–117. [CrossRef]
26. Wei, Q.; Zhang, J.; Scherpbier, R.; Zhao, C.; Luo, S.; Wang, X.; Guo, S. High prevalence of developmental delay among children under three years of age in poverty-stricken areas of China. *Public Health* **2015**, *129*, 1610–1617. [CrossRef] [PubMed]
27. Garroway, C.; De Laiglesia, J.R. *On the Relevance of Relative Poverty for Developing Countries*; OECD: Paris, France, 2012. [CrossRef]
28. Zhao, X.; Chen, Y. Research on Dynamic Measurement and Poverty-causing Factors of Multidimension-al-interventional Poverty of Rural Reservoir Resettlers. *J. Huazhong Agric. Univ.* **2021**, *3*, 128–137+189–190. [CrossRef]
29. Xu, X.; Teng, X.; Wen, C. The Multi-Dimensional Poverty Measurement of Migrants in Large-Scale Water Conservancy and Hydroelectric Projects—The Resarch Data Analysis Based on the Reservoir Area of Three Gorges. *J. Yangtze Univ.* **2019**, *42*, 52–57. Available online: https://kns.cnki.net/kcms/detail/detail.aspx?FileName=JZSZ201904009&DbName=CJFQ2019 (accessed on 25 April 2023).
30. Wang, Z.; Li, G.; Liu, F. Identification of Relative Poverty Characteristics and Poverty-Causing Factors of Emigrants in Hydropower Projects. *J. Huazhong Agric. Univ.* **2021**, *2*, 23–31+175. [CrossRef]
31. Grosh, M.; Glewwe, P. *Designing Household Survey Questionnaires for Developing Countries*; World Bank: Washington, DC, USA, 2000.
32. Foster, J.E. Absolute versus relative poverty. *Am. Econ. Rev.* **1998**, *88*, 335–341.
33. Yang, L. A comparative study on the calculation method of poverty line and its adjustment mechanism. *Comp. Econ. Soc. Syst.* **2010**, *5*, 52–62. Available online: https://kns.cnki.net/kcms/detail/detail.aspx?FileName=JJSH201005010&DbName=CJFQ2010 (accessed on 25 April 2023).
34. Shen, J.; Shi, G.; Shang, K. Research on Reservoir Resettlement Efficiency Based on Composite DEA Model. *Yellow River* **2009**, *31*, 9–11. Available online: https://kns.cnki.net/kcms/detail/detail.aspx?FileName=RMHH200903004&DbName=CJFQ2009 (accessed on 25 April 2023).
35. Wang, F.; Ke, Z. Objective of Settling Migrants from Reservoir Area: A Analysis Based on Welfare Economics. *Popul. Econ.* **2009**, *6*, 73–78. Available online: https://kns.cnki.net/kcms/detail/detail.aspx?FileName=RKJJ200906013&DbName=CJFQ2009 (accessed on 25 November 2009).
36. Wilmsen, B. After the Deluge: A longitudinal study of resettlement at the Three Gorges Dam, China. *World Dev.* **2016**, *84*, 41–54. [CrossRef]
37. Wilmsen, B.; Webber, M.; Duan, Y. Involuntary rural resettlement: Resources, strategies, and outcomes at the Three Gorges Dam, China. *J. Environ. Dev.* **2011**, *20*, 355–380. [CrossRef]
38. Alkire, S.; Seth, S. *Identifying Destitution through Linked Subsets of Multidimensionality Poor: An Ordinal Approach*; OPHI Working Paper 99; University of Oxford: Oxford, UK, 2016. [CrossRef]
39. Shangguan, Z.; Wang, M.Y.; Huang, J.; Shi, G.; Song, L.; Sun, Z. Study on Social Integration Identification and Characteristics of Migrants from "Yangtze River to Huaihe River" Project: A Time-Driven Perspective. *Sustainability* **2019**, *12*, 211. [CrossRef]
40. Alkire, S.; Shen, Y. Exploring Multidimensional Poverty in China: 2010 to 2014. In *Research on Economic Inequality*; Emerald Publishing Limited: Bradford, UK, 2017; pp. 161–228. [CrossRef]
41. Zeng, H. Analysis on Trend of Rural Residents' Gap in Level of Education Degree. *J. Chang. Univ.* **2018**, *28*, 92–97+102. Available online: https://kns.cnki.net/kcms/detail/detail.aspx?FileName=CDXB201806021&DbName=CJFQ2018 (accessed on 25 April 2023).
42. Zhou, B. Predictive values of body mass index and waist circumference to risk factors of related diseases in Chinese adult population. *Chin. J. Epidemiol.* **2002**, *23*, 5–10.
43. Peng, L.; Xu, D.; Wang, X. Vulnerability of rural household livelihood to climate variability and adaptive strategies in landslide-threatened western mountainous regions of the Three Gorges Reservoir Area, China. *Clim. Dev.* **2018**, *11*, 469–484. [CrossRef]
44. Wilmsen, B. Damming China's rivers to expand its cities: The urban livelihoods of rural people displaced by the Three Gorges Dam. *Urban Geogr.* **2017**, *39*, 345–366. [CrossRef]
45. Xu, D.-D.; Zhang, J.-F.; Xie, F.-T.; Liu, S.-Q.; Cao, M.-T.; Liu, E.-L. Influential factors in employment location selection based on "push-pull" migration theory—A case study in Three Gorges Reservoir area in China. *J. Mt. Sci.* **2015**, *12*, 1562–1581. [CrossRef]
46. Foster, J.; Greer, J.; Thorbecke, E. A Class of Decomposable Poverty Measures. *Econometrica* **1984**, *52*, 761. [CrossRef]
47. Foster, J.; Greer, J.; Thorbecke, E. The Foster–Greer–Thorbecke (FGT) poverty measures: 25 years later. *J. Econ. Inequal.* **2010**, *8*, 491–524. [CrossRef]
48. Villaseñor, J.; Arnold, B.C. Elliptical Lorenz curves. *J. Econ.* **1989**, *40*, 327–338. [CrossRef]
49. Alkire, S.; Foster, J. Counting and multidimensional poverty measurement. *J. Public Econ.* **2011**, *95*, 476–487. [CrossRef]
50. Xie, Y.; Zhang, X.; Xu, Q.; Zhang, C. Short-term trends in China's income inequality and poverty: Evidence from a longitudinal household survey. *China Econ. J.* **2015**, *8*, 235–251. [CrossRef] [PubMed]

51. Anhui Province Sets the 2018 Poverty Reduction Target: More Than 700,000 Poor People Out of Poverty. Available online: http://www.ceweekly.cn/2018/0126/217583.shtml (accessed on 26 January 2018).
52. Shangguan, Z.; Liu, J.; Wang, M.Y.; Chen, S.; Zhang, R. Have Water Conservancy Project Resettlers in Contemporary China Really Been Lifted Out of Poverty? Re-Measurement Based on Relative Poverty and Consumption Poverty. *Land* **2023**, *12*, 169. [CrossRef]
53. Kapustina, L.; Lipková, L'.; Silin, Y.; Drevalev, A. US-China trade war: Causes and outcomes. *SHS Web Conf.* **2020**, *73*, 01012. [CrossRef]
54. UNDP. Socioeconomic Impact Assessment of the COVID-19 Pandemic in 5 Poverty Counties in China-CN. 2021; p. 6. Available online: https://www.cn.undp.org/content/china/zh/home/library/crisis_prevention_and_recovery/socioeconomic-impact-of-the-covid-19-pandemic-in-five-poverty-co.html (accessed on 6 January 2021).
55. Li, Y. Industry Income Gap, Industrial Structure Upgrading and Regional Industry Positioning. *J. Shanxi Univ. Financ. Econ.* **2021**, *43*, 27–41. [CrossRef]
56. Zhang, X.; Jin, Y.; Wang, Y. Study on the Effect of Doscrepancy of Human Capital on Industrial Structure Optimization and Upgrading. *Sci. Decis. Mak.* **2020**, *2*, 24–41. Available online: https://kns.cnki.net/kcms/detail/detail.aspx?FileName=KXJC202002003&DbName=CJFQ2020 (accessed on 20 February 2020).
57. Li, T.; Barwick, P.J.; Deng, Y.; Huang, X.; Li, S. COVID-19 Pandemic and Unemployment: Evidence from Mobile Phone Data in China 1. *Statistics* **2020**, *134*, 135. [CrossRef]
58. Ong, L.H. State-Led Urbanization in China: Skyscrapers, Land Revenue and "Concentrated Villages". *China Q.* **2014**, *217*, 162–179. [CrossRef]
59. Rao, J.; Ye, J. From a virtuous cycle of rural-urban education to urban-oriented rural basic education in China: An explanation of the failure of China's Rural School Mapping Adjustment policy. *J. Rural. Stud.* **2016**, *47*, 601–611. [CrossRef]
60. Nielsen, I.; Nyland, C.; Smyth, R.; Zhang, M.; Zhu, C.J. Which Rural Migrants Receive Social Insurance in Chinese Cities? Evidence from Jiangsu survey data. *Glob. Soc. Policy* **2005**, *5*, 353–381. [CrossRef]
61. Shan, L.; Yu, A.T.W.; Wu, Y. Strategies for risk management in urban–rural conflict: Two case studies of land ac-quisition in urbanising China. *Habitat Int.* **2017**, *59*, 90–100. [CrossRef]

MDPI
St. Alban-Anlage 66
4052 Basel
Switzerland
www.mdpi.com

Sustainability Editorial Office
E-mail: sustainability@mdpi.com
www.mdpi.com/journal/sustainability

www.ingramcontent.com/pod-product-compliance
Lightning Source LLC
Chambersburg PA
CBHW040814250926
51850CB00007BA/1169
9783039285990